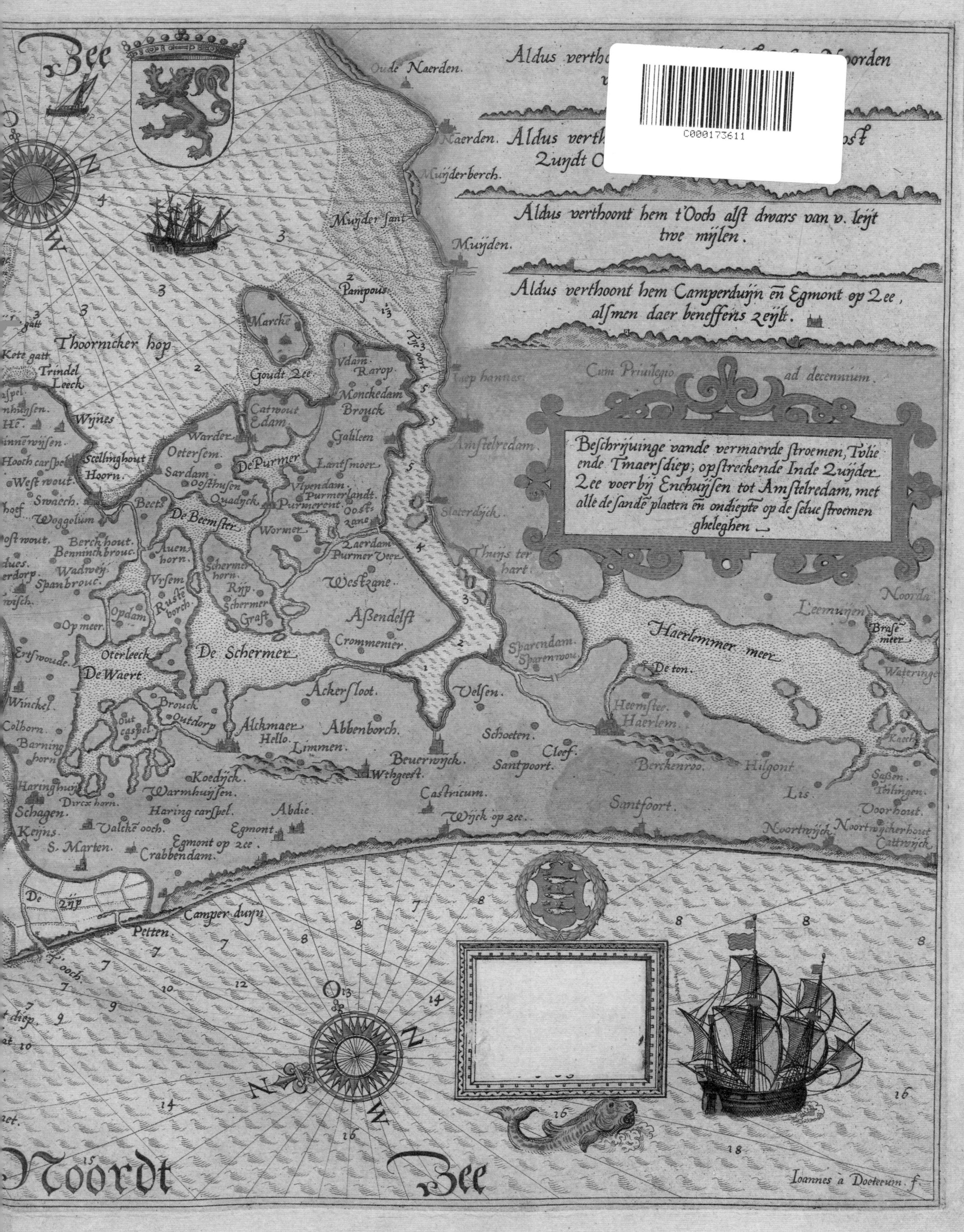

Aldus verthoont hem t'Ooch alst dwars van v. leijt
twe mijlen .
Aldus verthoont hem Camperduijn en Egmont op Zee ,
alsmen daer beneffens zeijlt .
Cum Priuilegio ad decennium .
Beschrijuinge vande vermaerde stroemen, Tvlie
ende Tmaersdiep; opstreckende Inde Zuijder
Zee voer bij Enchuijsen tot Amstelredam, met
alle de sande plaeten en ondiepte op de selue stroemen
gheleghen .
Oude Naerden.
Naerden.
Muijderberch.
Muijder sant
Muijden.
Pampous.
Marcke
Thoornicker hop
Trindel
Leeck
Wijnes
Goudt Zee.
Vdam
Rarop
Monckedam
Brouck
Catwout
Edam
Warder
Gatileen
Amstelredam.
Scellinghout
Hoorn.
Oetersem.
De Purmer
Lantsmoer
Sardam
Oosthusen
Quadyck
Purmerlandt.
Purmerent
Slooterdijck.
Beets
De Beemster
Wormer
Zaerdam
Purmer Veer
Thuijs ter
hart
Woggelum
Berch hout.
Benninckbrouc.
Auen
horn.
Schermer
horn.
Westzane
Wadwey.
Spanbrouc.
Vrsem
Ruste
borch.
Rijp
Schermer
Graft
Opdam
Op meer.
Aßendelft
Crommenier
Oterleeck
De Schermer
De Waert
Sparendam
Sparenwou.
Haerlemmer meer
Leemuijen
Noorda
Brasse
meer
Wateringe
De ton.
Ackersloot.
Velsen.
Winckel
Colhorn
Brouck
Outdorp
Out
caspel
Alckmaer
Hello.
Abbenborch.
Heemstee.
Haerlem.
Barning
horn
Limmen.
Schoeten.
Cloef.
Beuerwijck.
Santpoort.
Berckenroo.
Hilgont
Koedijck.
Warmhuijsen.
Wthgeest.
Haringhuijs
Dircx horn.
Castricum.
Santfoort.
Lis.
Sassen
Teilingen
Voorhout.
Schagen
Haring carspel.
Abdie
Wijck op zee.
Valckē ooch.
Egmont
Noortwijck.
Noortwijcherhout
Keijns.
Egmont op zee
S. Marten.
Crabbendam.
Cattwijck.
De Zijp
Camper duijn
Petten.
Noordt Zee
Ioannes a Doetecum. f.

Published by Times Books
An imprint of HarperCollins Publishers
Westerhill Road
Bishopbriggs
Glasgow G64 2QT
www.harpercollins.co.uk

First edition 2014

This edition 2015

A catalogue record for this book is available from the British Library

ISBN 978 0 00 814779 2

10 9 8 7 6 5 4 3 2 1

Printed and bound in Hong Kong

If you would like to comment on any aspect of this book, please contact us at the above address or online.
www.timesatlas.com
e-mail: timesatlases@harpercollins.co.uk

 @timesatlas

Facebook.com/thetimesatlas

THE TIMES

HISTORY OF THE WORLD IN MAPS

THE RISE AND FALL OF EMPIRES, COUNTRIES AND CITIES

Contents

Introduction

Maps are one of the oldest forms of human communication. The graphic representation of landscapes, whether real, imaginary or mystical, arguably dates back to pre-literate times, when Neolithic peoples scratched marks on rocks to symbolize the terrain in which they farmed or hunted. Mapping has certainly been a feature of most cultures since Babylonian scribes first drew plans of property boundaries and of temples around 2300 BC. The succeeding four millennia have brought many advances in the science of cartography and, in parallel, a widening in the geographic scope and purposes to which maps have been put. The introduction of the compass rose in fourteenth century portolan charts (or mariners' maps) may have been revolutionary, but so was the trigonometric survey of France undertaken by the Cassini family in the eighteenth century which made possible the first modern map of a nation, and the application of mapping to human demography, when the Victorian philanthropist Charles Booth produced maps of poverty in London in the 1890s.

Maps, though, have another important – and often neglected – role as records of critical developments in human history. A map has often been a crucial milestone, recording the expansion of an empire, the discovery of a new land, the imposition of a new political order or the agreement of a peace treaty.

The maps selected for this volume fall into three general categories. Firstly, there are those which show a growing knowledge of the world (no matter how imperfect), from the very first Babylonian 'world map' around 600 BC to Spanish conquistadors' maps of Mexico in the 1520s, and the first maps of the ocean floor in the 1870s. Secondly, and often allied to the first, there are maps which show advances in cartography or a widening of the application of mapping, such as the itinerary maps of the Roman Empire (most notably the *Peutinger Table*), Mercator's 1569 world map – which pioneered the map projection which has, ever since, been the most commonly used for world maps – and, in the twenty-first century, the appearance of 'virtual maps' which exist only in the digital realm but which impart to cartographers (and indeed would-be cartographers) an ability to manipulate data in ways which their predecessors could never have imagined.

Finally, there are maps which document great political developments – where the maps themselves have been actors in world events. The truce lines on the map of the Dayton Agreement which ended the Bosnian Civil War in 1995, and the complex mosaic of areas of control between Palestinians and Israelis in the West Bank shown on the Oslo Peace Agreement Map in the same year, were the subject of prolonged, and often bitter negotiations, in which each side literally argued over the placing of lines on a map. Similarly the power of the map as propaganda is demonstrated in a range of maps, from Reynolds' graphic warning of the potential spread of slavery in the United States in the 1850s, to maps produced in Nazi Germany in the 1930s showing the enlargement of the 'Greater German Reich' by its absorption of Austria and Czechoslovakia.

Mapping has never been so ubiquitous and powerful as in the twenty-first century, yet, as the maps in this book remind us, it has for nearly 5,000 years both reflected and shaped our view of the world and been a mirror of the political concerns, military conflict and cultural preconceptions which have given birth to the world we know today.

BABYLONIAN TOWN PLAN

The world's first cities emerged in Mesopotamia around 6,000 years ago. The needs of the centralized administrations of these city-states led to the development of formal written scripts and these were soon followed by plans and maps to document land boundaries and sales. Before long, they were also used to depict the shape of the cities themselves and the regions that lay beyond them.

The increasingly complex administrative requirements of the urban society which appeared in Mesopotamia from c.4000 BC were met by the development around 800 years later of a full script, known as cuneiform, incised on clay tablets. By the mid-third millennium BC, scribes of the Sumerian culture were producing lists of place-names, including rivers, mountains and towns as far distant as Ebla in Syria. They also compiled itineraries, detailing routes which extended even into Central Anatolia (Asia Minor).

While these may have been intended to assist or document military expeditions (or the trading voyages which took Mesopotamian traders as far afield as the southern Gulf and even to the cities of the Indus Valley), the earliest actual maps we possess from the region are much smaller-scale affairs, and were probably intended to document land-holdings. One of the earliest, from c.2300 BC, comes from Yorghan Tepe, north of Kirkuk (in modern-day Iraq). It shows a river, possibly the Euphrates, and indicates the name of the owner of the central plot, a certain Azana. It also includes written marks showing the direction of north, east and west, the earliest known examples of the indication of orientation on a map.

Somewhat later tablets, of other parts of Iraq, show part of the city of Uruk and a section of Babylon, but the most complete is this remarkable tablet from c.1500 BC incised with a plan of the town of Nippur. It shows the Euphrates, marked as wavy lines, two canals and buildings (labelled in cuneiform), including the Temple of Enlil, and seven gates in the city walls, each one of them named. That the map was based on an actual survey is suggested by the measurements in cubits given for several buildings, and though its purpose is not clear, it may have been compiled in connection with repairs to the city's defences.

Whatever the reason for its production, the Nippur plan is the earliest near complete town plan that we have and a sign of the broadening uses to which cartography was being put.

Terracotta fragment with map of the city of Nippur from Tell Telloh, Iraq.
Hilprecht-Sammlung Friedrich-Schiller-Universitat, Jena, Germany.

BABYLONIAN WORLD MAP

Mesopotamian cartography became more sophisticated over time, driven by the needs of land-surveying. A plan inscribed on a statue of Gudea of Lagash (*c.*2141-2122 BC), which depicts part of a temple enclosure wall, includes a ruler with gradated divisions, the first evidence of scale in mapping.

By the time of the neo-Babylonian empire in the seventh century BC, a sophisticated understanding of mathematics and astronomy had developed, together with an interest in celestial geography, which had led to the creation of the first star maps in the twelfth century BC. The city of Babylon was the largest the world had known, extending to over 10 sq km (4 sq miles) and with a population of some 200,000 by 600 BC. As Mesopotamia's cultural horizons grew, so the mapping of wider areas and the relationships between regions began and it is from this period that we possess the very first attempt at a world map.

Dating from between 700 and 500 BC, the Babylonian world map portrays the earth as a flat disk, surrounded by an ocean – the 'Bitter River'. Beyond this, shown as triangles, lie legendary lands populated by mythical beasts; the cuneiform labels name a land 'brighter than sunset or stars' and another as being veiled in perpetual darkness. The centre of the earth is occupied by a rectangle representing Babylon, with Assyria named to the right, and what may be Urartu (modern Armenia) beside it. Eight other circles indicate named cities, while the lines running vertically through the map probably represent the Euphrates river, with mountains to the north and a marsh marked at the south.

This is far from a topographically accurate map and, unsurprisingly, represents the world from a very Babylonian point of view, setting down their conception of how their territory related to the remotest (and possibly legendary) parts of the earth. Nonetheless, in the scope of its ambition, it represents a revolutionary advance in cartography.

World map carved on a Babylonian clay tablet, measuring just 122 x 82 mm (4.8 x 3.2 in), from Sippar, Babylonia (now southern Iraq). British Museum, London, UK.

EGYPTIAN PAPYRUS MAP

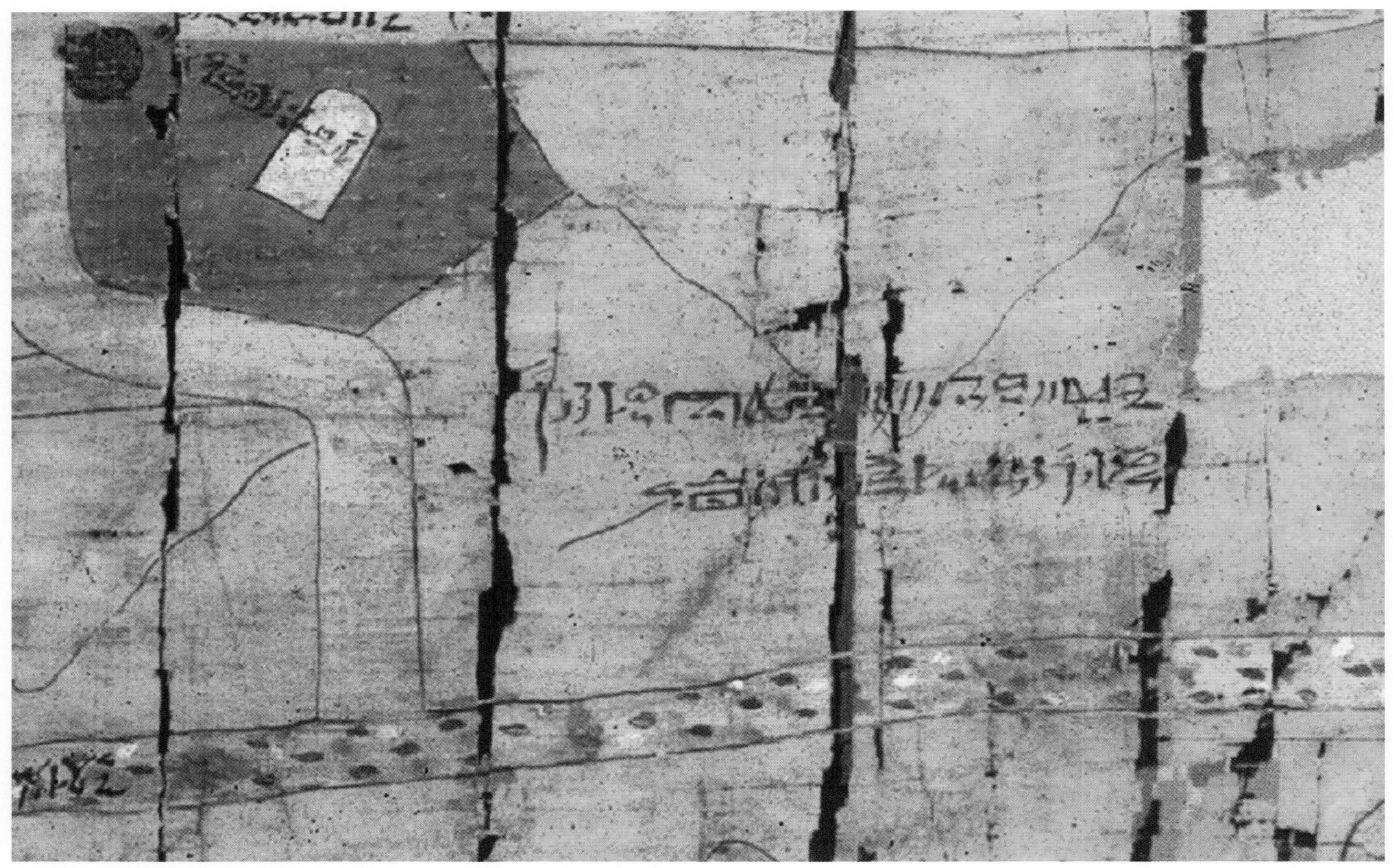

Reconstruction by James Harrell of the Turin papyrus map of Wadi al Hammamat, Egypt. Museo Egizio, Turin, Italy.

The Turin Papyrus, a painted papyrus scroll dating from c.1150 BC, contains a map of an area of Egypt's Eastern Desert between the Nile and the Red Sea. Probably intended as an aid to an expedition in search of sandstone for monumental statues of the pharaoh, it may be the world's earliest geological map.

Cartographic depictions of a sort survive from the very earliest periods of Egyptian history, including primitive topographic sketches of Nile boats and desert hills on pottery of the Nagada II period (c.3700–3100 BC). However, these mainly form part of a tradition of mystical or religious geography, showing the route a soul might navigate on its way into the afterlife, or the plots of land the deceased were said to work in the realm of Osiris, the god of the Underworld.

After the foundation of the New Kingdom around 1640 BC, a new Egyptian empire arose in the Levant and more intensive exploitation of resources in the Eastern Desert got underway. Maps of a sort were used at this time for political propaganda, such as the depiction of Seti I (1318–1304 BC) on a royal progress past frontier forts on his way to Canaan, or for astronomical purposes such as the star map on the ceiling of the tomb of Senmut, a minister of Hatshepsut (c.1470 BC).

Yet despite the patent administrative requirements of the pharaonic administration in governing a large and complex realm, there is only one surviving map which appears to have served those needs. The Turin Papyrus was discovered in the early 1820s by agents acting for Bernardino Drovetti, the French consul in Egypt, and its true significance was not recognized for some decades. The two fragments of a scroll that was originally 41 cm (16 inches) wide and 2.8 metres (9 ft 2 inches) long are painted with what are clearly depictions of *wadis* or dried-up river-beds and tracks running through one of New Kingdom Egypt's main gold and stone-quarrying regions.

The main route, the Wadi al Hammamat, is shown as a bold line crossing the lower portion of the map (which is oriented to the south, towards the source of the Nile), and is speckled with dots, which may represent the rocks and boulders in the dry watercourse. It is continued by another wadi which leads to the east, identified as the Wadi Atalla which connects with an ancient gold mining station at Bir Umm Fawkahir. Hills are shown on the papyrus by conical forms with wavy slopes, coloured in pink and black, while a number of other features are depicted, including a well, a cistern (probably used to separate out gold from pulverised quartz), a stela of the nineteenth-Dynasty pharaoh Seti I (1306–1290 BC), a temple of Amun and the houses of the gold-working settlement itself.

A series of inscriptions in hieratic (an everyday form of hieroglyphic script) give clues to the purpose of the papyrus. As well as labelling the features themselves, such as 'mountains of gold' and giving distances in *khets* (around 50 m/55 yds), they indicate that the map was drawn by Amenakhte, son of Ipuy, the Scribe of the Tomb during the reign of Ramesses IV (1151–1145 BC), in connection with an expedition to the Eastern Desert to find stone for building monumental statues of the pharaoh. This stone was probably *bekhen*, a type of greenish sandstone known as greywacke, the quarries for which are marked on the papyrus. In the first three years of his reign Ramesses IV mounted no fewer than six expeditions in search of *bekhen*, the last of these being a truly monumental enterprise involving over 8,000 men.

The shade of the hills on the map may represent the colours of the Eastern Desert mountains seen from afar: black for schist, pink for granites such as Atillah serpentinite. Even if it was not explicitly designed to be a geological map of the Wadi al Hammamat region, the Turin Papyrus is the earliest map we have which depicts the geology of an area.

Whether it was intended as a geological map or not, as the second oldest topographic map which survives (after the Babylonian town plan of Nippur) and the only major piece of large-scale cartography we have from the ancient Egyptians, the Turin Papyrus is a truly unique relic.

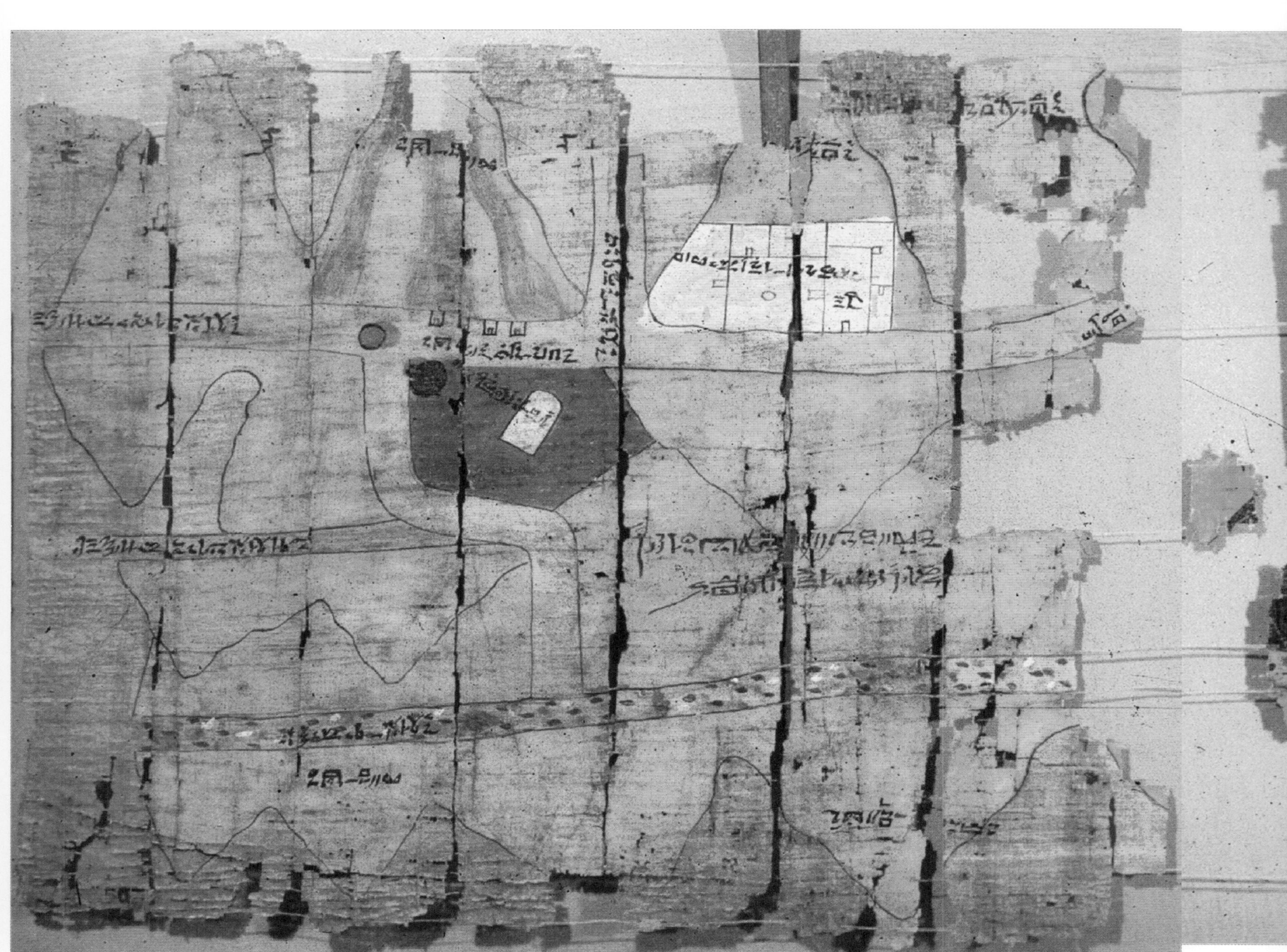

Ptolemy WORLD MAP

World map from a manuscript of Ptolemy's Geography.
Biblioteca Nazionale Vittorio Emanuele III, Naples, Italy.

The *Geography* of Claudius Ptolemy represented the highest development of ancient geography and cartography, incorporating the latest information from explorers in an elegant mathematical framework. As fate would have it, the book had its most important influence and widest diffusion more than a thousand years after it was written, and was crucial in the development of the modern atlas.

Ptolemy was a Roman citizen who wrote in Greek and lived in Alexandria, Egypt, in the second century AD. He was an amazingly versatile scientist and mathematician, writing about astronomy, astrology, music, optics, and geography. In his *Geography*, he brought Greek geography to its highest state of development, incorporating the latest information from recent voyages, and organizing his data mathematically, on the basis of latitudes and longitudes. The book consists of instructions for making maps, and a list of about 8,000 places, each with its coordinates: it provides both the framework for making maps, and the details for filling them.

There is some debate about whether Ptolemy himself made maps, though it is reasonable to think that someone who was providing instructions for making maps would try them out himself. It is more widely agreed that there was no transmission of Ptolemy's maps from antiquity to the Middle Ages – what survived was just the text of the *Geography*. That text was discovered by Maximos Planoudes (*c.*1260–*c.*1305), a Greek scholar in Constantinople, and in about 1300 Planoudes reconstructed Ptolemy's maps according to his instructions and using his data for the latitude and longitude of cities, mountains, and rivers. A Byzantine manuscript of the *Geography*, with maps, was taken to Italy in 1397 and the work proved very popular there. The Greek text was translated into Latin in 1409, we have reports of Florentine humanists enthusiastically studying a manuscript of the work in 1435, and by the 1460s a small industry producing lavish Latin manuscripts of the *Geography* had arisen in Florence. The first printed edition with maps was published in Rome in 1477, and on average new editions appeared more than once every three years until 1550.

The rediscovery of the *Geography*, in short, caused a sensation. The Renaissance represented a rebirth of interest in the culture of classical antiquity, and the enthusiastic reception of Ptolemy was certainly part of that movement, but it was much more. Ptolemy's mathematical system of mapping according to latitude and longitude, which seems obvious and inevitable to us today, had been out of use for centuries, and its virtues were quickly perceived. Its mathematical foundation made it easier for errors to be realized and corrected, and as new lands were discovered – an issue of great importance in the latter part of the fifteenth century – it was relatively easy to incorporate them into Ptolemy's system. We can see a prime example of that in Martin Waldseemüller's world map of 1507 (see page 64), which is based on Ptolemy but includes the New World. Also, Ptolemy's underestimate of the earth's circumference helped convince Columbus that he could reach Asia by sailing to the west.

Typical manuscripts and printed editions of the *Geography* contained one world map and twenty-six regional maps. As we can see from the world map illustrated here, Ptolemy's knowledge of the world's geography was far from perfect: for example, he believed that a huge land-bridge joined southern Africa with southern Asia, making of the Indian Ocean a huge lake. Renaissance scholars realized Ptolemy's errors, and also came to the conclusion that Ptolemy's maps were not practical for use in a world in which the name of almost every city, river, and mountain had changed over the centuries. But even while Ptolemy's geographical data was rejected over time, the format in which he had displayed his data, in a book with a world map and regional maps, was the basis of the first atlas in the modern sense of the word, Abraham Ortelius's *Theatrum Orbis Terrarum*, first published in 1570 (see page 84).

CAVRVS
CIRCIVS
SEPTEN
ZEPHIRVS
occeanus occidentalis
OCCEANVS OCCIDENTALIS
iberinia
britannia
Mediterraneum
Pontus euxinus
HESPERIVS SINVS
LIBIA INTERIOR
ETHIOPIA INTERIOR
ETHIOPIA SVB EGIPTO
TERRA INCOGNITA
ARABIA FELIX
Sinus Persicus
Rubrum mare
SINVS BARBARICVS
AFFRICVS
CIRCVLVS

AQVILO

WLTVRNVS

Sithia extra imaum montem

Serium regio

SINARVM REGIO

Sinus magnus

Sinus gangeticus

INDICVM

Taprobana insula

MARE

CIRCVLVS EQVINOCTIALIS

CIRCVLVS ZODIACI

TERRA INCOGNITA

EVRVS

Peutinger Table

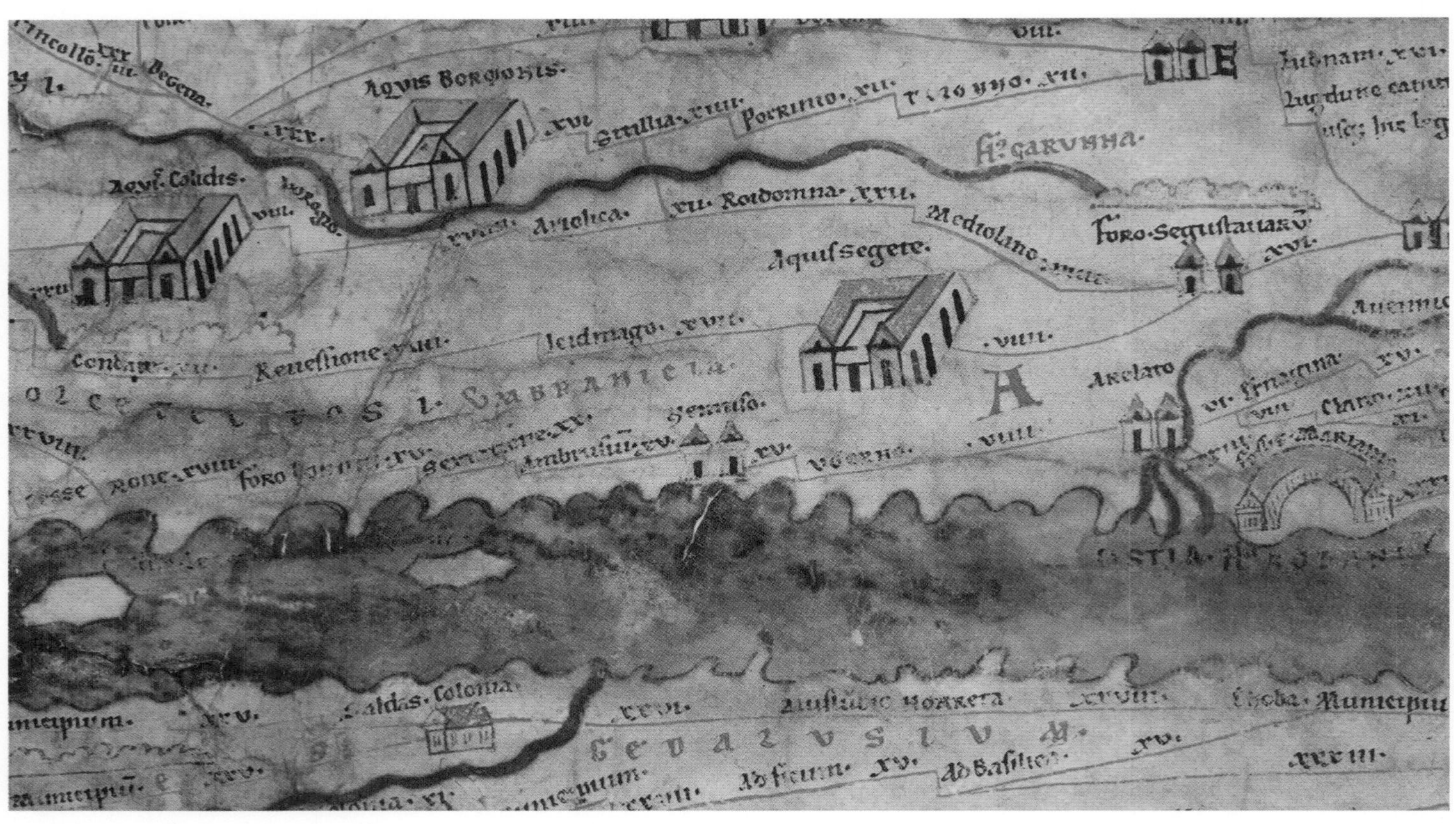

Section of the Peutinger Table Roman route map showing Gaul (France).
Major towns are shown as square fortifications.Österreichische Nationalbibliothek, Vienna, Austria.

ROMAN ROUTE MAP

Eleven sheets of parchment provide a fascinating glimpse into how the Romans saw their world, and a unique insight into the network of roads which bound it together.

Although the Roman empire was the largest the world had known, occupying an area of around 6.5 million sq km (2.5 million sq miles) at its height in the second century AD, and it possessed a long tradition of both civil and military surveying, it left us precious little in the way of maps. Maps are mentioned in Roman sources as early as 174 BC (a map of Sardinia allegedly once found in the Temple of Mater Matuta in Rome) and Augustus, the first emperor (27 BC – AD 14), commissioned a survey of the entire world (by which he really meant the empire). The resulting map is said to have been set up in the Porticus Vipsania in Rome, but of this, and its predecessors, not a trace survives.

Excluding much later medieval and renaissance reconstructions of the world map of the Greco-Roman geographer Ptolemy (see page 16), all we have are a few fragments of cadastral (or land survey) maps from France, two small pieces of a once monumental city map of Rome (the *Forma Urbis Romae*), which was probably completed around AD 205, a decorated third-century shield from Dura Europos in Syria which includes a map of part of the Black Sea coast and, more important than any of these, the *Tabula Peutingeriana*, or *Peutinger Table*.

The 'table' is a long, narrow parchment roll, some 6.8 m (22 ft 4 in) long and 33 cm (13 in) wide, composed of eleven strips – originally there were twelve but one, which covered most of Britain and Gaul, has been lost – which show major roads and routes throughout the Roman empire (and a few areas beyond it, such as Persia and parts of India). The map is not an original, probably representing a twelfth or thirteenth century transcription of a mid-fourth century template, which was then lost until its discovery around 1500 by Konrad Celtes, an Austrian humanist. He then bequeathed it to his friend Konrad Peutinger after whom it was named. It does not try to give a topographically accurate portrayal of the landscape (it distorts proportions so that north-south distances appear far less than in reality), nor to give any sense of scale. What it does provide is distances between locations – in Roman miles for the most part, but in Gaul switching to leagues, and in Persia employing *parasangs*, a local measure equivalent to around 5.6 km (3½ miles).

The map is decorated with symbols for towns, as well as signs for harbours, lighthouses, granaries and other important landmarks which might help orient a traveller. Its essential purpose seems to be a guide to the *cursus publicus*, the imperial Roman transportation system which maintained roads and operated way-stations for the despatch of official mail and to ensure the swifter movement of official traffic. It also forms part of the later Roman tradition of written itineraries, which similarly gave the distances between key cities and staging posts. The most important of these, *The Antonine Itinerary*, which was finalized around AD 290, but may originally have been compiled as an aid to planning long-distance journeys by the Emperor Caracalla (198–217), includes maritime as well as land routes.

The *Peutinger Table* is on a grander scale and its inclusion of over 4,000 place names, a few of them tiny places whose existence has only subsequently been confirmed by archaeologists, gives us an invaluable snapshot of the Roman empire in the fourth century. As perhaps the earliest surviving map with a very definite purpose, and the distant prototype of much later road atlases, it also occupies a unique role in the history of mapping.

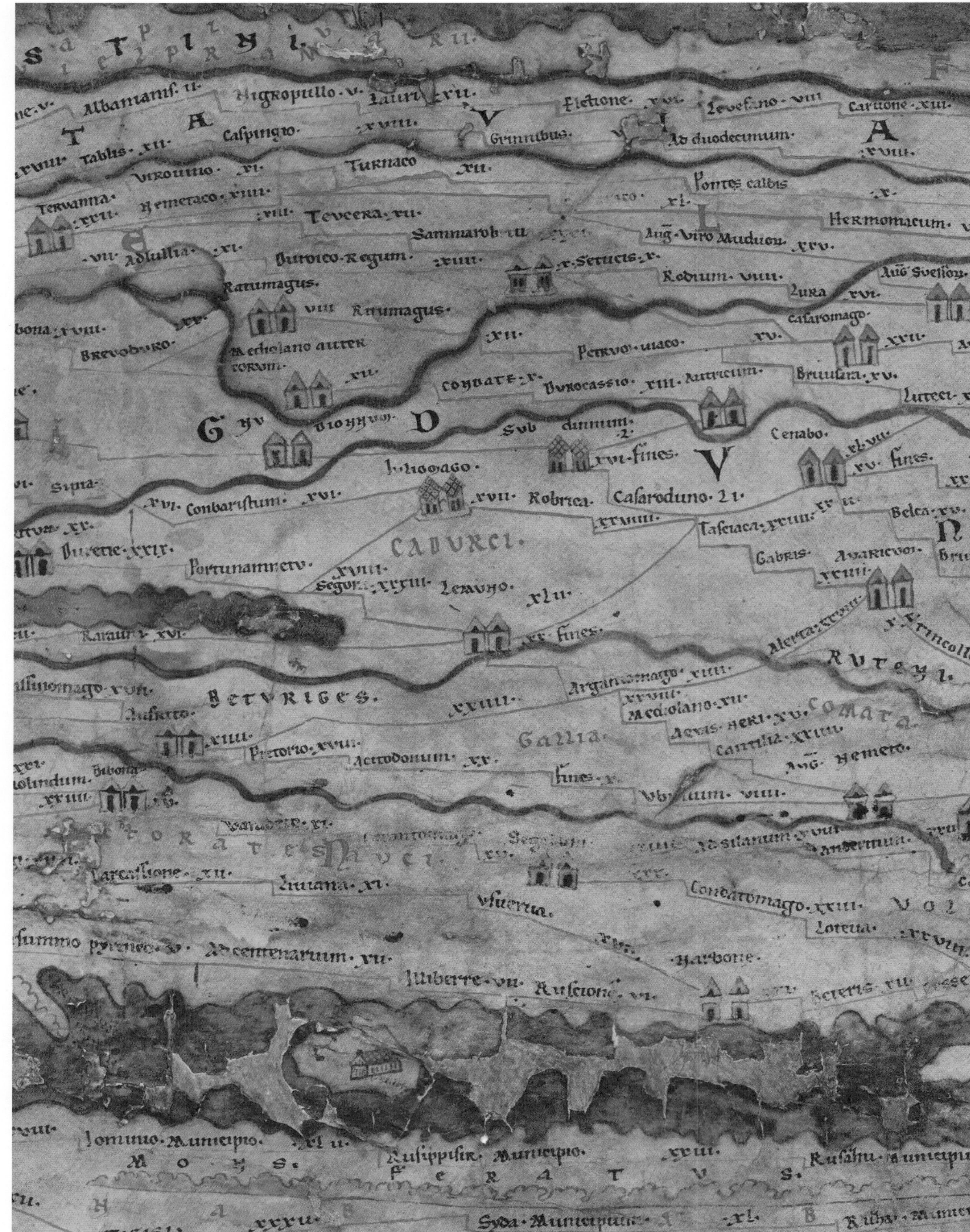
Albamanis. ii.
Nigropullo. v.
Caspingio.
Grinnibus.
Ad duodecimum.
Carvone. xiii.
Tablis. xii.
Turnaco xii.
Pontes caldis
Tervanna.
Tevecra. xii.
Hermomacum.
Aug. Viro Muduor. xxv.
Ratumagus.
Rodium. viiii.
Casaromago.
Brevodiro.
Condate. x.
Durocassio. xiii.
Briuisara. xv.
Cenabo.
Iuliomago.
Conbaristum. xvi.
Robrica.
Casaroduno. 21.
CADURCI.
Belca. xv.
Gabris.
Lemuno.
xlii.
fines.
BETURIGES.
Mediolano. xii.
Gallia
Pretorio. xvii.
Cantilia. xviii.
Aug. Nemeto.
fines.
Carcasione. xii.
Liviana. xi.
Condatomago. xviii.
Usuerva.
Loteva.
Narbone.
Ad centenarium. xii.
Illiberre. vii.
Ruscione.
Beteris. xii.
Tomimo Municipio. xlii.
Rusippisir Municipio. xviii.
MONS
FERRATUS.
Syda Municipium. xl.
Tigisi.

viii. Rotomagi.
x. Arenatio.
vi. Burginatio. v.
PARISI.
Baca conervio.
Ceuclum
Blariaco. xii.
Feresne. xvi.
Atuaca.
xvi. Cortovallio. xii.
xii. Vogo Borgiaco. xvi.
xi. Duronum. x
Gemmico vico. xl iii.
Pernaco. xvi.
Mose. viii.
Vironum. viii.
G
REMIGES.
I
Auxenna. x.
Durocortoro.
xii. Rotomagus. xxv.
Calagum. xxx.
NITIOBRO
Bibe. xxii.
xix. Tanomia. xxv.
Riobe.
Aug. Bona. xviii. O.
Corobilium.
Mose.
Metleglo. xv. Condate.
Segessera. xvi.
Andemantunno.
Eburobriga.
xi. x.
Varcia. vi.
XX vi.
fl. Riger.
Filena. xix.
Vidubia. xx.
Autesshio.
Duro. xxii.
E N
Cabillione.
Aballo. xvi.
fl. Arar.
xiiii. Ponte.
CAMBIOVICENSES.
Aquis Hisinei.
S
Sidoloco. xviii.
Tenurcio. xii.
xiiii.
xxii.
viiii.
Borvo
viii.
E
Degena.
Aquis Bormonis.
xvi.
Stillia. xiiii.
Pocrinio. xii.
xii.
fl. GARUNNA.
Aquis Calidis.
viii.
xii. Rodomna. xxii.
Mediolano.
Foro Segustavarv.
xvi.
Aquis Segete.
Icidmago. xvii.
Revessione. viiii.
viiii.
A
Arelato
vi.
SI. UMBRANICIA
Gerriso.
xv.
viiii
Foro
Ambrusio. xv.
OSTIA
Saldas. Colonia.
xxv.
xxvii.
Ausilicio Horreta
xviiii.
Choba. Municip
SEVALUSIUM
xv.
xxviii.
Ad fletum. xv.
Ad Basilica
xxiii.
fines affrice et mauritanie.

Palestine

Section centred on Jerusalem of the Madaba mosaic map. The 'Old Church', Madaba, Jordan.

MADABA MOSAIC MAP

The Madaba Mosaic, a sixth-century Byzantine map of the eastern Mediterranean found in a church in Madaba, Jordan, shows a new development: the use of cartography in the service of religion. It depicts a fundamentally Christian landscape, a long way from the functional surveying and administrative purposes of earlier Roman mapping.

As Rome's western provinces were overrun by Germanic barbarians in the fourth and fifth centuries AD, so the tradition of cartography in the remaining eastern portion of the empire (which historians call Byzantium) went into decline. The Emperor Theodosius I is recorded as having commissioned a map of the empire in AD 435, but it has not survived and Byzantine *periploi* (books with navigational directions) and land itineraries have no maps to match the *Peutinger Table* (see page 20).

When maps did re-emerge, in the sixth-century *Christian Topography* of Cosmas Indicopleustes, an Alexandrian merchant and traveller, they were a curious hybrid of classical geographical notions and Christian theology. Cosmas rejected the long-standing orthodoxy that the earth was spherical, theorizing instead that it was flat and that the universe was like a rectangular vaulted box with the upper part being the realm of God and the angels, and the lower part the domain of mankind. In maps accompanying his manuscript, the world is shown as a large land-mass with four gulfs (the Caspian and Mediterranean Seas and the Arabian and Persian Gulfs), surrounded by an ocean, into which drain four rivers (the Nile, Tigris, Euphrates and "Pheison"). Paradise is to be found at the easternmost extremity of the earth.

This Christianization of cartography, which was to be a feature of mapping well into the Middle Ages, was put to a more practical use in the Madaba Mosaic. Its surviving portions, on the floor of the church in Madaba, show a section of Palestine centred on Jerusalem and a part of Egypt, but the map probably originally encompassed the entire Holy Land. There are other Byzantine mosaic maps, notably one at Nicopolis, near Preveza in Epirus (western Greece), which shows the world as a rectangle surrounded by an ocean (and which may depict the earth before the creation of mankind), but the Madaba map is notable for its attempt to show something like real geographical detail. It does not do this by including much in the way of topographical features, confining itself to the Dead Sea, the Mediterranean, the Nile and Jordan rivers and a scattering of wadis. The map instead presents a dense network of cities, towns and villages, many of them with accompanying annotations setting places within a Christian context, such as 'Selo, where the ark once was' and 'Aliamon, where the moon stood still in the time of Joshua, the son of Nun.'

As befits the mosaic's preoccupation with Christian topography, Jerusalem is given pride of place in the centre of the map, and is shown at a far larger scale (1:1,600, as opposed to the 1:15,000 of parts of Judaea). The whole map originally probably consisted of some two million mosaic *tesserae* picking out the architectural highlights of the towns in greens, blues, reds, black, white, brown, violet and grey. Buildings such as the Church of the Holy Sepulchre in Jerusalem are clearly identifiable and the mosaic's attention to detail in this most critical section can be gauged by its showing of one road leading to the city's main street which remained unknown until archaeologists found it while conducting a rescue excavation in 2010.

The Madaba map, with its wealth of scriptural annotations, probably served as an instructional device, giving cartographic life to episodes from the Bible. Yet it is also symbolic of a greater trend, that for the moment, mapping was more about making sense of God's divine purpose and how that was reflected on earth than in creating an accurate representation of the world for its own sake.

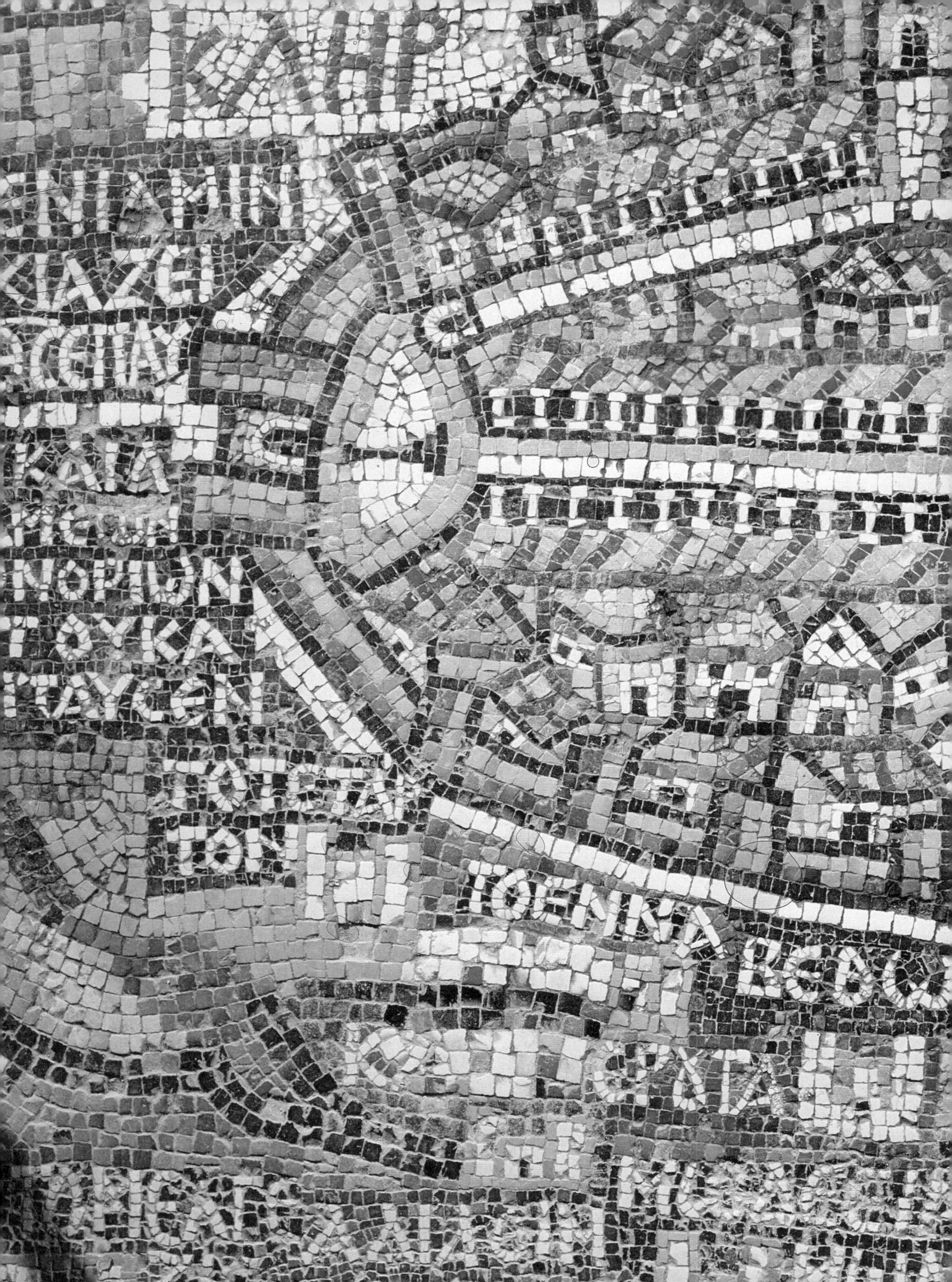

CHINESE STONE MAPS

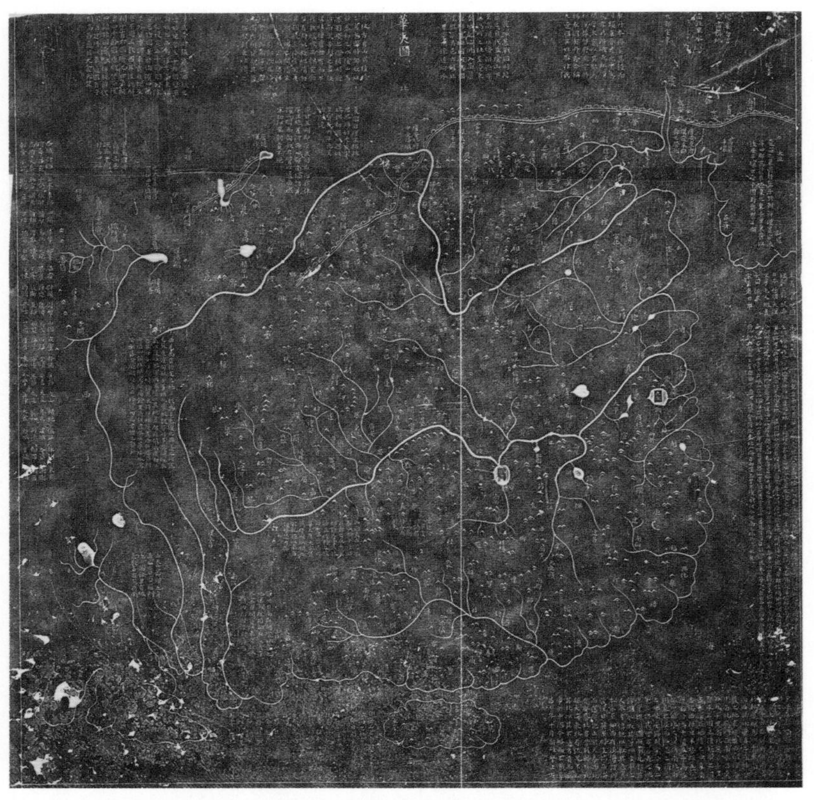

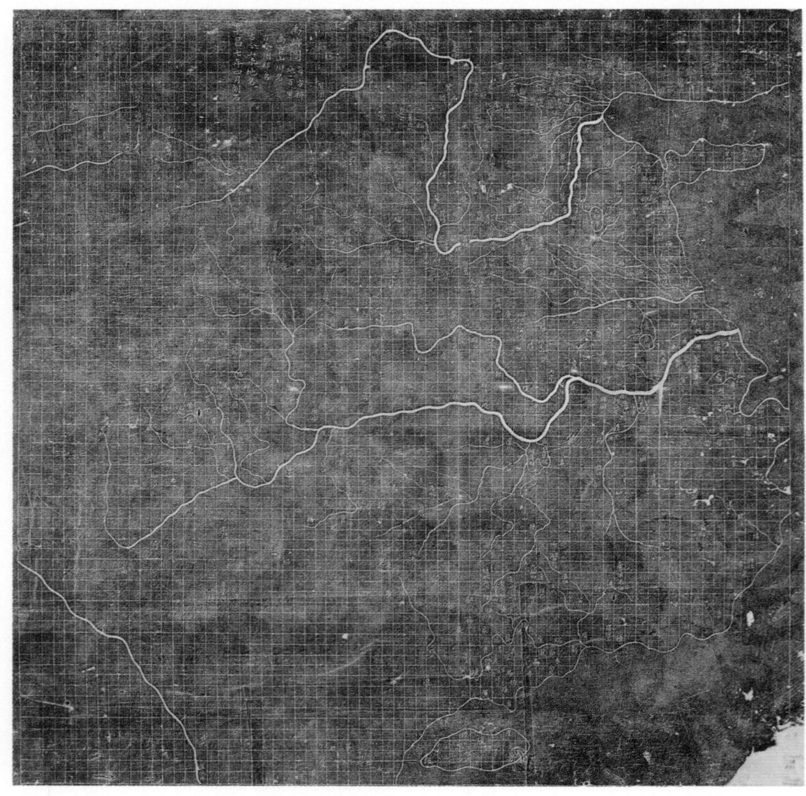

Rubbings of Chinese maps carved in stone: *Hua yi tu* (top) and *Yu ji tu* (bottom). Original carved stones are in the Forest of Stone Steles Museum, Xi'an, China. Library of Congress, Washington D.C., USA.

Hua yi tu AND *Yu ji tu*

A twelfth-century Chinese map carved on a stone slab (or stele) in Sichuan province contains the earliest evidence of grid squares to indicate scale. It is the culmination of a cartographic and mathematical tradition which dates back at least 1,500 years in a country where mapping was deeply embedded in bureaucratic and administrative practice.

References to maps, and their use in government, are relatively common in early Chinese texts, with the earliest survey of the country recorded in the sixth century BC. The Han dynasty *Zhou li* (Ritual forms of Zhou), which was composed in the second century BC, but purports to detail bureaucratic practices several centuries earlier, refers to *Qi Fangshi* or Directors of the Regions who were charged with producing maps of the empire, and even to a geographer-royal tasked with overseeing maps of a ruler's inspection circuits of his domains. Even at this very early stage, maps were clearly integrated into the Chinese administrative mind-set.

It is from around 310 BC, during the era of the Warring State which succeeded the Zhou, that we have the earliest surviving Chinese map, a bronze plaque found in the tomb of King Cuo of the Zhongshan state which bears a plan depicting a series of royal sacrifice halls. A further set of seven maps found on thin wooden boards in a tomb in Gansu province date to about 239 BC, and are notable for their use of black lines for rivers and tributaries. Yet it was the unification of China under the Qin in 221 BC, and the long period of rule by the Han over a united China (from 206 BC to AD 220) which must have given an impetus to cartography, driven by the needs of the government to comprehend and organize its vastly increased realm. They performed a military function also; Li Ling, who led a campaign against the Xiongnu nomads beyond China's northern borders in 99 BC, is recorded as having produced maps of the expedition.

Even so, few Han maps have survived, the most notable being a set on silk from Mawangdui in Hunan province which appear to be garrison maps showing military installations and the topography of a key border area. They use a colour-coded system for geographical features (light blue for rivers, red for roads) and bear annotations giving the distance between key points. The increasing sophistication of Chinese cartography is indicated in the works of Pei Xiu, Minister of Works in the 260s (just after the fall of the Han), which laid out six principles of map-making, including the importance of scale, location, distance, elevation, gradient and precision in measurement. He also created a map of China, the *Yu gong diyu tu*, ('Regional Maps of the Yu Gong') which like all maps of the era is long lost.

After the Mawangdui maps, there is a curious near-millennium-long gap in the cartographic record, and not a single map survives from the Tang dynasty (618–907) which projected Chinese power far into central Asia (and fought and lost a decisive battle in 751 with the Arabs expanding in the opposite direction). Only under the Song dynasty in the twelfth century do a series of maps engraved on stone tablets show that cartography, far from having died out in China, had achieved new levels of technical excellence. The earliest of these, the *Jiu you shouling tu* ('Map of the Prefectures and Counties of the Empire') which was carved in 1121, names more than 1,400 administrative units, together with depictions of mountains, trees and rivers.

The slightly later *Hua yi tu* (*Map of China and Foreign Lands*) was engraved on a stone stele in 1136, but is slightly less sophisticated, including only about 500 place-names and depicting the sources of the Yangtze and Yellow Rivers incorrectly. On the reverse side of the stele is the most famous of the stone maps of China, the *Yu ji tu* (*Map of the tracks of Yu*), which was carved the same year, although evidence from the place names included indicate that it may be based on an original from about 1100. The *Yu ji tu* is remarkably accurate for a medieval map, particularly in its depiction of rivers and coastline, but what is most extraordinary about it is the grid pattern of equal-sized squares superimposed on it, each – according to an annotation on the map – representing about 100 *li* (around 32.5 km/20 miles). This is the earliest known appearance of grid-lines and their use to achieve consistency of scale is a remarkable tribute to the technical skills of the Song map-makers. It is also a testament to political expediency, for it depicts Song control as it was before a treaty with the Jurchen Jin nomads, who occupied the capital Kaifeng and took over much of northern China in 1127.

The stone maps were clearly intended to be permanent and were set up in prefectural headquarters and schools where they could be studied and copied by producing rubbings from them. The *Yu ji tu* formed the basis of an 1142 stone map set up at Zhenjiang, but thereafter the practice appears to have fallen into disuse. Gridded maps next appear on paper during the Ming dynasty in the sixteenth century, with the works of Luo Hongxian (1504–1564). In keeping with the Chinese reverence for tradition, Luo based his work on the fourteenth-century cartographer Zhu Siben, who in turn stated that in composing his own maps, he had consulted the *Yu ji tu*.

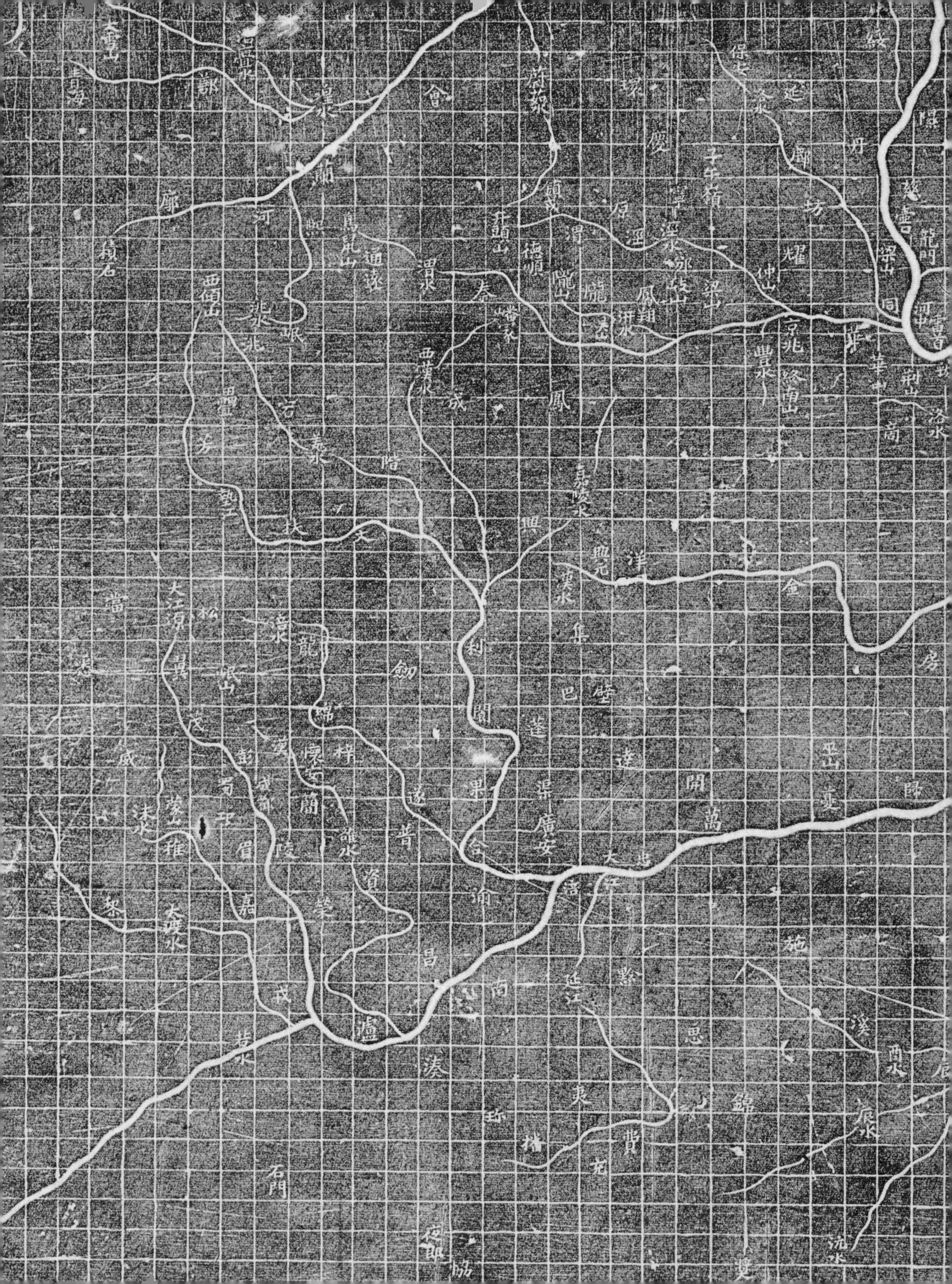

大雪山
青海
鄯
湟水
會
環
保安
延
慶
子午嶺
鄜
坊
丹
隰
絳
龍門
梁山
同
河中
晉
蘭
廓
積石
河
熙
馬銜山
通遠
渭水
秦
鎮戎
原
涇
涇水
邠
岐山
耀
仲山
渭
德順
隴山
隴
鳳翔
汧水
西傾山
洮水
岷
洮
疊
宕
芳
階
嘉陵水
鳳
成
西漢水
豐水
京兆
終南山
華
華山
商
洛水
文
扶
興
興元
洋
金
漢水
集
巴
壁
房
利
閬
劍
綿
梓
潼水
龍
松
大江源
茂
岷山
當
威
汶
彭
漢
成都
蜀
懷安
簡
遂
果
蓬
達
渠
廣安
開
萬
平山
夔
歸
邛
雅
沫水
蒙山
眉
陵
雒水
普
資
合
榮
渝
大江
涪
忠
黎
嘉
大渡水
昌
戎
瀘
南
黔
延江
施
溪
思
酉水
辰水
錦
夷
費
珍
播
溱
黃水
石門
夜郎
協
沅水

太原
汾
汾水
太岳
羊頭山
潞水
太行山
析城
王屋山
沇水
濟源
孟
懷
大河
西京
熊耳山
陸渾山
伊水
嵩山
汝
洛水
趙
邢
洺
磁
滏水
漳水
澤
衛
相
大伾
濬
大陸
冀
恩
博
濮
棣
德
濟水
齊
鄆
漢
陶丘
曹
大野澤
濟
泗水
兗
泰山
原山
淄水
淄
青
胸山
濰
濰水
沂
沂水
蒙山
密
萊
膠水
膠山
琅邪山
登
成山
之罘山
文登山
羽山
嶧山
單
南京
滎澤
鄭
東京
潁昌
宋
亳
睢水
宿
汴水
徐
沭水
淮陽
泗水
漣水
海
楚
潁水
陳
蔡水
潁
渦水
蔡
淮水
濠
壽
方城山
唐
光化
鄧
荊山
襄
桐柏山
淮源
信陽
光
霍山
盧
肥水
巢湖
滁
高郵
泰
揚
通
真
和
無為
隨
陪尾山
安
黃
皖水
舒
潤
江陰
常
茅山
太平
江寧
宣
廣德
太湖
蘇
秀
湖
松江
池
郢
荊門
江陵
復
內方山
漢陽
鄂
蘄
彭蠡
南康
江州
敷淺原
廬山
興國
歙
杭
睦
越
四明山
會稽山
天台山
饒
洪
贛水
筠
信
衢
婺
明
處
溫
澧
澧水
岳
洞庭湖
青草湖
袁
鼎
沅水
湘水
潭
臨江
吉
建昌
撫
邵武
建
鯉溪
福
南劍
資水
衡山
衡
邵
郴
虔

al-Idrisi WORLD MAP

Copy of the world map from al-Idrisi's *Recreation of Journeys into Distant Lands.* South is to the top, following Islamic tradition. Bodleian Library, Oxford, UK.

The twelfth-century World Map produced by the Islamic geographer al-Idrisi is a synthesis of Muslim, Greek and Latin cartographic scholarship. Its compilation in Sicily, at a time the island was experiencing a brief period of tolerance which embraced all of these cultures, is symbolic of the meeting of three traditions which were soon to diverge.

The armies of Islam which swept through the Middle East and North Africa in the seventh century absorbed lands with deep-rooted scientific traditions and much of this Greek and Persian learning was preserved and transmitted by Arab scholars. Ptolemy's influence in particular was passed on by Islamic astronomers and geographers, who tried to harmonize the seven celestial spheres of Aristotle's cosmology with the seven climatic zones of the earth set out by Ptolemy.

Islamic map-making emerged under the Abbasid caliphs, whose capital, Baghdad, became a notable centre of learning after its foundation in 762. The first recorded Islamic world map was the *al-Surat al-mamuniyya* (*Map of al-Mamun*) commissioned by the caliph al-Mamun who reigned from 813 to 833. Although it has not survived, the tenth-century geographer al-Masudi said that it was based on Ptolemy's seven geographical regions. Combining with a burgeoning interest in geographical portrayal, Islamic geographers concerned themselves with compiling tables of latitude and longitude, seeking to perfect and enlarge those which they found in Ptolemy. A particularly important example by al-Khwarizmi (780–850), who made a compilation of 2,042 places, formed a platform for Islamic scholarship for centuries to come.

At the height of Abbasid, power in 846, Ibn Khurradadhbih, director of posts and intelligence at the caliphal court in Baghdad, composed his *Kitab al-masalik wa al-mamalik* (*Book of Routes and Provinces*) which showed the main trade routes within the Islamic empire. This concern to portray the extent of the Muslim lands, the *Dar al-Islam*, only grew as the caliphate frayed and then fractured over the tenth century. By the middle of that century, a thriving cartographic tradition, the Balkhi school, had developed, whose works generally included a world map, together with sectional maps of the seventeen provinces (or *iqlim*) of Islam. These characteristic round maps were oriented to the south (the direction of prayer towards Mecca in much of the Islamic world), and showed the world as a single mass punctuated by gulfs and seas, surrounded by an ocean.

The greatest of Islamic cartographers, al-Sharif al-Idrisi (*c*.1099–*c*.1166), both developed and deviated from the Balkh tradition – a sign of his vast experience in both the Islamic and Christian worlds. Born in Ceuta into the Moroccan royal nobility of the Alawi Idrisids, he was educated in Cordoba, Spain, whose vast royal library was said to hold over 400,000 volumes. His desire for broader horizons and the vulnerability of the Muslim emirates of al-Andalus to Christian expansion, led him to travel for almost a decade through France, England, Asia Minor and Morocco, finally arriving in Sicily in 1138.

Here he took service with Roger II of Sicily, whose reign saw a period of exceptional tolerance in which Islam was not persecuted and its tradition of scholarship was both valued and promoted. Roger desired that 'he should know accurately the details of his land and master them with precise knowledge', and to this end he commissioned scholars to extract what information they could from Ptolemy and other ancient authors and to combine this with information gleaned from contemporary travellers across the inhabited world. Al-Idrisi's experience was ideally suited to this task, and he came to be at the heart of the project, working for fifteen years in a painstaking exercise which first saw the gathering and sifting of information and then its transference onto a *lawh al-tarsim* or 'drawing board' before the final stage, the engraving of an enormous silver disk (some 135 kg/298 lb in weight) which contained the finished world map.

The precious silver disk disappeared in time, but what survived was a geographical compilation, the *Nuzhat al-mushtaq* (*Recreation of Journeys into Distant Lands*) – also known as *The Book of Roger* after its patron – intended as a gloss to the main map 'adding whatever they had missed as to the conditions of lands and countries'. Most of the surviving manuscripts of the *Nuzhat al-mushtaq* themselves contain a world map oriented like the Balkhi maps to the south. These depict the land-mass of the earth as an island and contain the traditional Ptolemaic division into seven climatic zones, with Baghdad and the core of the Islamic world situated in the fourth of these. As a novel refinement, al-Idrisi further partitioned these into ten sub-zones, each of which makes up one of seventy sectional maps which also appear in many manuscripts.

Al-Idrisi's map is clearly heavily influenced by Ptolemy, prominently showing the Mountains of the Moon (top right of the map) which were believed to lie at the source of the Nile, and the southern portion of the earth beyond Africa covered by a large continent (for which explorers would still fruitlessly search in the eighteenth century). But in common with most Islamic maps it shows no cities, country names and precious few rivers (all those were covered in the compendium). It was completed in January 1154, just weeks before Roger's death. Just as al-Idrisi's map represented a fleeting moment of the joining of western and Islamic cartographic conditions, so the tolerant atmosphere of Sicily ebbed away. Al-Idrisi himself slipped away to Ceuta where he died around 1166, while under Roger's nephew Frederick Barbarossa, Sicilian Muslims were forcibly converted or exiled, until by 1229, when Barbarossa died, there were virtually none left at all.

الواق واق
سفاله
الزنج
الصين
الهند
كرمان
فارس
خراسان
الشاس
بلغار
ياجوج
ماجوج

جبل القمر وهو منبع النيل
صحارى وتلال خلف وسط الارض
الحلبشه
النوبه
التاجرين
كانم
كواز
الواحات
بلاد لملم
بلاد مغزاره
قزان
بلاد غانه
صنهاجه
صحاري برنيق
الجريد
افريقيه
السوس
المغرب الاقصي

Matthew Paris

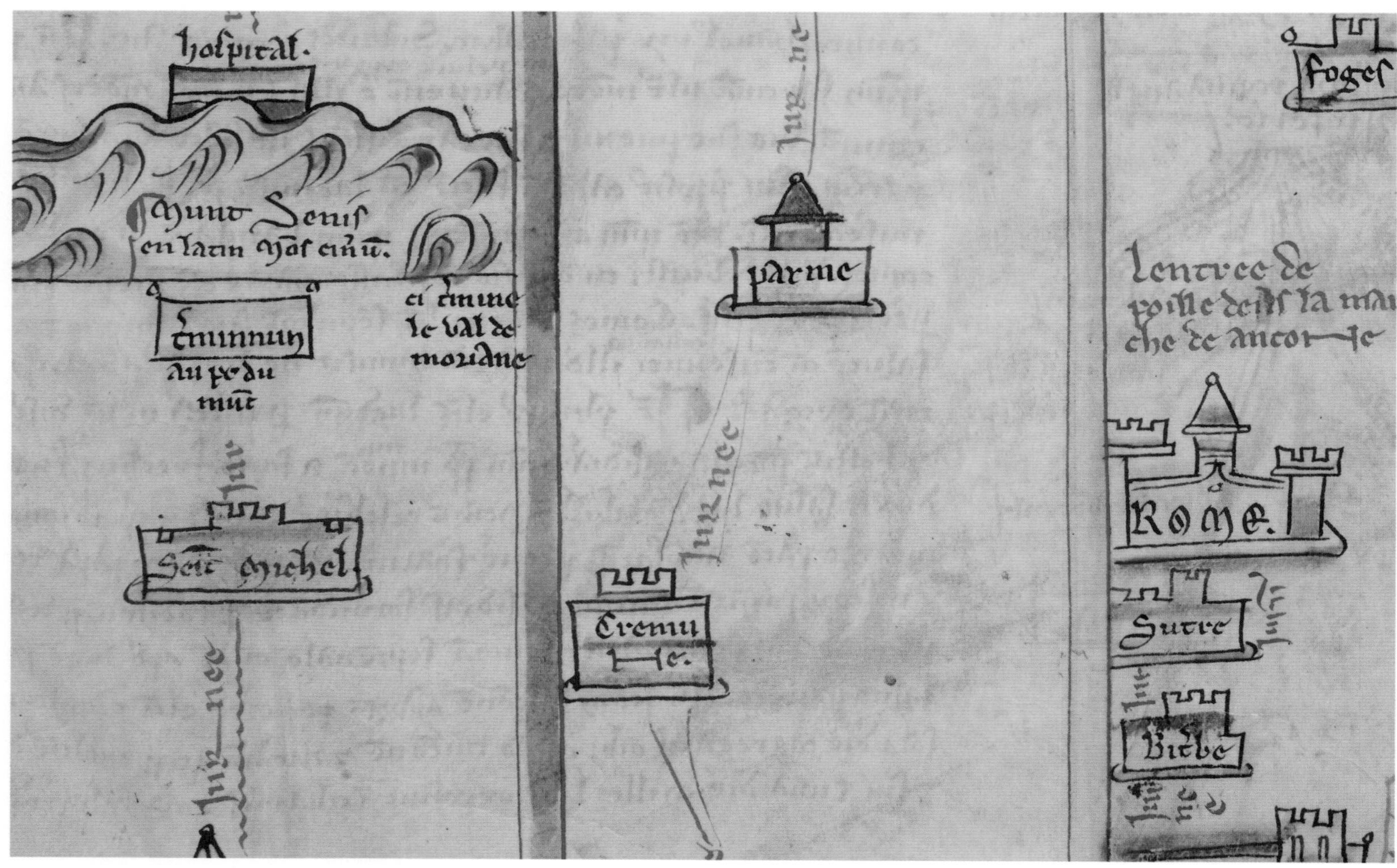

Part of Matthew Paris's itinerary map for a surrogate journey from London to the Holy Land. The full section shown overleaf covers the route from London (bottom left) to the Apulian port of Otranto in southern Italy (top right). British Library, London, UK.

ITINERARY MAP

At a time when maps were not used at all for way-finding or route planning, the sole travel aid was an itinerary – a list in correct sequence of the names of places between the point of departure and the desired destination, the only pair of names known with absolute certainty. Matthew Paris, a monk at St Albans, took this idea to another level with a beautifully illustrated itinerary map depicting the route from London to the Holy Land.

Ideally, one would know not only the names of the start and end points for a planned journey, but also the name of every intermediate place, large or small, along the route which could be counted on to offer safe accommodation. This was impractical, however, and in all but major towns it was more likely that travellers would count on being directed every morning to the right road and on checking with local people and other travellers as they went. Before the arrival of the printed road book in the sixteenth century, such lists usually consisted of informal lists of place names with the distance between them jotted down on a spare leaf at the beginning or end of a manuscript book or in the margins of a letter. Apart from, perhaps, a note about frontiers or political disturbances, nothing would be said about forests, mountains, even rivers; the variable and occasionally challenging geography of the route was a normal condition of travel.

All the more puzzling, then, is the artistic detail and elaborateness of this itinerary map by the Benedictine monk Matthew Paris which he included in the chronicle he was charged with keeping up to date at his monastery in St Albans, Hertfordshire. The itinerary starts in London, continues through Canterbury to Dover, across the Channel, on to St Denis and Paris, south to Lyons, then either on to the Languedoc port of St Gilles or over the Alps by Mont Cenis into northern Italy, down the western side of the peninsula to Rome, and across the Apennines via Beneventum to the Adriatic coast of Apulia and the port of Otranto. The map following the one shown here depicts Acre, main port of entry for the Holy Land, confirming that the terminus of the route was Jerusalem, which is also marked.

Instead, however, of the standard dry list of place-names, Matthew Paris – undoubtedly the most multi-talented of English medieval chroniclers – portrayed the entire route between London and Apulia as a linear map. Starting in the bottom left corner with London and reading up each column in turn across the page, some 2,700 km (1,700 miles) are represented by vertical lines. A note (in French) gives the distance in units of a day's journey (*journées*). Alternative routes are also indicated. Places are displayed pictorially, in more or less correct order, with some fascinating vignettes.

Clearly, Matthew Paris's unique creation was never intended to be a map for an actual journey, despite his remarkable geographical knowledge of both the main route and the alternatives he sometimes indicated. Clearly, too, it was not produced as a master itinerary for merchants or pilgrims to consult before setting out. Instead, Matthew Paris compiled it for his fellow monks at St Albans, barred by the rules of the order from leaving the monastery, to encourage them in their prayers and meditations and as an aid to their *surrogate* travel to Rome and the Holy Land. Matthew Paris's itinerary map was for spiritual, not physical, pilgrimage.

Nothing resembling Matthew Paris's graphic itinerary is known from the 400 years following Paris's work. In 1675, however, John Ogilby published his *Britannia*, a volume containing 100 maps of the post roads of England and Wales arranged in scroll-like columns up and down the page. It may have been pure chance that Ogilby was consulting his acquaintances in London over his work exactly at the time when Robert Cotton's library, which contained a copy of Matthew Paris's itinerary map, was being catalogued. It was not chance, though, that the catalogue – with its reference to a 'geographical guide ... depicted in columns on the page' – reached Robert Hooke, a Fellow of the Royal Society and Ogilby's main advisor for the presentation of the post-road maps. And surely it was not accidental that Ogilby's well-known, and much copied, road maps imitate so closely the strip layout of Matthew Paris's itinerary.

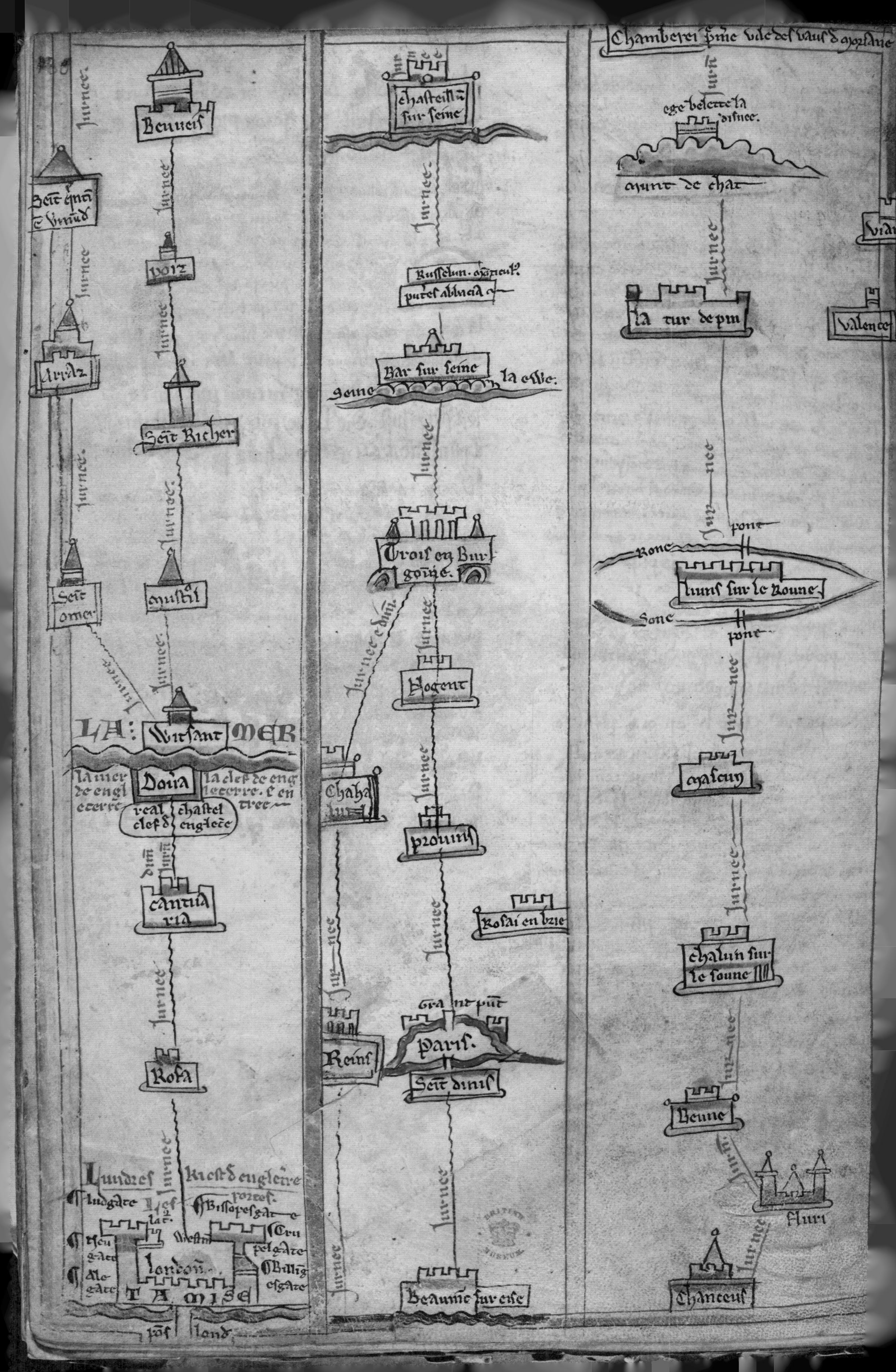

Beuueis
Arraz
Seit Richer
Seit omer
Mustil
LA
Witsant
MER
Doura
la mer de engleterre
la clef de engleterre e l entree
real chastel clef d engleterre
Cantuaria
Rosa
Lundres ki est d engleterre
Ludgate
Bissopesgate
London
TAMISE
Chasteilun sur seine
Bar sur seine
Seine
Trois en Burgoine
Nogent
Prouins
Rosai en brie
Grant punt
Paris
Reins
Seit dinis
Beaumunt sur eise
Munt de chat
La tur de pin
Valence
Rone
Liuns sur le Roune
Sone
pont
Chalun sur le soune
Beune
Fluri
Chanceus

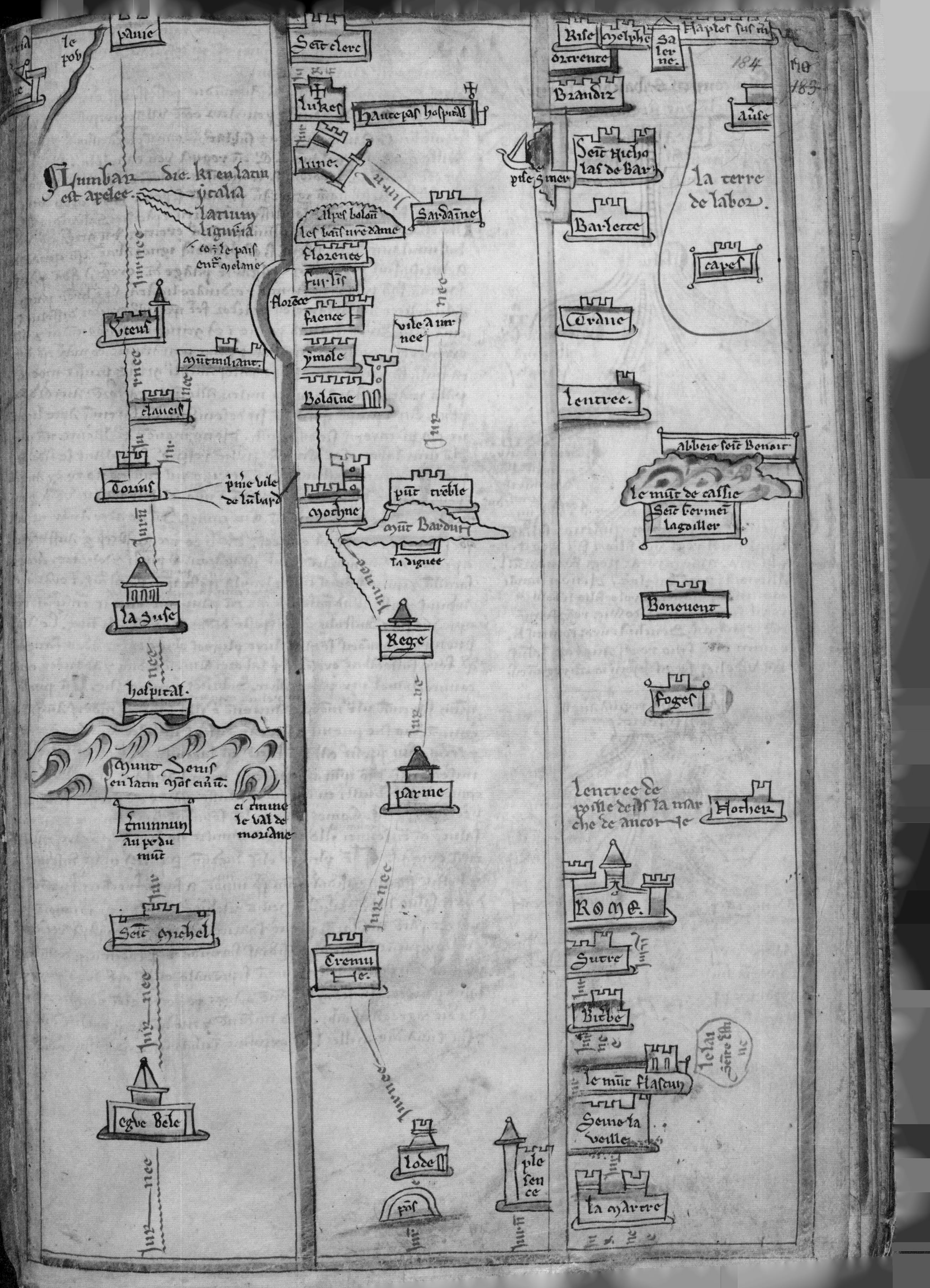

le pou
Pavie
Lumbardie ki en latin est apelee
ytalia
latium
liguria
Verceus
Mutmiliant
Turins
Pme vile de lumbard
la Suse
hospital
Munt Senis en latin mont cinis
ci comence le val de moriane
Seint michel
Seint clere
Lukes
Haute pas hospital
Lune
Sardaine
Alpes bolon
les bains nre dame
Florence
Furlis
Faence
Florece
Ymole
vile a iurnee
Bolonne
Mothne
Punt treble
Munt Bardun
Rege
Parme
Cremune
Lode
Pos
Plesence
iurnee
Rise
Melphe
Salerne
Naples sus mer
Otrente
Brandiz
Seint Nicholas de Bar
Barlette
Trane
Lentree
Ause
La terre de labor
Capes
Abbeie seint Beneit
le munt de cassie
Bonouent
Foges
Lentree de poille deuers la marche de ancone
Rome
Sutre
le munt flascun
Seine la veille
la Martre
184
185

Carte Pisane

The *Carte Pisane*, the oldest surviving example of a portolan chart. It shows much of Europe, with a very detailed coastline of the Mediterranean Sea. Bibliothèque national de France, Paris, France.

PORTOLAN SEA CHART

The *Carte Pisane* is a manuscript sea chart drawn on vellum dating from the end of the thirteenth century and is significant because it is the oldest surviving example of a genre called portolan charts. This chart, because of its commitment to geographical accuracy, represents a first step towards a more scientific approach to cartography and a significant development in navigational science, opening up new possibilities for seafarers. As one of the first of its kind, it set the format and style for navigational charts for the next four hundred years.

The name 'portolan', which actually came into use much later, relates to a chart's purpose: navigating between seaports along the coast of the Mediterranean Sea. Portolan charts differ dramatically from the medieval *mappae mundi* (see page 44) which were intended to offer a religious perspective of the world with little interest in spatial accuracy.

Neither the authorship, the place of production, nor the exact date of the *Carte Pisane* are known, but we do know that it must have been produced sometime after the introduction of the magnetic compass into the Mediterranean area and was probably compiled in Genoa. Its name derives from the fact that it was discovered in Pisa, on the Tuscan coast of Italy, down the coast from Genoa.

Mariners – Phoenicians, Greeks, Romans – had been routinely sailing these waters since the classical period but no navigational charts are known to have survived from these times. The *Carte Pisane* is unlikely to have been the very first of its kind, but whatever chart may have served as the initial pattern, the style and cartographic content were repetitively copied, and this model remained the standard style of chart for 400 years. In that period, Genoa and Venice were leading trading states in the Mediterranean Sea and surrounding waters and they took the lead in marine science. This included ship design and construction as well as the development of their chart-making enterprises.

Thanks to the compass and the aggregated information from years of practical sailing experience, this chart exhibits unprecedented accuracy in depicting these waters. Although the coverage also includes parts of the Atlantic coast and England, information on this area is sketchy at best. However, analysis of later charts shows a gradually improvement in coverage of these coasts through the fourteenth and fifteenth centuries. Primitive magnetic compasses were introduced into Mediterranean Sea navigation sometime around 1200, and the development of the portolan charts – along with the compass upon which they were based – represented a significant improvement in navigational science. Prior to this, sailing was typically limited to coastal itineraries between adjacent ports. With this advancement, longer itineraries, and wintertime and overnight sailing became possible.

Physically, the chart was made by drawing in ink on a prepared animal skin. Two points were selected to be the centres of tangent circles. From each of these two central points sixteen straight lines – rhumb lines – radiated. They indicated the major cardinal points – North, South, East and West – and also intermediate directions. Against this background network, the coastlines of land masses and islands were drawn in. The rhumb lines were used by the mariner to determine the correct compass course to be sailed.

This chart introduced several features not specifically seen on earlier maps which became the standard on sea charts for centuries: the circle which defines the radiating rhumb lines, the scale bars, place names set on the land side perpendicular to the coast. Unlike later examples of this genre, the *Carte Pisane* contains no compass rose. However, the cardinal points are indicated on the circumference of the circles using archaic names: *tramontana* for North, *mex jorno* for South, *levante* for East and *ponente* for West. Another notable aspect of the chart is the complete absence of any detail within the landmasses. The *Carte Pisane* was intended for navigating at sea, and details of the interior would have been seen to be unnecessary.

From this very basic early example, the genre of portolan charts expanded during the next 400 years in terms of place of production and the areas they covered. By the fifteenth century production centres sprung up well beyond the Mediterranean Sea: in Portugal, Netherlands, England and the overseas colonies of these seafaring powers. Before their dominance would be challenged by printed charts in the seventeenth century, the coverage of portolan charts would eventually include nearly the entire globe.

nermandia
bertaigna
rocella
sca maria de sar
bordelos
sco nicolau
gascoma
bolona
sco sabastiano
sco andrea
porto colombo

A CHRISTIAN WORLD VIEW

The Hereford *Mappa Mundi* – a T-O map centred on Jerusalem with east at the top. The continents of Asia (top), Africa (lower right) and Europe (lower left) are separated by the Mediterranean Sea and the Nile and Don rivers. Hereford Cathedral, Hereford, UK.

Hereford Mappa Mundi

As knowledge of Roman cartography withered away in medieval Europe, maps became more than ever a way of presenting a Christian vision of the world in geographic form. A series of increasingly complex *mappae mundi* (maps of the world) culminated in the late thirteenth-century Hereford map, which drew deeply on a combination of classical scholarship, the Bible itself and the works of early Christian writers.

The main maps which emerged in Late Antique and Early Christian Europe were concerned with the portrayal of the topography of the Holy Land, where Christ's story had unfolded, or with a broader attempt to frame maps of the world in a Christian context. The latter type, of which around 1,100 have survived, came to be referred to as *mappae mundi*, from the Latin *mappa* (or 'cloth') and *mundus* (meaning 'world').

Some maps drew on the tradition of seven climatic zones espoused by Ptolemy (see page 16), particularly those illustrating the *Commentary on Scipio* by the fifth-century Roman scholar Macrobius. The majority, however, are of a type known as 'T-O' maps, which portray the world as a flat circular disk, surrounded by an ocean, with the body of the T dividing the land into the continents of Asia (generally the largest, taking up the whole top half of the map), Europe and Africa, separated by great waterways (the Tanais, or Don, the Nile and the Mediterranean) represented by the 'T' itself. Orientation of the maps to the east allowed Jerusalem to take centre stage in most of the *mappae mundi*, a sign of their fundamentally theological rather than geographical concerns. It is no coincidence that the T-shape was later compared to the form of the cross on which Christ was crucified.

Such maps became gradually more complex as time went on. Reflecting this development are the Albi map, one of the oldest, from the eighth century, with its almost stylized land masses; the maps from the *Commentary on the Apocalypse* by the eighth-century Spanish cleric Beatus of Liebana; and the eleventh-century Cotton *Mappa Mundi* which shows signs, in its accurate depiction of coastlines, of being based on a much-copied Roman original.

The Hereford *Mappa Mundi* is the summit of this development, compiled around 1290, one of a group of thirteenth-century maps which are associated with an English-based tradition, which include the largest ever found, from Ebstorf in Germany, which at approximately 3.56 m (11½ ft) square covered thirty sheets of calf vellum, but was destroyed by an Allied bombing raid on Hanover in October 1943. The Hereford map is thus the largest surviving mappa mundi, its surface – about 1.58 m x 1.33 m (5 ft x 4½ ft) packed with diagrammatic representations of towns, biblical scenes, fabulous monsters and 420 place names, including the Indus, Ganges, and Hereford itself, nestled in a corner on the River Wye. That accuracy was not always the strongest point of such maps is amply indicated by the transposition of the labels for Europe and Africa, neatly picked out in bold gilded letters.

It is clear that the map's intention is the presentation of Christian history. At the top, the figure of Christ sits in judgement over virtuous souls, who are led upwards to heaven by the angels, and a cohort of the damned, who are dragged by demons to the burning mouth of hell. Christian topography is further privileged by the setting of the Earthly Paradise – from which Adam and Eve were expelled – at the top (far east) of the map and the disproportionately large drawing of Jerusalem at its very centre.

Yet the map's classical antecedents are also clear, drawing from scholars such as Pliny the Elder, and the fifth-century historian Orosius, as well as encyclopaedists such as Isidore of Seville. The manticores, unicorns and mandrakes which adorn the map, together with the unlikely figure of the Sciapodes, whose single enlarged foot they were said to use as a living parasol, are figures far more of fable than of theological erudition. The author of the map, who has been variously identified as Richard of Laffingham or Richard de Bello, also shows his concern for drawing upon Roman heritage by depicting the Emperor Augustus in the bottom left of the map despatching four surveyors to compile a map of the known world.

In its presentation and concerns, the map is typical of medieval European scholarship as a whole, a careful synthesis of surviving classical knowledge and Biblical tradition to produce a uniquely Christian world-view. Its role was a didactic one, to present a visual encyclopedia of faith in an age where relatively few could read, but all could wonder at such a marvel.

Cresques

The Catalan Atlas of 1375 represents one of the most important milestones in the history of cartography and provides a fascinating glimpse into how the world was seen in the late Middle Ages. It is a beautiful and complex work which offers a spectacle to the eye and an almost inexhaustible range of features, concepts and information to challenge the mind. The generally accepted attribution is to a Jewish resident of Mallorca – variously referred to as Cresques Abraham or Abraham Cresques – and his son, Jehuda. But there must have been several collaborators in the production of this work, given its many dimensions, which include cartography, history, anthropology, astrology, mythology and more.

Because the records from a time over six hundred years ago are scanty, the origins of this mapping project are unclear and there are several competing versions. The most credible involves a commission by Prince John of Aragon, (future King John I) who wanted to offer a map as a gift for his cousin, Charles (future King Charles VI of France). At that time, there was no significant centre of mapmaking on continental Spain, so John looked to Mallorca where a nascent industry in manuscript sea charts had begun a few decades earlier. John selected Cresques Abraham and his son, Jehuda, of Palma de Mallorca for the work, and the price was set at 150 gold florins of Aragon for the father and with sixty Mallorcan pounds for the son. Abraham was not a cartographer, nor pilot, nor astronomer; rather he was a well known illuminator of books. He was a Jew and had good access to cartographic information from Arabic as well as western sources which are fully reflected in the map and the accompanying texts. There is also some uncertainty on the date, but on panel number two is found a perpetual calendar constructed with the reference date of 1375, and there is good information that the map was recorded in Charles's inventory of 1380.

The work has gone through some conservations over the centuries and is now in a somewhat different form than the original. Physically, it now consists of twelve narrow panels – four of which cover cosmographical, navigational data with the remaining eight panels containing the actual map. Laid together, they form a single work of about 65 cm x 300 cm (25½ in. x 9 ft 10 in.).

The four non-cartographic panels together with the numerous text boxes embedded in the map itself indicate that a significant amount of research from medieval and classical sources went into the production of this work. These classical sources include Herodotus, Homer and early versions of the lives of the Christian saints. But there are also reports of western travellers to the East in the thirteenth and fourteenth centuries: the Morrocan Ibn Battuta, the Franciscan friar Odorico da Pordenone and the Venetian merchant Marco Polo.

The map itself is built around a standard portolan chart (see page 40) of the Mediterranean Sea, Black Sea, Baltic Sea and adjacent lands. The western part of the Atlas is likely to have been taken directly from a portolan chart produced in Palma de Mallorca prior to 1375, possibly the chart of Angelino Dulcert (1339), which is the first chart known for certain to have been produced in the Balearics. The scale of this part is almost exactly the same as the portolan charts of the time – about 1:6,000,000 – and there are also some decorative and stylistic similarities.

While the four panels which make up the western part of the map reflect the rigid precision of the portolan chart style, the eastern part (which is illustrated here) is necessarily much more speculative, abstract and philosophical. By 1375 there had been no seaborne exploration east of the Mediterranean by western mariners, so Abraham and his son were limited to what geography they could deduce from the textual reports from the very few travellers to the faraway lands to the East. Even discounting the natural tendency for exaggeration, the sheer duration of these expeditions – twenty-nine years for Inb Battuta and twenty-four years for Marco Polo – would test the memory of any traveller as reporter.

There are a few recognizable geographical features in this eastern part of the map. The Persian Gulf, the Caspian Sea and the South China Sea more or less delineate this vast continent, but there was not much detail available to Cresques on the interior. Nonetheless, the space is filled with beautiful iconography and informative text. We see the Three Wise Men on horseback travelling west through the desert, the Tower of Babel in the extreme southwest corner of the panel, and a caravan travelling east on the Silk Road. But probably the most informative aspect of this eastern part of the map is the many text boxes which provide a good picture of the understanding of the mysterious east in the minds of educated Mediterranean scholars.

Because there are so few surviving fourteenth-century portolan charts, the Catalan Atlas would be considered a very important piece even if the eastern part and the four panels of cosmographical information were excluded. But the inclusion of the East makes this truly one of the most remarkable cartographic works of all time.

CATALAN ATLAS

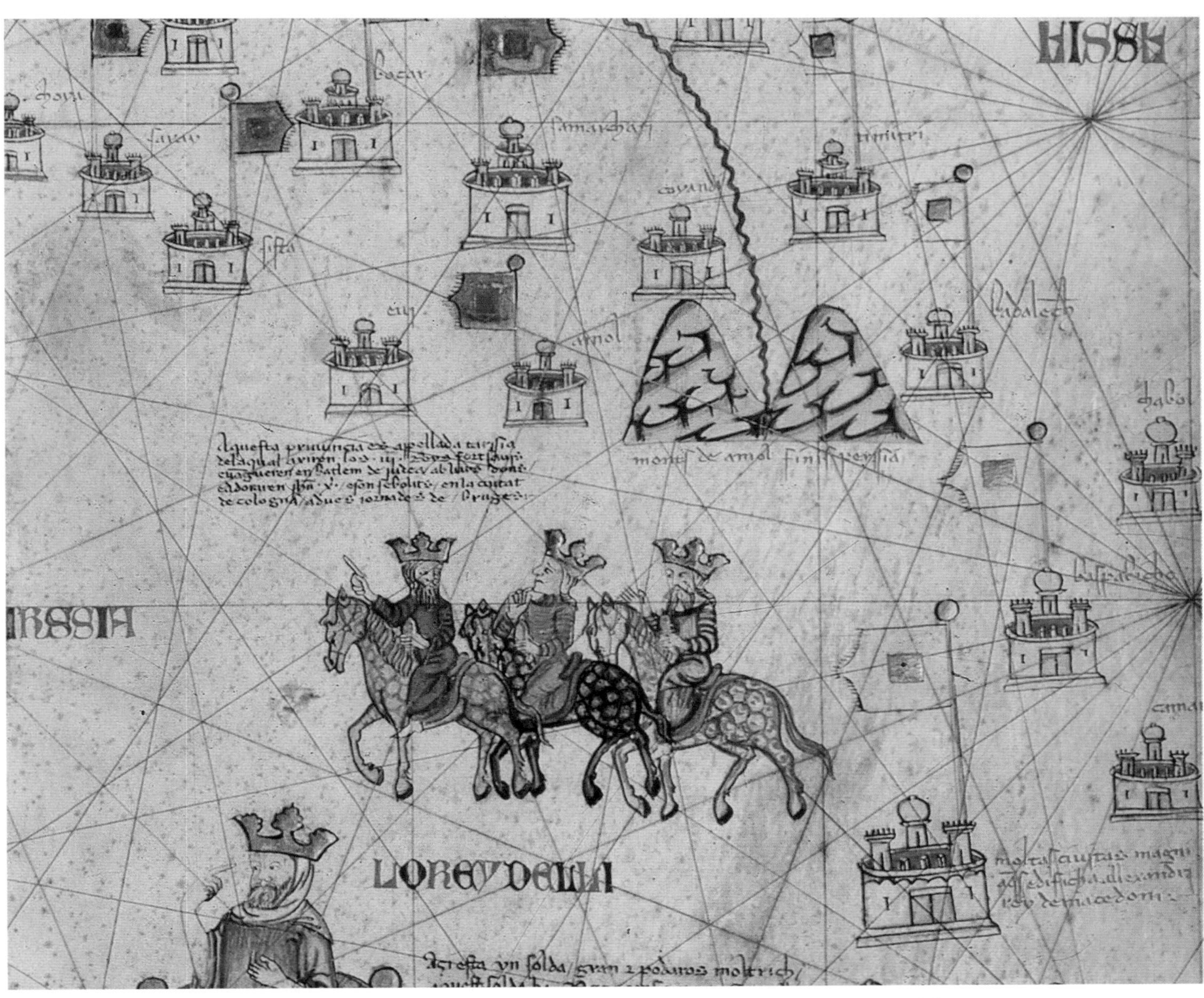

Extracts covering Asia from the Catalan Atlas. The panels shown overleaf stretch from the Caspian Sea and the Persian Gulf in the west to China and the South China Sea in the east. Illustrations depict cities, Bible stories and mythological figures. Bibliothèque national de France, Paris, France.

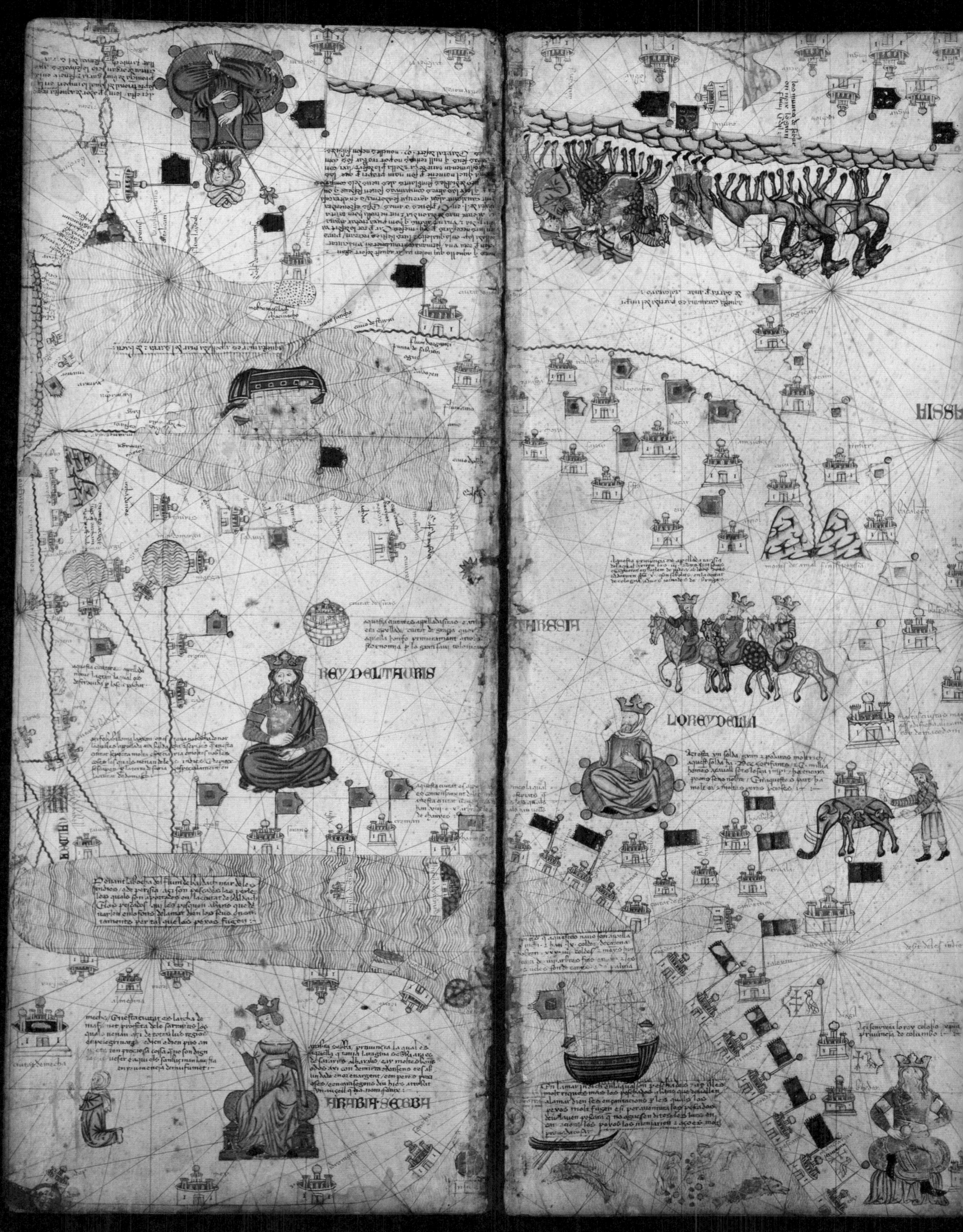
REY DEL TAURIS
ARABIA SEBBA

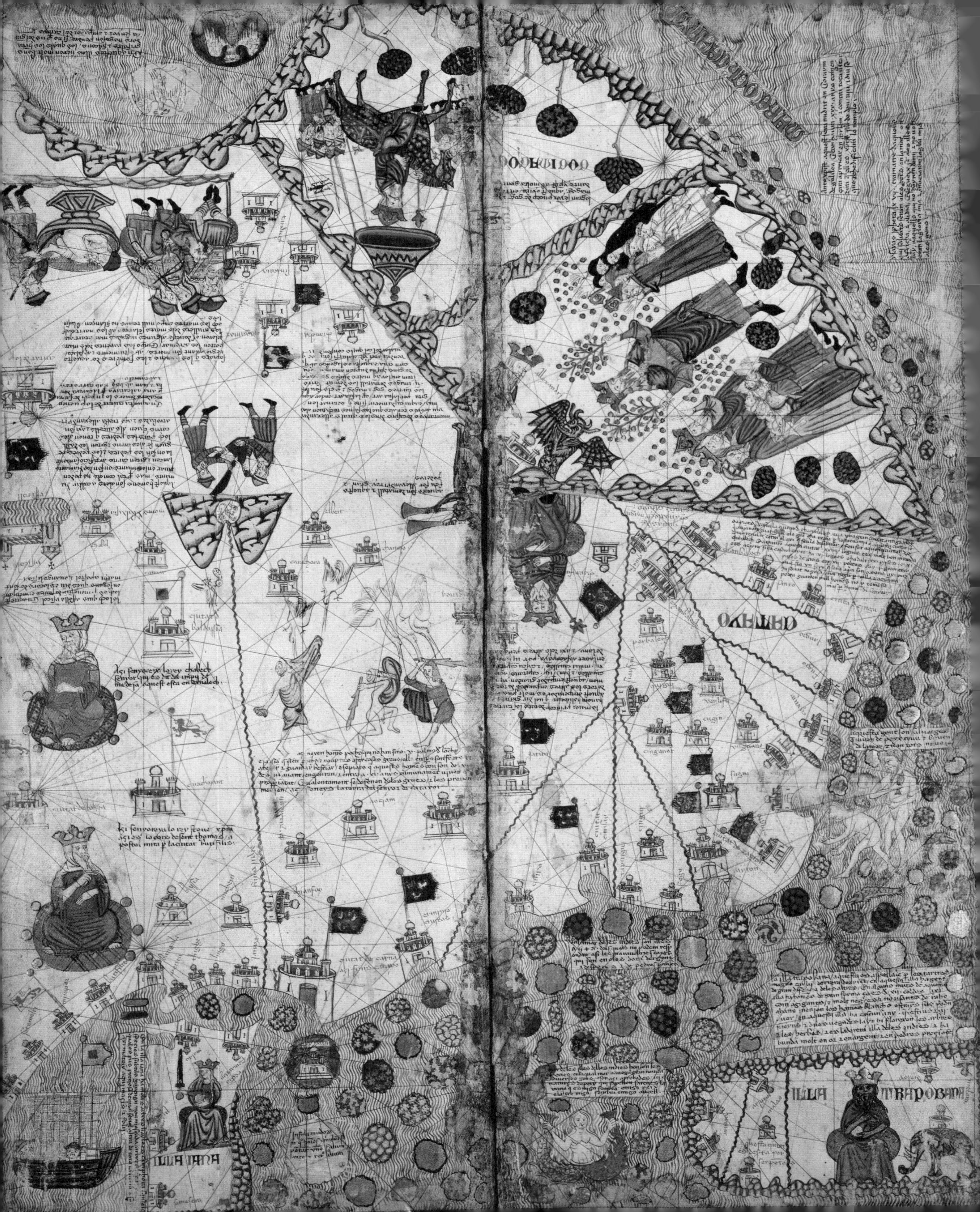

Constantinople

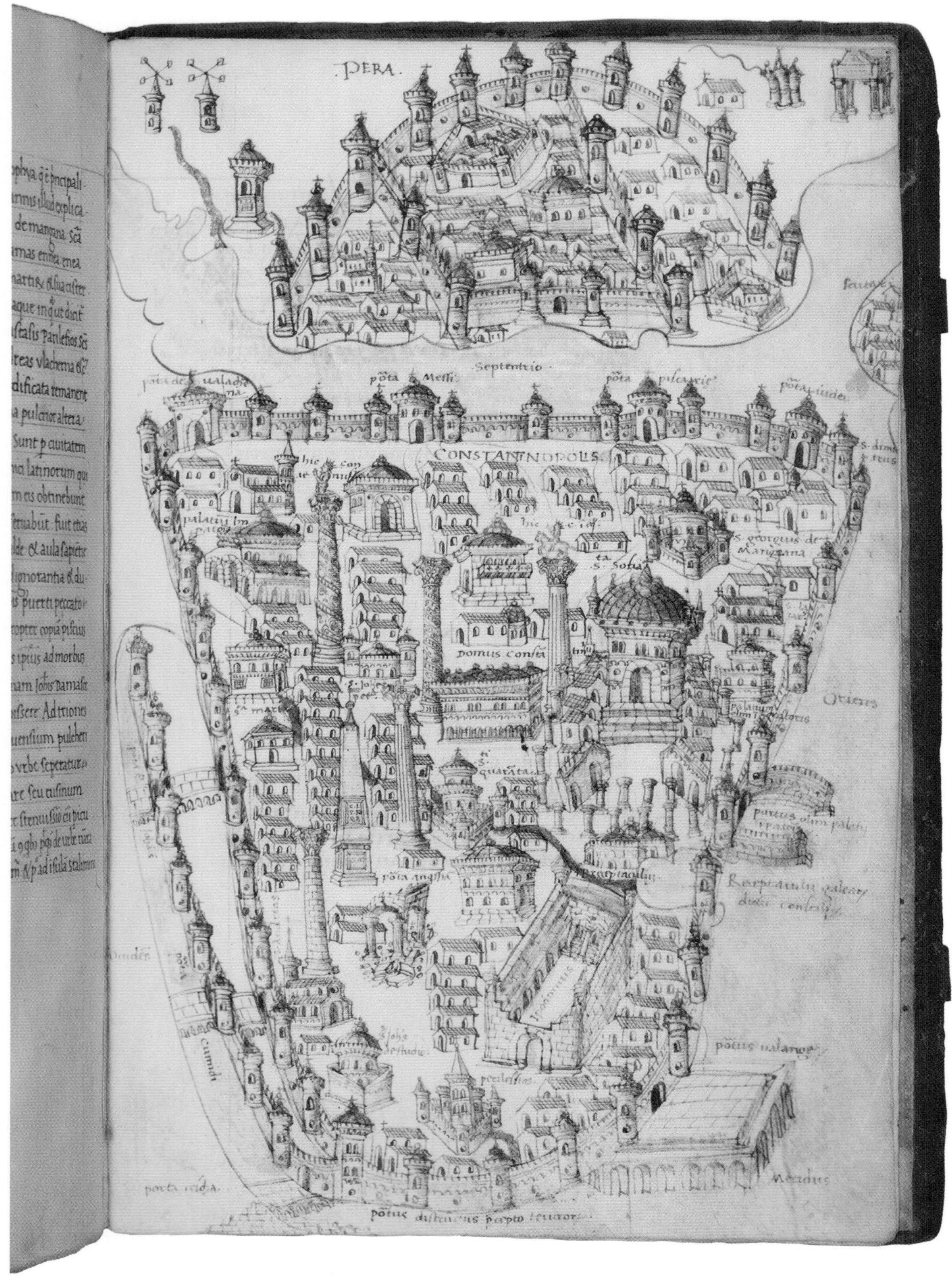

Map of the city of Constantinople (modern-day Istanbul), produced on vellum by Cristoforo Buondelmonti. From a private collection.

CITY MAP

Although world maps had become relatively common in Europe by the later Middle Ages, maps of regions and local areas from that time are extremely rare. One exception is the tradition of *isolarii* or collections of island maps begun by the Florentine nobleman Cristoforo Buondelmonti in the early fifteenth century, among which can be counted his map of Constantinople – an especially fine example.

Large-scale mapping, including the drawing of city maps, had almost died out in Europe after the fall of the Roman Empire. The single notable exception is a series of maps of Jerusalem, beginning with one that accompanied an account of Arculf's pilgrimage to Jerusalem around 670. These became more sophisticated and after the Crusader occupation of the city in 1099 show signs of being drawn on the basis of local knowledge. Local maps began to appear again in Italy in the twelfth century, including one showing an aerial view of Rome. Elsewhere, evidence for them is very sparse, although a 1422 map of Vienna, which was provided with a scale, survives.

The rediscovery of classical learning that fuelled the Renaissance also played a role in this revival in local mapping. The Italian humanist Leon Battista Alberti produced his *Descriptio Urbis Romae* (*Description of the City of Rome*) in the 1440s, outlining a procedure for drawing a scale map of Rome through the use of surveying devices. The appearance of *isolarii* ('books of islands') was an offshoot of the study of classical geographers and interest in the topography of the Mediterranean world of classical antiquity. Aristotle had noted in his *De Mundo* that many geographers dealt with the study of islands as a separate discipline and some of these, such as the works of Dionysius Periegetes, survived into the fourteenth century to inspire works such as the *De Insulis* (*On Islands*), a dictionary of islands compiled by Domenico Silvestri around 1400.

Cristofero Buondelmonti, born into a Florentine noble family, grew up in this milieu – one of his teachers, Domenico Bandini, wrote the *Fons Memorabilium*, which includes sections on various geographical features of the Mediterranean world, including mountains and islands. He travelled to Rhodes in 1415 and spent some years wandering around the Aegean islands, gathering local historical and geographical information. As a priest he was able to travel freely and his service with Jacopo Crispi, the Italian Duke of the Archipelago, further smoothed his way.

The result of those journeys was two books, the first a detailed *Description of Crete*, completed by 1418 and the second a *Book of Islands* finished around 1420, which includes descriptions of seventy-nine locations (principally islands, but with a few other coastal locations, such as Gallipoli, Mount Athos, Athens and Constantinople). The outlines of the islands are reasonably accurate and may have been derived at least in part from nautical charts, although the additional information included which Buondelmonti refers to as 'an illustrated book of the Cyclades and the various other islands surrounding them' doubtless derived from his own travels in the region.

The map of Constantinople is of a different order. It depicts the city in the early fifteenth century, just a few decades before it finally fell to the Ottoman Turks in 1453. By then, the diminutive domain of the Byzantine emperors had shrunk to contain little more than the area inside the city walls, but Buondelmonti portrays it at its height, and not the neglected field of ruins to which in many places it had decayed. He made his own measurements of key buildings, concluding that the great church of Hagia Sophia (which can be seen to the right of the city centre) was 134 cubits long and 120 cubits from one corner to the other, while the Hippodrome, the ancient horse-racing track, he determined to be 356 cubits long and 124 cubits wide. Among the other landmarks he includes are the distinctive Galata Tower in the district of Pera on the Asian side of the Bosphorus and the imposing city walls, which had resisted all attempts to storm them for over a thousand years.

A combination of map and travel guide, Buondelmonti's *Book of Islands* spawned other *isolarii*. By the early seventeenth century there were more than twenty, many of them directly using his work and adding additional maps (such as for Crete, Sicily and other Mediterranean islands). They act as a testament to the growing breadth of mapping in Europe at a time when it was still poised between scholarship grounded in the classical world and a description of the new horizons opened up by the Renaissance and the Age of Exploration.

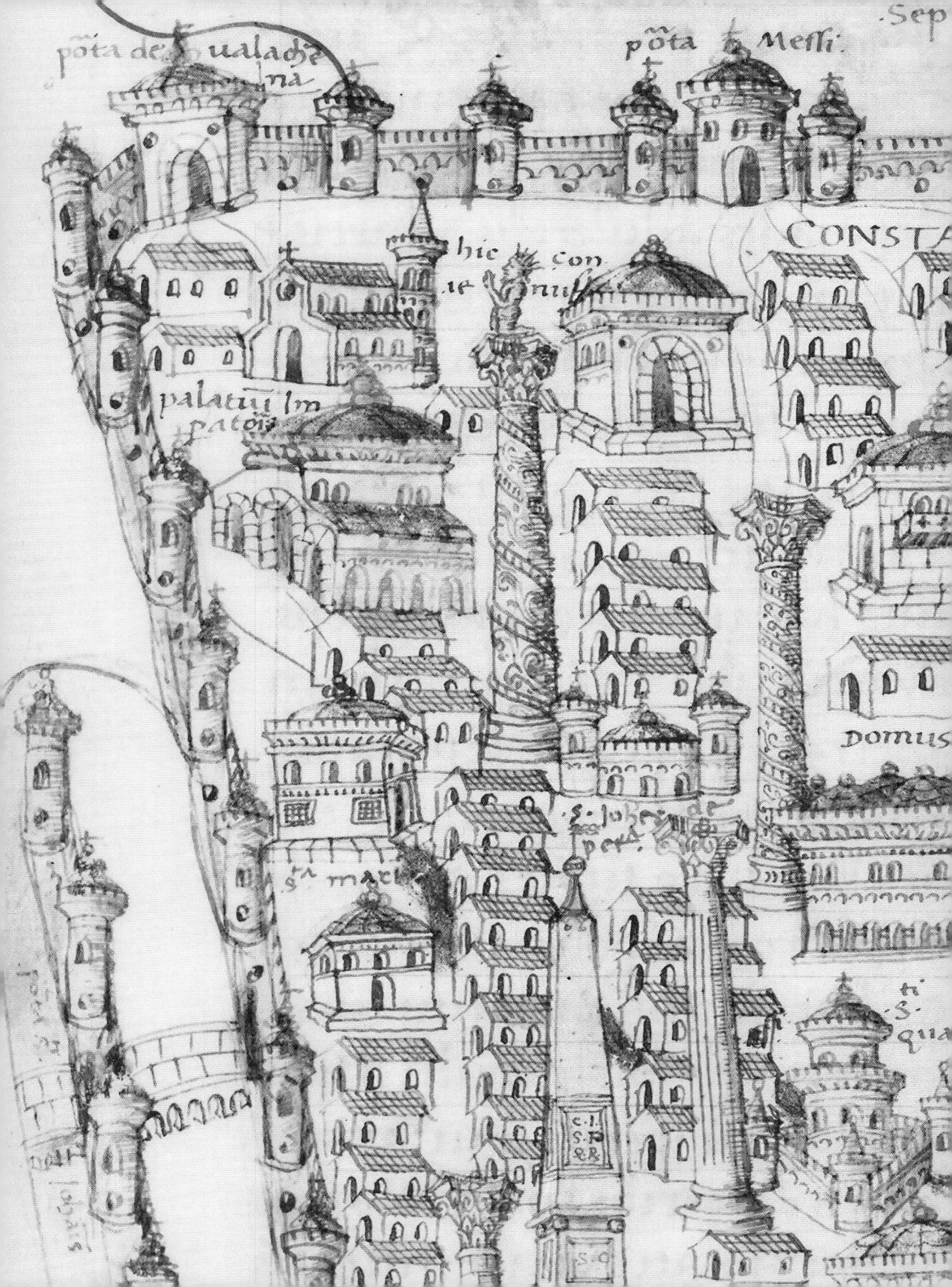

Sep
pôta Messi
CONSTA
palatiũ Im patoꝝ
DOMUS

pota piscarie
pota iudei
S. dimi trius
TINOPOLIS.
S. georgius de Mangana
S. Sofia
S. lazar
Oriens
palatium olim
portus olim palatii

The Vinland Map

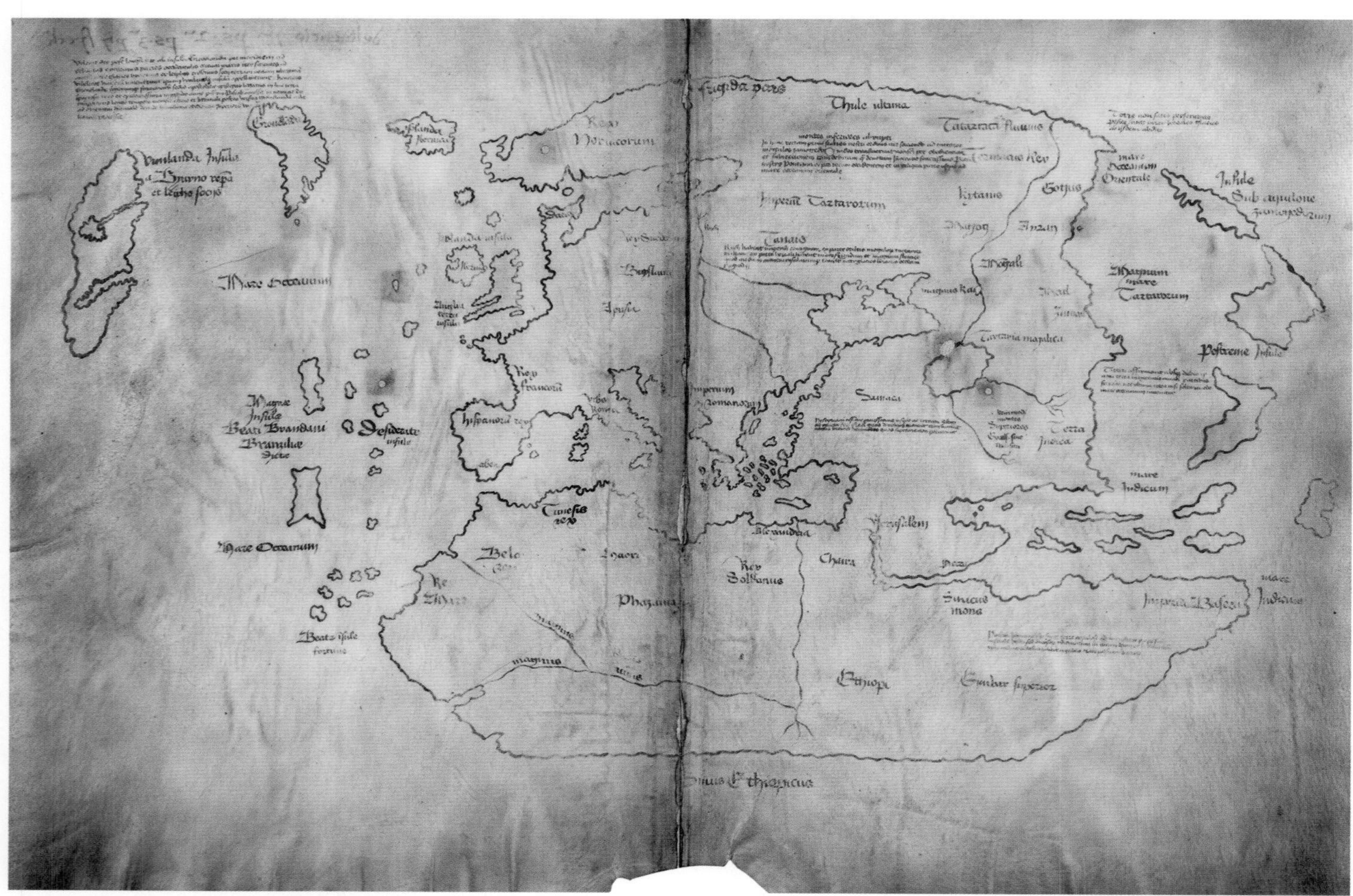

Debate has raged over the authenticity of this map. It was originally claimed that it proved the discovery of North America by the Vikings, decades ahead of Columbus. Beinecke Rare Book and Manuscript Library, New Haven, USA.

A VIKING DISCOVERY?

The announcement in 1965 that a map had been found proving that the Vikings had travelled to North America and which contained the earliest representation of its coastline provoked astonishment. Known as the Vinland Map, the authenticity of this single sheet of vellum was for decades one of the most debated subjects in cartographic history.

Until 1961, and the discovery by archaeologists of a Viking site at L'Anse aux Meadows on Newfoundland, there was no undisputed evidence that the Vikings had reached North America. A series of medieval sagas, the *Saga of Eirik the Red* and the *Saga of the Greenlanders* recounted voyages by Viking explorers setting out from Greenland around 1000. These expeditions under Bjarni Herjolfsson, Leif Eiriksson and others came across lands they named Helluland, Markland and Vinland, where the Vikings established a short-lived colony. But despite efforts by scholars to identify the places named in the sagas or to identify proof that the stories were true, no definitive proof had ever been found.

A considerable stir, therefore, was caused in 1957 when an antiquarian bookseller in New Haven showed a map he had bought in Barcelona to the curator of Medieval and Italian Literature at Yale University Library. The map, drawn on vellum, was bound together with the *Tartar Relation*, a manuscript telling the account of a diplomatic mission in the 1250s by the Franciscan friar Giovanni del Piano Carpini to the court of a Mongol Khan. It appeared to be a fifteenth-century world map which included the north-eastern coastline of America, as well as an astonishingly accurate outline of Greenland. It bore a number of Latin inscriptions, which seemed to lend it authenticity, including one that read 'By God's will, after a long voyage from the island of Greenland to the south toward the most distant remaining parts of the western ocean sea, sailing southward among the ice, the companions Bjarni and Leif Eiriksson discovered a new land, which was extremely fertile and even possessed vines, and they name that island Vinland.'

The map was bought by Yale for $1 million and intense study of the manuscript began, culminating in the public announcement in October 1965 of the map's existence and an exhibition with the Vinland Map as its centrepiece. It was said to date to around 1440, but almost immediately doubts began to surface about its authenticity. The subsequent debate has involved much polemic, centring mainly around the composition of the ink used on the map and whether the Vikings could have known that Greenland was an island, given that this was not proven until the late nineteenth century and there had been no actual circumnavigation of it even in 1965. Other points of contention included the fact that none of the surviving sagas mention Bjarni and Leif Eiriksson taking part in the same voyage, despite the label on the Vinland Map which states that they did.

In 1972 chemical analysis of the ink on the map concluded that it contained titanium anatase, a chemical compound which was not synthesized until the 1920s (though it occurs in traces in nature) and in 1974 the map was declared a forgery at a symposium of scholars on the subject. Although further analysis in 1985 seemed to contradict those findings, a fresh study in 2002 reconfirmed them and showed that the ink on the Vinland Map was different to that on the accompanying *Tartar Relation*, another indication it might have been forged.

Efforts to prove a connection between the Vinland Map and some lost portolan chart or to establish that the Vikings could have known Greenland was an island have proved inconclusive. It remains likely, therefore that the map is one of the most celebrated and accomplished forgeries in cartographic history. Other attempts to create or label artefacts as Viking, such as the Kensington Runestone (an alleged Viking inscription from the 1360s unearthed in Minnesota in 1898) or the Newport Tower in Newport, Rhode Island (claimed to be a Viking construction), were based on hoaxes or misreadings of the evidence. Together with the Vinland Map, these show how great the desire was to find proof of the Viking discovery of North America. Once real archaeological evidence emerged, this imperative subsided, leaving the Vinland map as a cartographic curiosity.

Vinlanda Insula
a Byarno repa
et leipho socijs

Gronlanda

Isolanda Ibernica

Rex Norwegorum

Islanda insula
Ibernica

Anglia terra insula

Rex Suedorum

Rex francorum

Mare Occeanum

Magnae Insulae Beati Brandani Branzilae dictae

Desiderate insule

hispanorum rex

Urbs Roma

Tunesis rex

Mare Occeanum

Beate isule fortune

Rex Marr

magnus

Phazania

cigida pars

Thule ultima

Tatartati fluuius

Rex

Imperiu Tartarorum

Tanais

kytanis

Gotfus

mare Occeanum Orientale

Insule Sub aquilone Zamoyedorum

Aruan

Moghali

magnus Kan

Magnum mare Tartarorum

Tartaria magalica

Postreme Insule

Samaca

Terra Indica

mare Indicum

Alexandria

Jerusalem

Chaira

Rex Soldanus

Sinicus mons

Imperiu Basser

mare Indicum

Ethiopi

Zinzibar superior

nus Ethiopicus

Cantino Planisphere

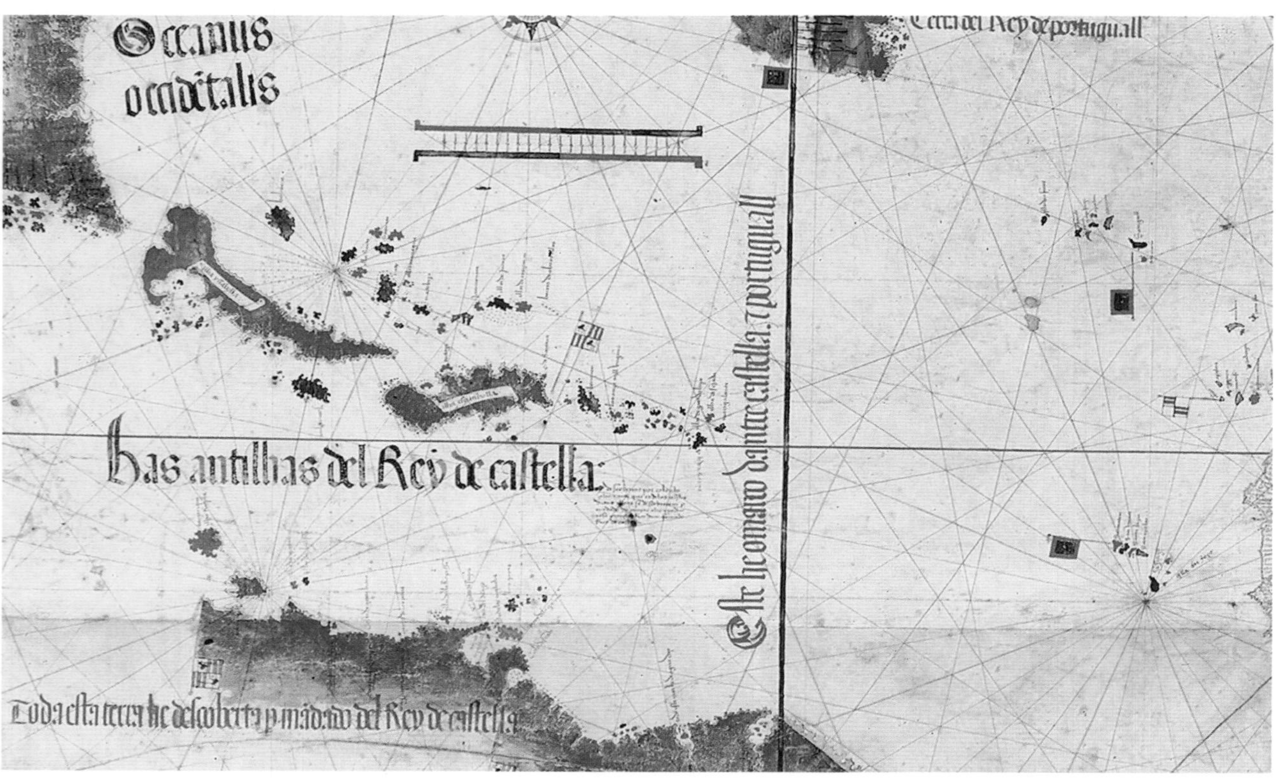

The Cantino Planisphere was one of the first charts to be compiled, at least in part, from measurements of latitude, and was highly influential in the mapping of the New World. Biblioteca Estense, Modena, Italy.

A NEW WORLD CHART

The Cantino map is one of the earliest surviving nautical charts to incorporate observations of latitude. The map incorporates the latest data from Portuguese explorers, and this same Portuguese information, which then appeared in a chart by Nicolo de Caverio made in c.1504, was very influential through its wider dissemination in maps by Martin Waldseemüller (see page 64).

Traditional nautical charts such as the Carte Pisane (see page 40) and the Catalan Atlas (see page 48) were based on compass directions and estimated sailing distances between places. The rediscovery of Ptolemy's *Geography* in the fourteenth and fifteenth centuries (see page 16) generated an appreciation of the value of latitude and longitude – of a grid to specify north-south and east-west location – in mapmaking. It was not until the eighteenth century that an accurate method for determining longitude (east-west location) was determined, but latitude was relatively easy to determine. King João II of Portugal began ordering explorers to determine latitude during their voyages in about 1485, and the Cantino map, which was smuggled out of Portugal by the Italian Alberto Cantino in 1502, incorporates some of those new data. A recent analysis of the chart by Joaquim Gaspar demonstrates that the map incorporates latitude data in some areas, such as southern Africa and a few points in Brazil, while in others, such as the Mediterranean, it uses traditional nautical chart data. The map thus represents an important step forward in the history of mapmaking: it was constructed using a new cartographic technique, though that technique was not employed throughout the map.

The Cantino map or 'planisphere' (a world map which represents the *sphere* of the world in a *plane* surface) is also a step forward in the long and difficult process of understanding the nature of the New World, and specifically its relationship with Asia, which is where Columbus believed he was during his New World explorations. An island on the Cantino map which from its shape and location must be Cuba, is instead labeled *Isabella*, and there is a landmass northwest of this island which looks like it might be intended to represent Florida – but Florida was not discovered until 1513, by Ponce de León. When Columbus was sailing along the coast of eastern Cuba, he was convinced that it was part of the mainland of Asia, and it seems that the cartographer of the Cantino map was mapping according to Columbus's conception. Remarkably though, the anonymous cartographer also incorporates the theory that the New World and Asia were separate. The landmass at the northwestern edge of the map apparently represents part of Asia – but at the eastern edge of the map, the eastern coast of Asia is depicted, but there is no land at the same latitude at the eastern edge of the map to join with that at the western edge.

As curious as this map's portrayal of the New World is, the advanced Portuguese data which the map incorporates was to prove very influential. Much the same data was used in a manuscript world map made by Nicolo de Caverio in about 1504, and this map was used as a model for the world maps printed by Martin Waldseemüller in 1507 (see page 64) and 1516. Waldseemüller's maps, both in their original editions and the various copies which were made of them, diffused this information all across Europe.

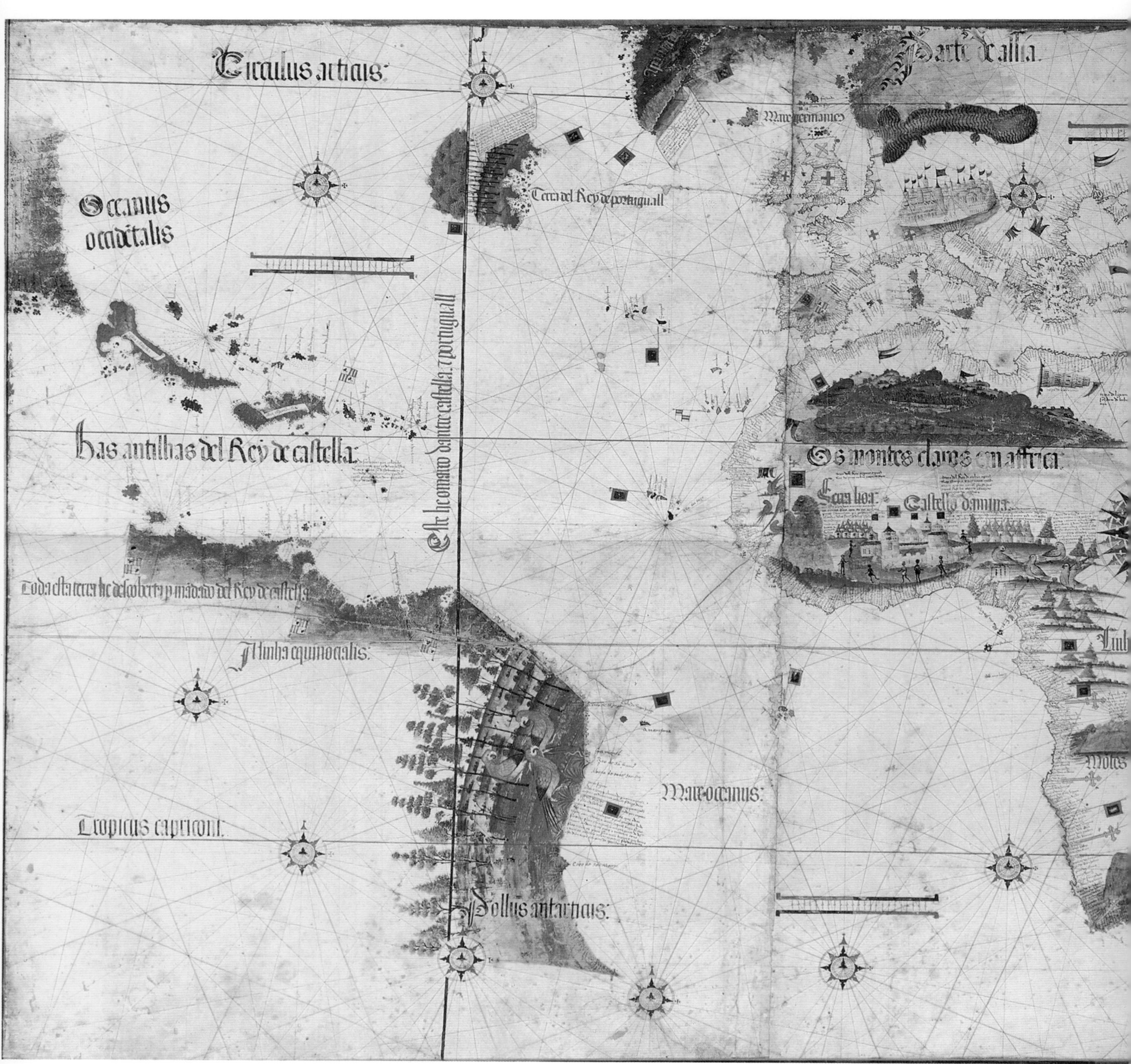
Circulus arcticus:
Parte de assia
Oceanus occidētalis
Terra del Rey de portuguall
Este he o marco dantre castella y portuguall
Has antilhas del Rey de castella:
Os montes claros em affrica
Serra lioa
Castello damina
Toda esta terra he descoberta p mãdado del Rey de castella
Linha equinocialis:
Mare oceanus:
Tropicus capricorni
Pollus antarticus:

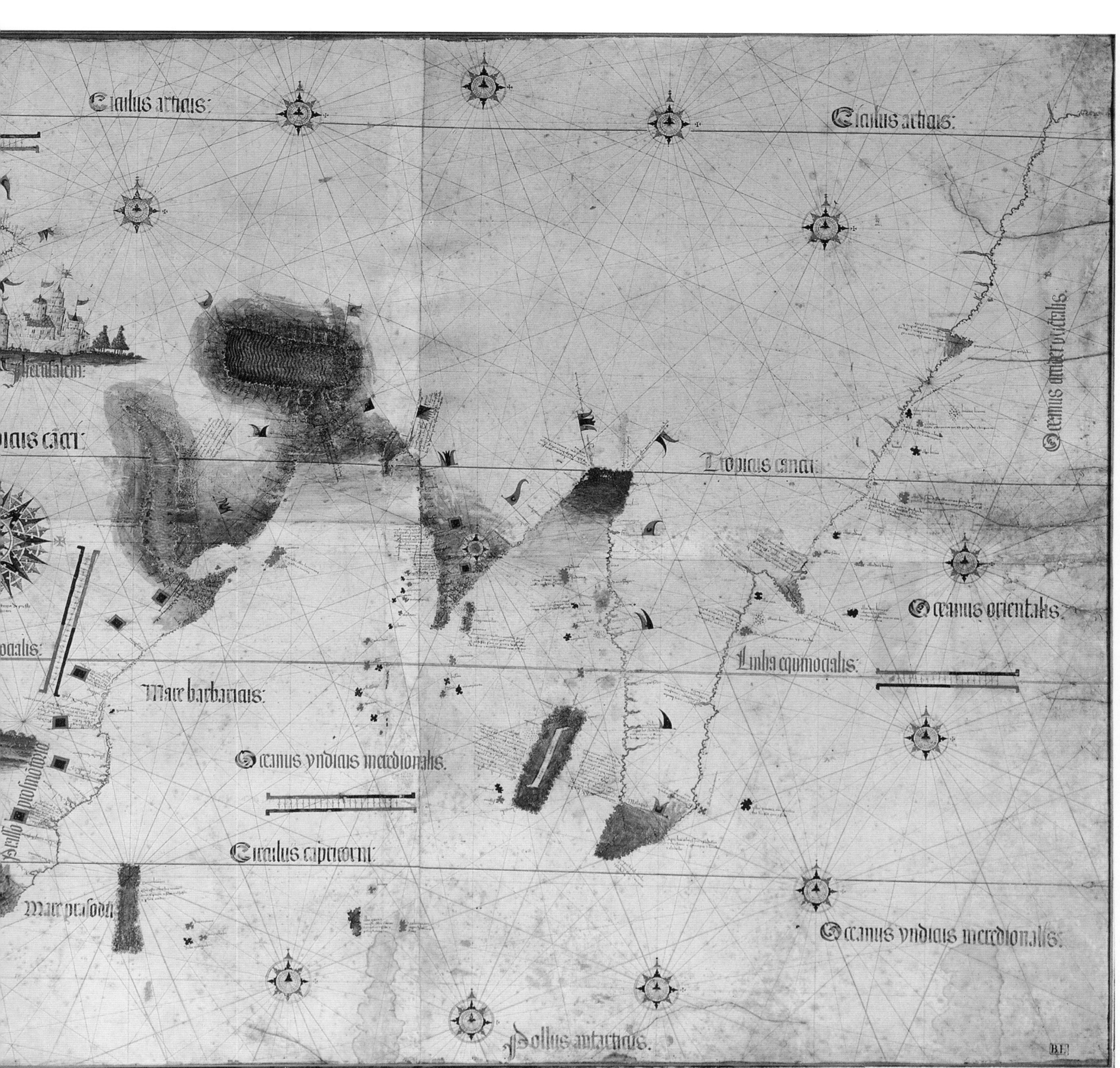
Circulus articus:
Circulus articus:
Tropicus cancri
Oceanus orientalis.
Linha equinocialis:
Mare barbaricus:
Oceanus yndicus meridionalis.
Circulus capricorni
Oceanus yndicus meridionalis:
Polus antarticus.
B.F.

Waldseemüller

Originally carved on twelve woodblocks, this was the first world map to apply the name 'America' to the New World. Library of Congress, Washington D.C., USA.

WORLD MAP

The audacious 1507 world map by Martin Waldseemüller was the first to apply the name 'America' to the New World, and it was also one of the first both to include the New World *and* to show all 360 degrees of the earth's circumference, placing the new discoveries in a global context.

It was in the small town of Saint-Dié, in the Vosges Mountains in what is now eastern France, that 'America', the modern name of a quarter of the earth's land area, was first devised and first applied to a map. The name comes from that of the Italian explorer Amerigo Vespucci (1454–1512), who voyaged to the New World some years after Columbus. It was proposed for the newly discovered lands in the west – initially just to what we know as South America – by the German humanist Matthias Ringmann (1482–1511) and his colleague the cartographer Martin Waldseemüller (*c.*1475–1520). Why did they choose Vespucci's name rather than Columbus's name of the 'Indies' – which Columbus used believing he had travelled to Asia? Simply because Vespucci had been more adept than Columbus at publicizing his voyages. Waldseemüller and his colleagues published the map which bears the name 'America' in 1507 as part of a set of three items: there was a short book entitled *Cosmographiae introductio* (*Introduction to Cosmography*), which includes an explanation of the name 'America'; a printed sheet intended to be pasted onto a ball to make a small globe; and twelve printed sheets which could be assembled into a wall map measuring 128 × 233 cm (4¼ × 7½ ft).

Waldseemüller's map is based on the cartographic system of Claudius Ptolemy (see page 16): he organizes space using Ptolemy's system of latitude and longitude, and uses a projection based on one of Ptolemy's. But he modified Ptolemy's projection so that he could include a much wider sweep from west to east, and to Ptolemy's image of the world he added large tracts of Asia in the east, and the New World in the west. In fact Waldseemüller boldly portrays all 360 degrees of the earth's circumference at a time when much of the world was still unknown to Europe. In particular, Europeans were still to encounter the Pacific Ocean, and yet the cartographer shows water to the west of America, thus indicating that the newly discovered lands were separate from Asia when many, including Columbus, still believed that they were part of that continent.

For much of his depiction of Europe, Africa, and Asia, Waldseemüller relied on a large world map made by Henricus Martellus in about 1491 which was based on Ptolemy and incorporated information from Marco Polo. He based his depiction of the New World on a map made by Nicolo Caverio *c.*1504 which is very similar to the Cantino Planisphere (see page 60), but he gave those lands the name 'America' and showed them surrounded by water based on his interpretation of Amerigo Vespucci's writings. At the top of the map Waldseemüller includes portraits of his two most important (albeit to some extent indirect) sources, Ptolemy and Vespucci, together with small maps of the parts of the world known to each of them.

The 1507 map was printed from twelve exquisitely carved woodblocks; the carving in the map's elaborate borders, with their decorative 'wind-heads', is particularly fine. Waldseemüller claimed to have printed 1000 copies of his 1507 map, so the project was a major effort to disseminate knowledge of the New World and of recent explorations of the coast of southern Africa. Today only one exemplar of the map survives, on permanent display at the Library of Congress in Washington D.C. Despite this, and the fact Waldseemüller himself did not use the name 'America' on his later maps, the 1507 map and its accompanying *Introduction to Cosmography* were influential enough in the crucial period of the early sixteenth century that the name 'America' prevailed against the competing candidates such as 'Terra sanctae crucis' (the Land of the Holy Cross), and 'Terra nova' (the New World).

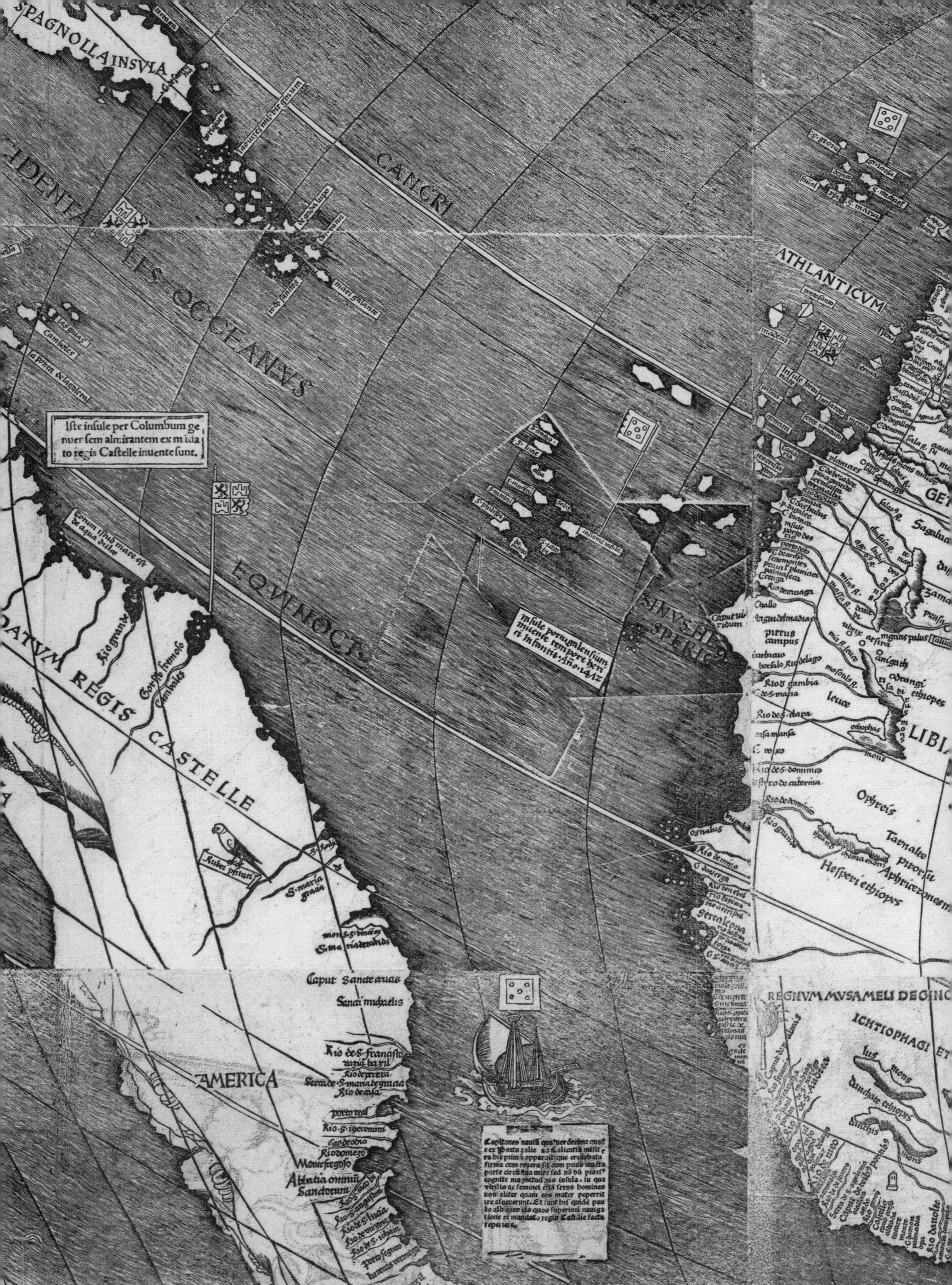

SPAGNOLLA INSULA
CANCRI
ATHLANTICUM
OCCEANUS
Iste insule per Columbum genuensem almirantem ex mandato regis Castelle inuente sunt.
EQUINOCT
Insule portugalensium inuente tempore henrici infantis.
DATUM REGIS CASTELLE
Rio grande
Gorffo fremoso
AMERICA
Abbatia omnium Sanctorum
LIBIA
Ophreis
Hesperi ethiopes
Serraleona
REGNUM MUSAMELI DE GINC
ICHTIOPHAGI

GERMANIA
EUROPA
GALLIA
HISPANIA
ITALIA
GRECIA
paludes Meotides
PONTUS EUXINUS
MARE MEDITERRANEUM
mare adriaticum
ligusticum mare
mare tireni
balearium
Sardonium
Sardinia
Sicilia
Candia
Cyprus
Sirtis magna
Affrica
Egiptus
Saxonia
Schlesia
Polonia
Prussia
Valachia
bulgaria
Austria
hungaria
Syria
IUDEA
ARABIA DESERTA
Cirenaica
marmarica
libia
nigrite ethiopes
ETHIOPIA EGIPTO
ETHIOPIA SUB
Chelom
Calate
meroe insula
Hic sunt magne solitudines z deserte in quibus sunt Leones pardi tigris des elephantes et multorum aliorum animalium.
Vncia est animal seu statura canis parum lõgius a quo si quis vulneratur mures concurrũt & mingunt super eum & inmediate moritur. Draco alatus mire magnitudinis volans os tidit cum cauda elephantes & leones.
Hic castor repitur est animal longum dentibus acutis cauda piscis pedibus posterioribus vt anser anterioribus vt canis ambulat gregatim
Cherisidal siue aliidras hic nascit et est serpens qui terram fumare facit
arange mons
AGANGINE ET HIOPES
REGNUM ORGUENE
AFFRICA
XILICEIS ETHIOPES
REGNUM NUBIE
ELEPHANTOPHAGI
CALCEI ETHIOPES
CINOMIE ERA REGIO
ETHIOPIA
AGIZIMBA
TROGLODICA REGIO
NINOTORA REGIO
SINUSALITES
MARE RUBRUM
Struthophagi ethiopes

Tenochtitlán

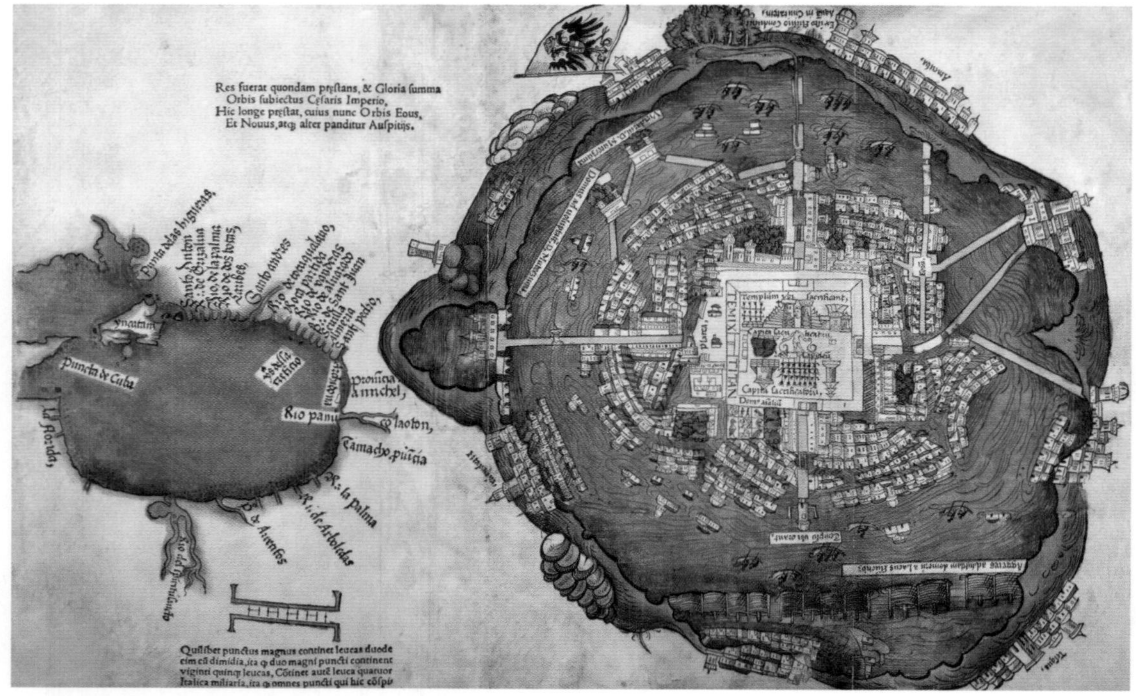

Two representations of the Aztec city of Tenochtitlán, central Mexico, one in the European cartographic style (the 'Cortés Map', left), the other following Aztec traditions (right). Newberry Library, Chicago, USA (left); Bodleian Library, Oxford, UK (right).

AZTEC CITY MAPS

When Spanish conquistadors reached the Aztec cities of Mexico in 1519, they were impressed by the size and complexity of the settlements they found there. The greatest of all was Tenochtitlán, home to around 200,000 people which was recorded, uniquely, in both native and European maps before its destruction by fire after its conquest by the Spanish in 1521.

The Aztec empire was a complex network of city-states owing allegiance to the *tlatoani* (emperor) based in the central Mexican city of Tenochtitlán. The Aztecs had migrated into the Valley of Mexico from a mythical homeland far to the north they called Aztlan, some time in the thirteenth century, and began a major expansion under Itzcoatl (1428–40) when they became the dominant power in the region. Motecuzoma II, the *tlatoani* when the Spanish arrived, was thrown off balance by the newcomers' appearance and allowed the small band led by Hernan Cortés time to acquire local allies who were unhappy with Aztec rule, and to reach Tenochtitlán. Although they were expelled in July 1520 with heavy losses (having murdered Motecuzoma), the Spanish returned the following summer and captured Tenochtitlán after a two-month siege which left most of the city in ruins.

These two maps belong to wholly separate cartographic traditions and present radically different visions of the city before its destruction. The 'Cortés Map' accompanied a letter sent by the conquistador to his master, the Emperor Charles V, in 1524, and was probably engraved and printed in Nuremberg in Germany. A combination of plan and aerial view, it accurately shows the layout of the city, situated on an island on Lake Texcoco, and accessible only by four causeways. In the centre is the large Sacred Precinct, with the main temple (dedicated to the rain god Tlaloc and the god of war, Huitzilopochtli) occupying the prime position. The densely clustered houses are arrayed along a series of canals in a manner which may have been intended to evoke images of Venice in a European audience.

The second map, taken from the *Codex Mendoza*, a traditional manuscript produced in 1541 (some time after the conquest, as many surviving manuscripts are) belongs to a long-standing Aztec tradition of mapping. The conquistador Diaz del Castillo recorded that the Spanish received from Motecuzoma maps drawn on cloth which showed 'all the pueblos we should pass on our way'. These may have been itinerary maps, a form of cartography used by Mesoamerican societies to show the migrations which had led to the foundation of their cities. One such, the *Map de Sigüenza* shows the peregrinations of the Culhua-Mexica (one of the Aztec peoples) from Aztlan to Culhuacan in the Valley of Mexico, the meandering route of their wanderings shown by a line of footprints.

The *Codex Mendoza* map belongs to an allied tradition of cartographic histories of local communities. In common with many Aztec maps it does not separate the depiction of space from that of time, and encapsulates much of its information in hieroglyphic form or through images, rather than through accurate representation of topography. Tenochtitlán is shown at the centre of the map identified by an eagle perched on a stone cactus (a reference to the foundation myth of the city, when the Aztecs were told they could establish their city when they came across such a combination). The map is replete with symbolic references. The four crossed waterways may signify the city's canals, while the skull rack just to the right of the cactus could refer to the rack situated beside Tenochtitlán's main temple. Just to the left of the eagle sits Tenoch, the eponymous founder of the city, surrounded by nine other Aztec clan leaders who may also symbolize the nine districts of the capital. The round shield in front of a bundle of arrows at the base of the cactus may well represent the martial prowess of the Aztecs in conquering the Valley of Mexico.

Although traditional mapping of the type shown here – including maps of the heavens which drew on the sophisticated Mesoamerican understanding of astronomy – went into a decline after 1600 and much of it (having been drawn on maguey fibre or cotton) perished, the production of traditional community maps survived even into modern times.

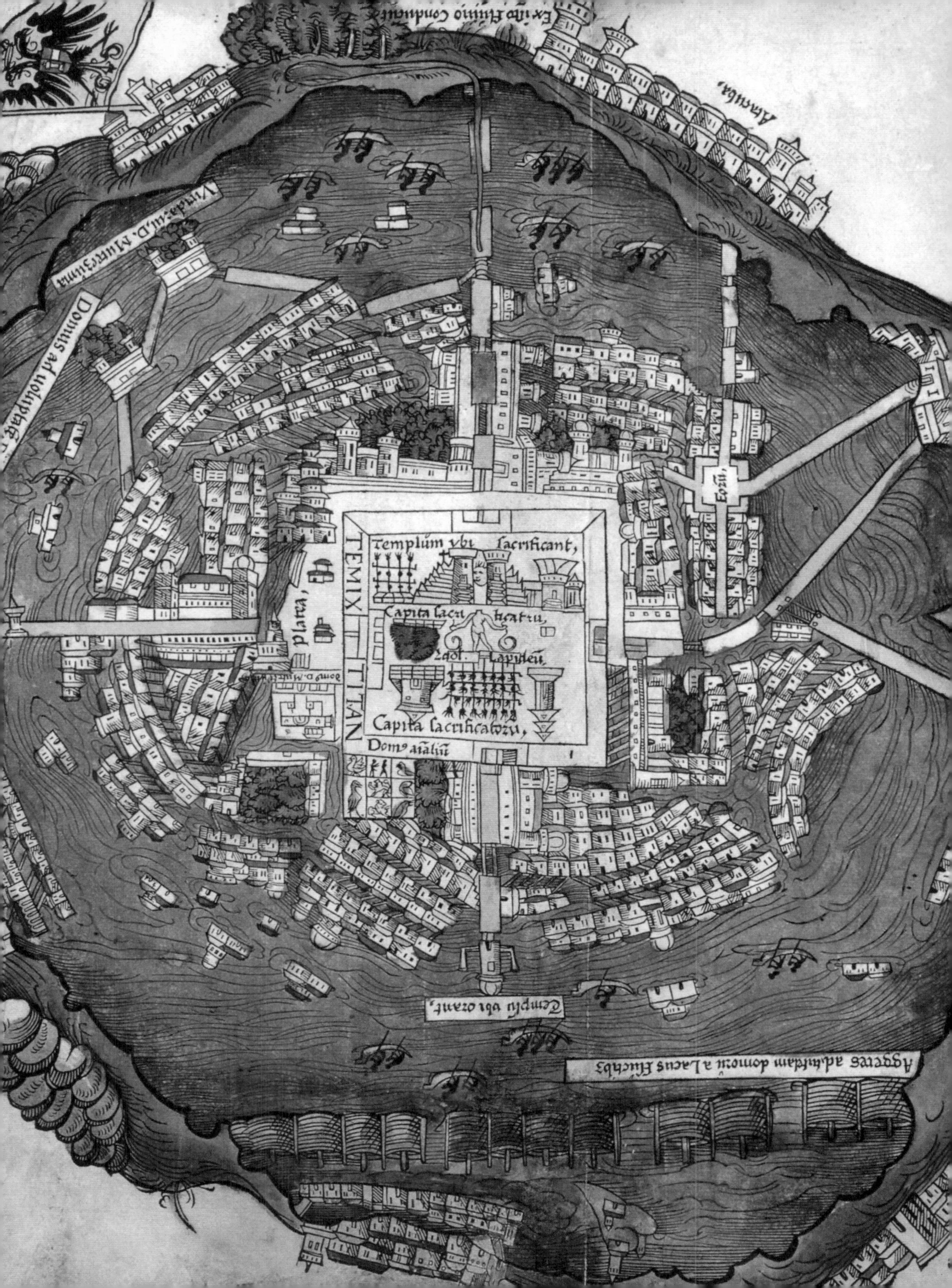

Templum ubi sacrificant,
TEMIXTITAN
plaꝛea,
Capita sacrificatoꝛu,
Dom⁹ aĩaliũ
Templũ ubi orant,
Aggeres ad tutelam domorũ a Lacus fluctibus
Domus ad uoluptat.

Acacitli
quapan
ocelopan
aguexotl
tecineuh
tenuch
xomimitl
xocoyol
tenochtitlan
xiuhcaqui
atototl
colhuacan. pueblo.
tenayucan. pueblo.

Reinel

During the last decade of the fifteenth century and the first two decades of the sixteenth century, the Atlantic Ocean provided the early Iberian explorers a conduit as they tentatively searched its shores for fertile targets for exploitation. By 1535, the estimated date of this Reinel sea chart, the role of the Atlantic had shifted and it had become a veritable highway of commerce between the Iberian ports and the newly established destinations in the Americas.

Bearing neither the author's signature nor date of production, the chart can nevertheless be attributed confidently to the Reinel family, who were pioneers of Portuguese cartography in the first four decades of the sixteenth century. In fact, Pedro, the father, enjoys the honor of being the author of the earliest signed Portuguese chart (c.1504). Pedro, together with his son, Jorge, produced nine charts which have survived. They worked most of their lives for the Portuguese crown and eventually in 1528 earned significant annuities from King João III for their service. But in 1519 an incident occurred which cast doubt on their loyalty to their native country.

Their fellow countryman Ferdinand Magellan, after failing to convince King João to support a project to find a westerly route to the Spice Islands (in modern-day Indonesia), turned to King Charles of Spain for support. An equally important objective of the Magellan project from King Charles's point of view was to demonstrate that the Spice Islands lay in the Spanish sphere of influence as defined by the Treaty of Tordesillas. In 1494 this treaty established a line in the mid-Atlantic with all lands and waters to the west of this line reserved for exploitation by Spain and those to the east reserved for Portugal. At the time, the Pacific Ocean was unknown and no thought was given to position of the line in the eastern hemisphere. The Magellan voyage would be crucial in determining the position of the islands vis-à-vis the line in the Pacific – the so-called Anti-Meridian.

While preparations for the great voyage progressed, Jorge Reinel accepted the offer to come to Seville to make the charts for the voyage, and Pedro is known to have followed Jorge there to assist his son. Specifically, it is claimed that Pedro set the location of the Spice Islands on the world chart his son had made for Magellan on the Spanish side of the Anti-Meridian, even though this was known by the Portuguese to be untrue and of course emphatically against Portuguese interest by inviting Spanish interference in an established Portuguese trading area. Nonetheless, father and son returned to Lisbon and resumed making charts for Portuguese voyagers until 1540. Because the careers of Pedro and Jorge were so intertwined, it is impossible to know if this particular chart is the work of father or son, or possibly even a joint effort. However, based on the calligraphy, most scholars favor the father, Pedro, as the more likely author.

The chart is in the classic Portuguese style – bold font for the nomenclature, twelve beautiful compass roses, thirty-seven flags, mostly Portuguese and Spanish, and strong coloured outlines on major coastlines. In addition to representations of Lisbon and the major Portuguese establishment of El Mina on the West African coast, there are also detailed vignettes of Venice, Genoa, Cairo and Jerusalem. The east-west coverage is from the Red Sea to the 'island' of Yucatan, and the latitude scale indicates south-north coverage from the Tropic of Capricorn nearly to the Arctic Circle. The latitude scale could have been placed at any convenient place on the chart, but interestingly its position is very near the Line of Demarcation established in 1494 at Tordesillas. There is a row of nail holes along the eastern edge of the chart, indicating that it had been nailed to a dowel, which was the typical manner of handling charts aboard ship. However, this chart is so large – 122 x 81 cm (4 ft x 2½ ft) – and so elaborately decorated, that it may have been intended primarily for use on land, possibly for planning trans-Atlantic voyages.

In 1534, King João established the system of colonization of Brazil called *capitanias hereditárias*, a kind of land grant to exploit the interior of the country. This system dramatically increased trans-Atlantic traffic as colonists and slaves travelled west from Europe and Africa, and commodities – mainly sugar and brazilwood – were sent back to Europe. The Reinel chart undoubtedly played a part in this seaborne traffic.

The earliest available information on this chart dates from 1928 when a dealer in Florence named Otto Lange listed it. After changing hands one or more times, the chart was acquired by the National Maritime Museum in 1934 where it remains to this day. As one of only nine charts by the Reinels, and one of only about two dozen Portuguese charts of this period, this beautiful chart of the Atlantic Ocean must be considered a rare and precious document of this time of European navigation to the Americas.

TRANS-ATLANTIC INFLUENCE

Early sea chart of the Atlantic Ocean in the classic Portuguese style, by Pedro or Jorge Reinel. National Maritime Museum, Greenwich, London, UK.

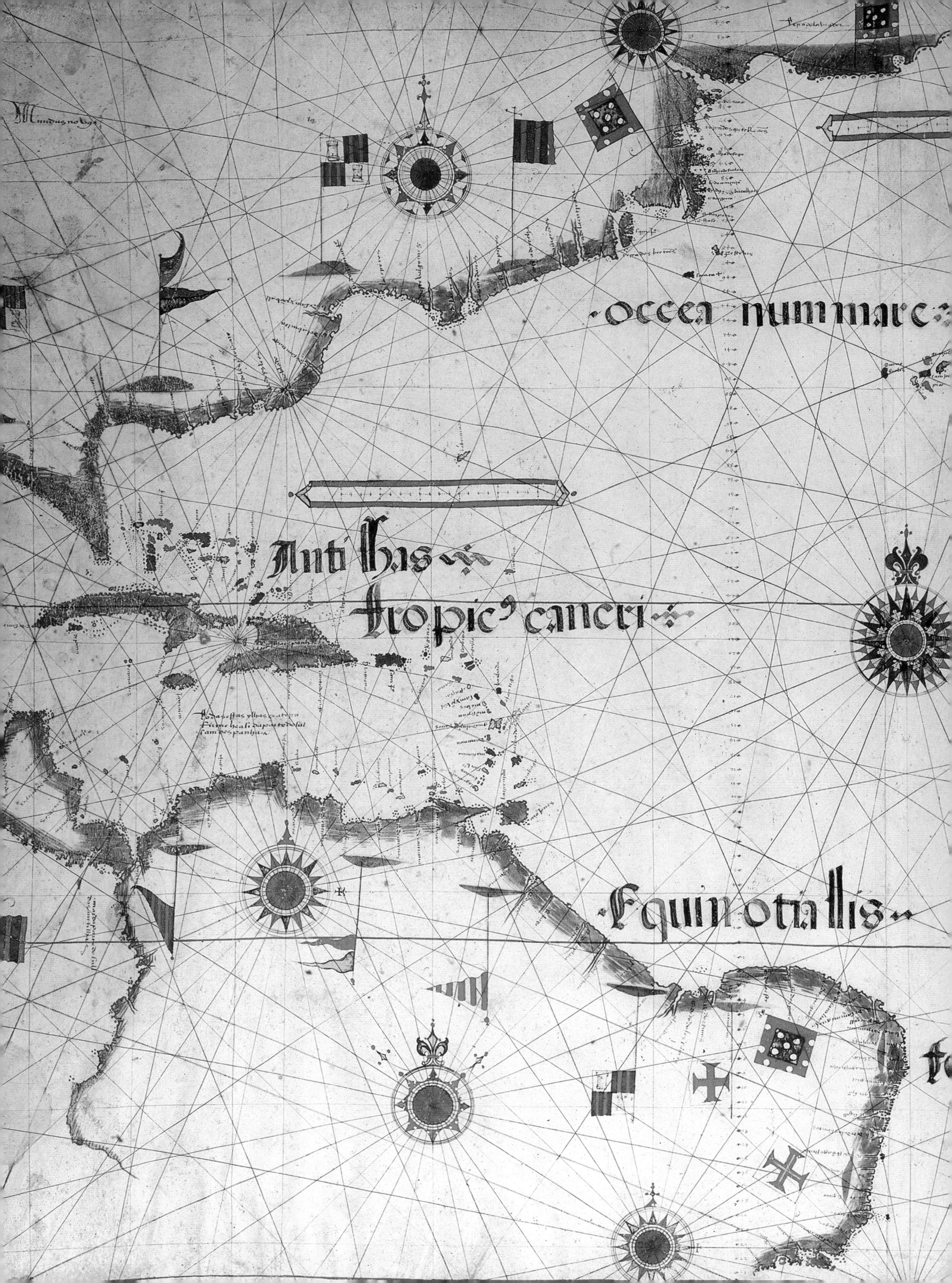

Mundus novus
Occea num mare
Antilhas
Tropic' cancri
Equinoctialis

Jhūs
Europa
Partes dafrica
tropic' capricorni

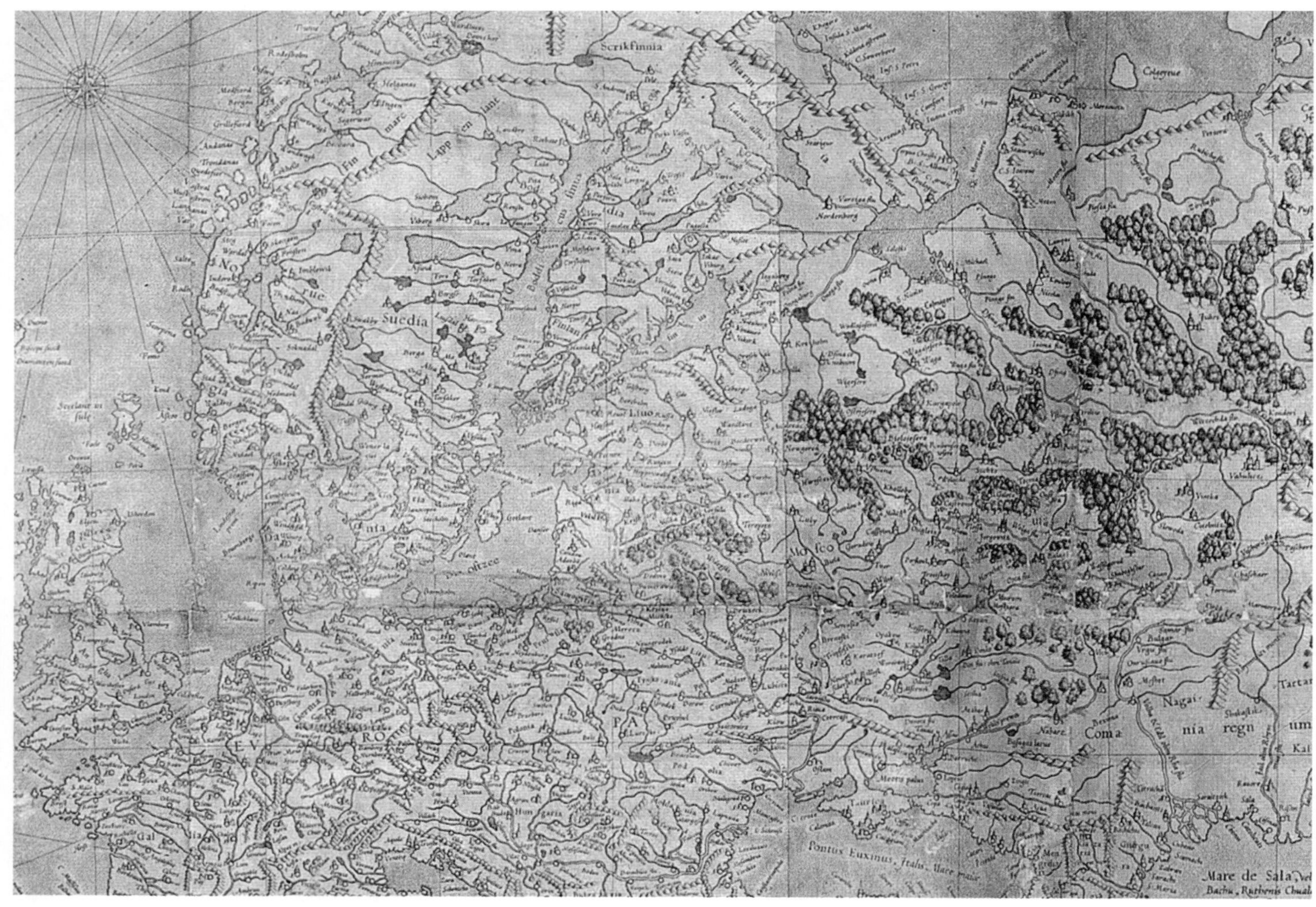

Mercator's world map, compiled on the map projection which bears his name – possibly the most influential projection of all time. Bibliothèque Nationale de Cartes et Plans, Paris, France.

Mercator

It is dangerous to claim that any single map should be the greatest cartographic breakthrough of all time. But Mercator's *Nova et aucta orbis terrae descriptio ad usum navigantium emendatè accommodata* was the Holy Grail of cartography, the map the world had been searching for. Its value lay in the idea it conveyed rather than the geography it described, but that idea was so fundamental to the way we relate to our planet that it's still being used today. Published at the height of the Age of Discovery by a cobbler's son from the Low Countries, this huge wall map revolutionized navigation and cartography.

Mercator was in his late fifties when his greatest single work was published in his hometown of Duisburg on the Rhine. He had already produced the most remarkable terrestrial globe of the sixteenth century, and his numerous cartographic labours included a wall map of Europe and another of the British Isles. Like so many of his contemporaries, his talents spanned many fields. He was a skilled instrument maker, he was the author of a world history and he wrote the first manual of italic handwriting to be published north of the Alps. It was the breadth of his learning and intensity of his geographical curiosity which provided the insights to think laterally and produce his world map of 1569.

Navigators on the high seas had always struggled to harmonize the compass bearings taken from the decks of their ships, with those suggested by the sea charts they were using. Many different map projections existed, but there were none in use that allowed a seafarer to lay a straight edge on a map, read off a bearing, and then apply that bearing to a compass which would take the ship to the required spot. The nub of the problem was the mathematical relationship between a sphere and a plane – between the earth and a map. Nobody had devised a map projection which would allow the spherical globe to be 'flattened' in such a way that a required compass bearing would be the same on both. Mercator used his world map of 1569 to demonstrate how he'd solved the problem.

At first glance, the 1569 map appears misleading. Lands closer to the pole are 'stretched' horizontally, making them far larger than their true area. North America is a bloated monster occupying nearly half of the northern hemisphere and the landmasses at each pole (it was a common belief at the time that there were islands filling the Arctic) reached across the full width of the map. The stretching is caused by the straightening of lines of longitude, which was necessary to make the projection work. To cover himself, Mercator explained in the map's title that this was a *New and more complete representation of the terrestrial globe properly adapted for use in navigation.* The map was not intended to convey the relative areas of the world's landmasses, but to help navigators find their way across oceans – and to make it possible for cartographers to transfer compass-derived coordinates of new-found lands onto a universal projection. Mercator referred to his breakthrough as 'the squaring of the circle'.

Whilst it became an invaluable aid to navigation, Mercator's projection was misunderstood and misused. In demonstrating his new navigational tool, Mercator had placed Europe at the top and centre of his map, a decision which reflected both the market for the map and the fact that, at the time, Europe was more accurately surveyed than any other continent. Aesthetically, the map worked too, with the exaggerated bulk of north America on the left, balancing the combined mass of Europe and Asia on the right. Unfortunately, Mercator's cartographic demonstration became the projection of choice for publishers whose interests were unrelated to navigation. Maps showing Europe at the centre of the world, and the northern hemisphere 'above' the southern, were a gift to colonial administrations, particularly the British, whose overseas possessions were exaggerated in their relative size on this projection. And with its neat, rectangular format, Mercator's world map was ideally suited to the spaces on schoolroom walls. Generations of young people absorbed the projection as if it was the only spatial truth. Publishers played to demand by reproducing a world view which consumers expected, rather than take commercial risks by introducing, for example, unfamiliar equal-area projections which had no navigational utility but did represent landmasses in their true proportions. A wider understanding of projections among consumers eventually led to the use in atlases and wallmaps of a variety of projections. Today, derivatives of Mercator's projection are used by leading mapping agencies, including the Ordnance Survey, the UK Hydrographic Office and NASA.

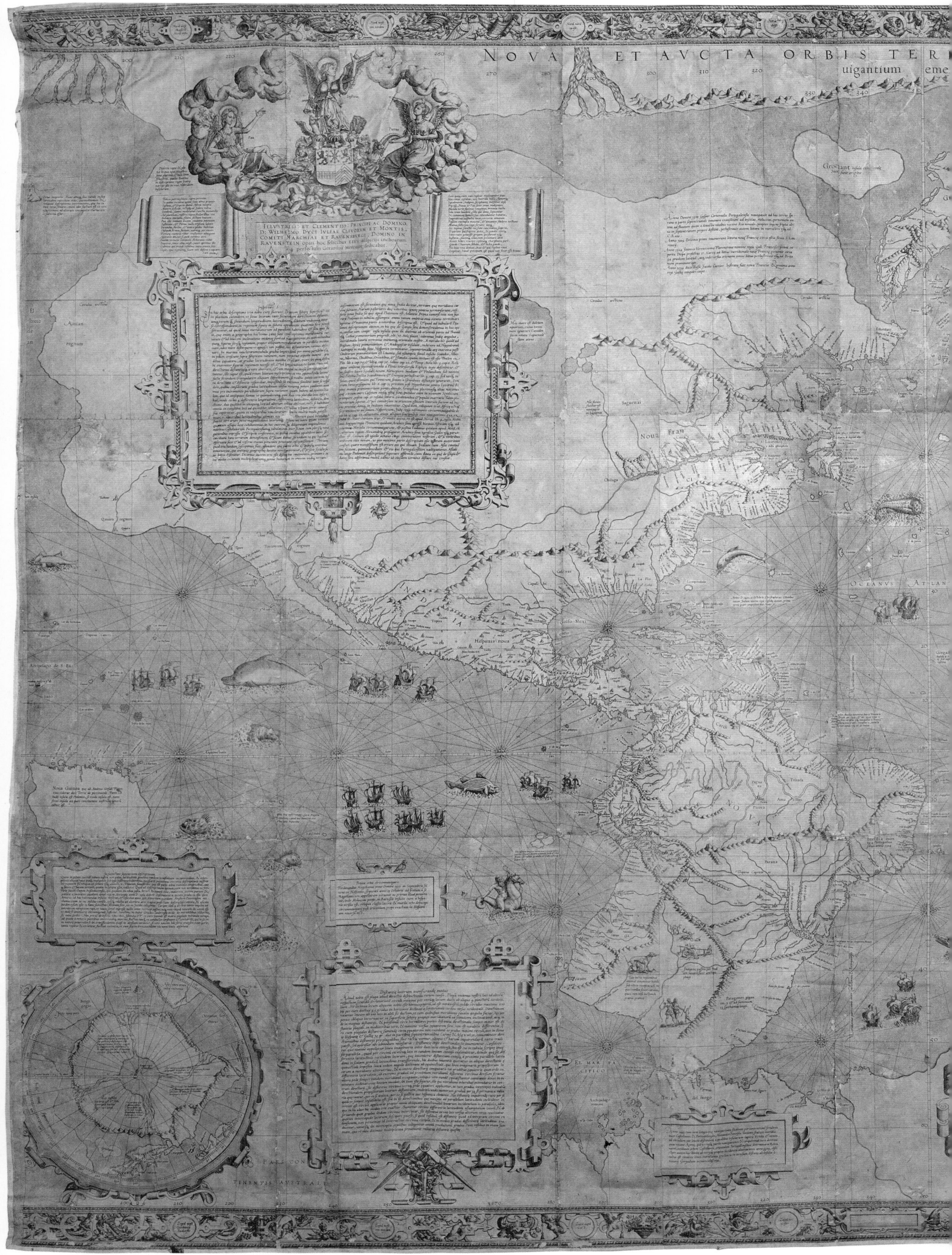

NOVA ET AVCTA ORBIS TER
uigantium eme
ILLVSTRISS: ET CLEMENTISS: PRINCIPI AC DOMINO, D: WILHELMO DVCI IVLIAE CLIVORVM ET MONTIS, COMITI MARCHIAE ET RAVENSBVRGI, DOMINO IN RAVENSTEIN opus hoc felicibus EIVS auspicijs inchoatum perfectum Gerardus Mercator dedicabat.
Groclant
Saguenai
Noua Fran
Chilaga
Golfo de Mexi
Oceanvs Atlan
Patagones gigan
Terra del fuego
Nova Guinea

DESCRIPTIO AD VSVM NA
accommodata
Organum directorium
PARS CON TINENTIS AV STRALIS

Waghenaer

Although recognized by few as a pillar of modern civilization, the nautical chart is what makes possible the movement of billions of tons of goods each year from port to port and nation to nation. This chart, one of a series made by Lucas Janszoon Waghenaer in 1584, set the standard for the development of nautical charts as we know them today.

Without nautical charts, the oil, automobiles, electronics, clothing, and other items of modern commerce could not be safely transported by the huge tankers, container ships, and bulk carriers of the modern maritime world. In time of war, they are indispensable aids to the fighting fleets of the world for defining tactical operating areas and for moving men and materiel. The concept of the nautical chart, however, did not even exist until late in the sixteenth century. Although portolan charts and textual pilot guides existed, they bore little resemblance to the modern nautical chart. Very little, if any, depth information or navigational information was included on these early charts other than the configuration of the coast and compass roses.

It was not until 1584 that the first atlas of modern sea charts was published, when an obscure Dutch mariner, Lucas Janszoon Waghenaer from the Dutch seaport of Enkhuizen, produced *Spieghel der Zeevaerdt*, an atlas of forty-four nautical charts extending from Scandinavia to Portugal. According to his biographer, Cornelis Koeman, the charts 'form a miraculous "first" which ... must have been a revelation for the seaman....'. These charts differed fundamentally from the portolan charts as they included depths, shoreline and terrestrial landmarks such as church steeples and castles, aids to navigation including buoys and daymarks, stippled areas for sand and mud shoals, crosses for known rocks, scale bars, compass roses, numerous coastal views, and various landmarks which could be used for navigational purposes. Like the portolan charts, they were beautifully drawn and engraved. The charts included fish, sea monsters, various types of cargo and naval vessels, cartouches, and artfully rendered wavelets in oceanic areas in addition to the basic navigational information.

Waghenaer was born in 1533 and sailed for nearly three decades, retiring from the sea in 1579. He described himself as 'a simple citizen and pilot at sea'. However, this is certainly an understatement as he had a vision to produce a new type of map for the mariner. He had spent his life sailing in northern European waters – an area of low-lying coastal areas, shifting shoals, large tidal ranges, strong currents, and much foul weather. Perhaps copious notes from this experience led to his book of charts which was published in 1584. He chose a master engraver, Van Deutecom, to prepare the plates for the charts. Much of their elegant appearance can be attributed to Van Deutecom's work. Waghenaer's charts were immediately embraced by much of the maritime community which made him moderately wealthy. He published a number of editions of his atlas and also published a traditional pilot book , *Thresoor der Zeevaerdt* in 1592, and another seaman's guide *Enchuyser Zeecaertboeck* in 1596, which incorporated both written text and charts. However, towards the end of his life he was beset by financial difficulties and was given a modest government pension until his death in 1606. He left little or no estate as the municipal authorities of Enkhuizen petitioned the Netherlands States General for continuation of his pension for one year 'in view of the poverty and miserable state of the widow of Lucas Jansz. Waghenaer...'.

Since the time of Waghenaer, there have been few philosophical changes to the nautical chart. Depth and shoreline are still the foundational information of the chart. Superimposed on these are various landmarks and aids to navigation such as buoys, lighthouses, and daymarks. Scales, compass roses, anchorage areas, stippling for sandy shoals, and crosses denoting rocks can still be found. What has changed is the technology behind the production of the modern nautical chart. The tools of modern sea surveyors could not have been dreamed of by Waghenaer and his colleagues – multi-beam sonars which cover the entire seabed of charted areas; sidescan sonars which give images akin to aerial photography; navigational systems which attain decimetre-level positioning on the surface of the earth for the location of individual soundings; and tidal observations which are accurate to the centimetre, are the capabilities of the modern hydrographer. Draftsman and engraver have given way to modern computer geographic information systems in producing the modern nautical chart and the computer screen and electronic chart are replacing the paper chart of the last four centuries. However, Waghenaer's spirit continues – for his invention of the nautical chart, he should be remembered as a true benefactor of mankind.

NAUTICAL CHARTS

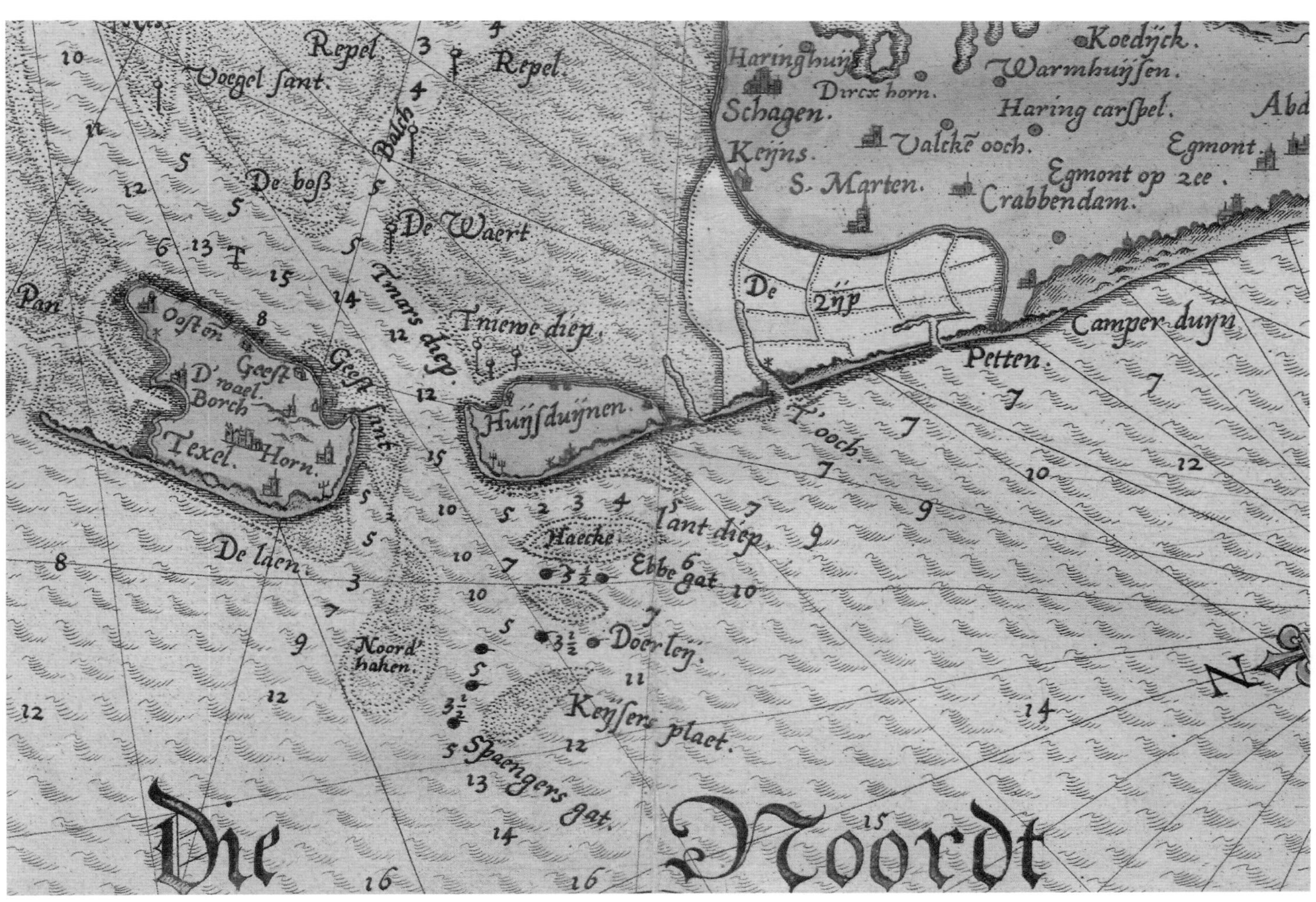

Plate of the Zuyderzee from Waghenaer's beautiful and innovative sea atlas *Spieghel der zeevaerdt*. Utrecht University Library, Utrecht, Netherlands.

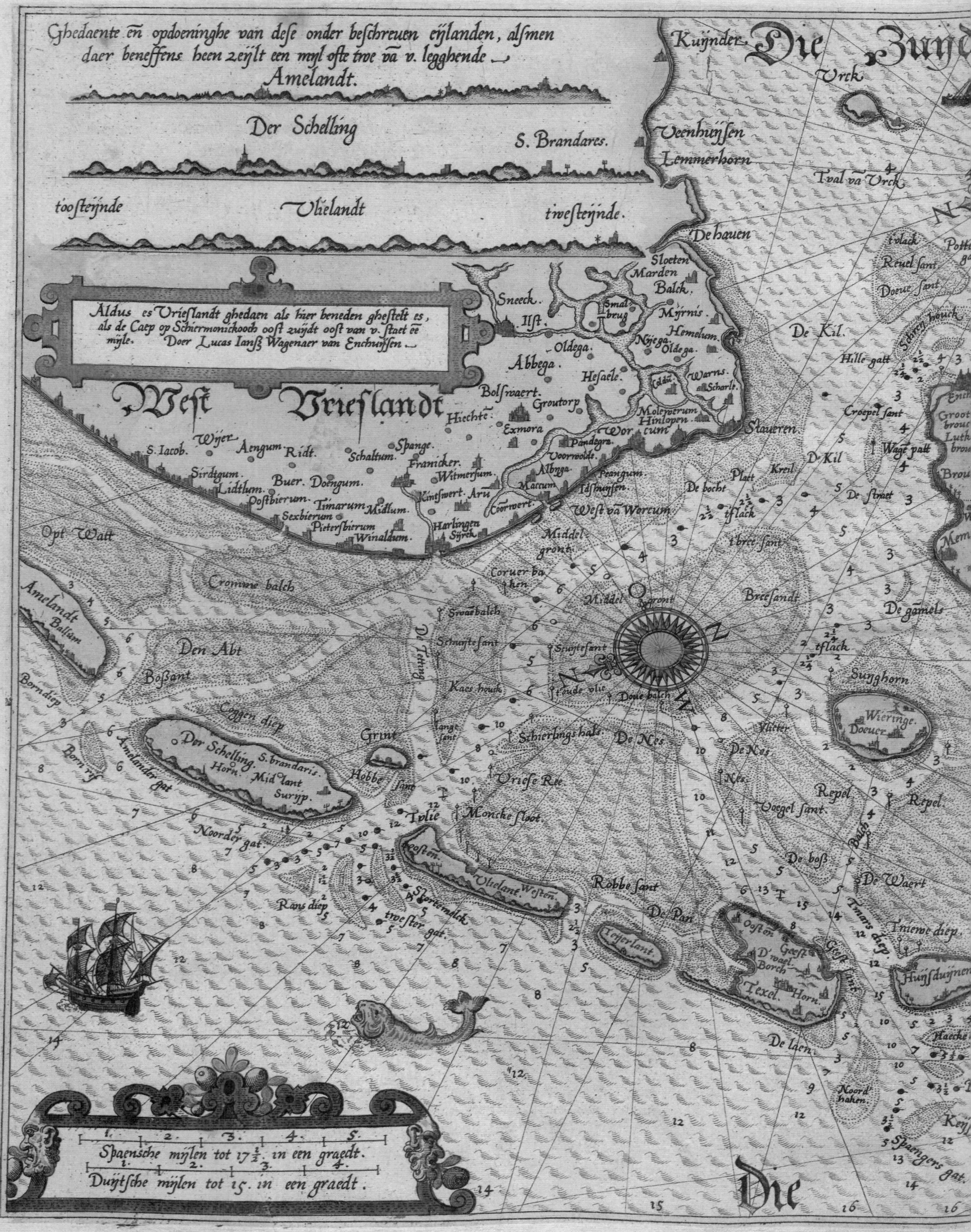

Ghedaente en opdoeninghe van dese onder beschreuen eylanden, alsmen daer beneffens heen zeylt een myl ofte twe va v. legghende
Amelandt.
Der Schelling
S. Brandares.
toosteynde
Vlielandt
twesteynde.
Aldus es Vrieslandt ghedaen als hier beneden ghestelt es, als de Caep op Schiermonickooch oost zuydt oost van v. staet ee myle. Doer Lucas Ianß Wagenaer van Enchuyssen
West Vrieslandt
Die Zuyd
Die
Spaensche mylen tot 17½. in een graedt.
Duytsche mylen tot 15. in een graedt.
Kuynder
Vrck
Veenhuyssen
Lemmerhorn
Tval va Vrck
De hauen
Sloeten
Marden
Balck
Sneeck.
Ilst.
Myrnis
Hemelum.
Oldega
Abbega.
Hesaele.
Bolswaert.
Groutorp
Warns.
Staueren
Hinlopen
Wortum
Exmora
Hiechte
S. Iacob.
Wyer
Aengum
Ridt.
Spange.
Schaltum.
Franicker.
Witmersum.
Sirdigum
Lidtlum.
Buer.
Doengum.
Oostbierum.
Timarum
Midlum.
Sexbierum
Pietersbierum
Winaldum.
Harlingen
Opt Watt
Cromme balch
Amelandt
Ballum
Den Abt
Boßant
Coggen diep
Der Schelling
S. brandaris.
Horn.
Midlant
Noorder gat.
Grint
Hobbe sant
Vlie
Monche sloot
Vriese Ree.
Vlielant
Robbe sant
De Pan
Texerlant
Texel.
Oosten
Horn.
De Kil
Hille gatt
Croepel sant
Wage gatt
De straet
Breesandt
De gamels
tflack
Sunghorn
Wieringe.
Doeuer
De Nes
Vlitter
Repel
Vogel sant
De boß
De Waert
Tnieue diep.
Huysduynen
De Laen
Noord haken.
Spaengers gat.
Middel gront
Schierlingshals
Doue balch
Reuel sant
Doeur sant
Ramsdiep
twester gat

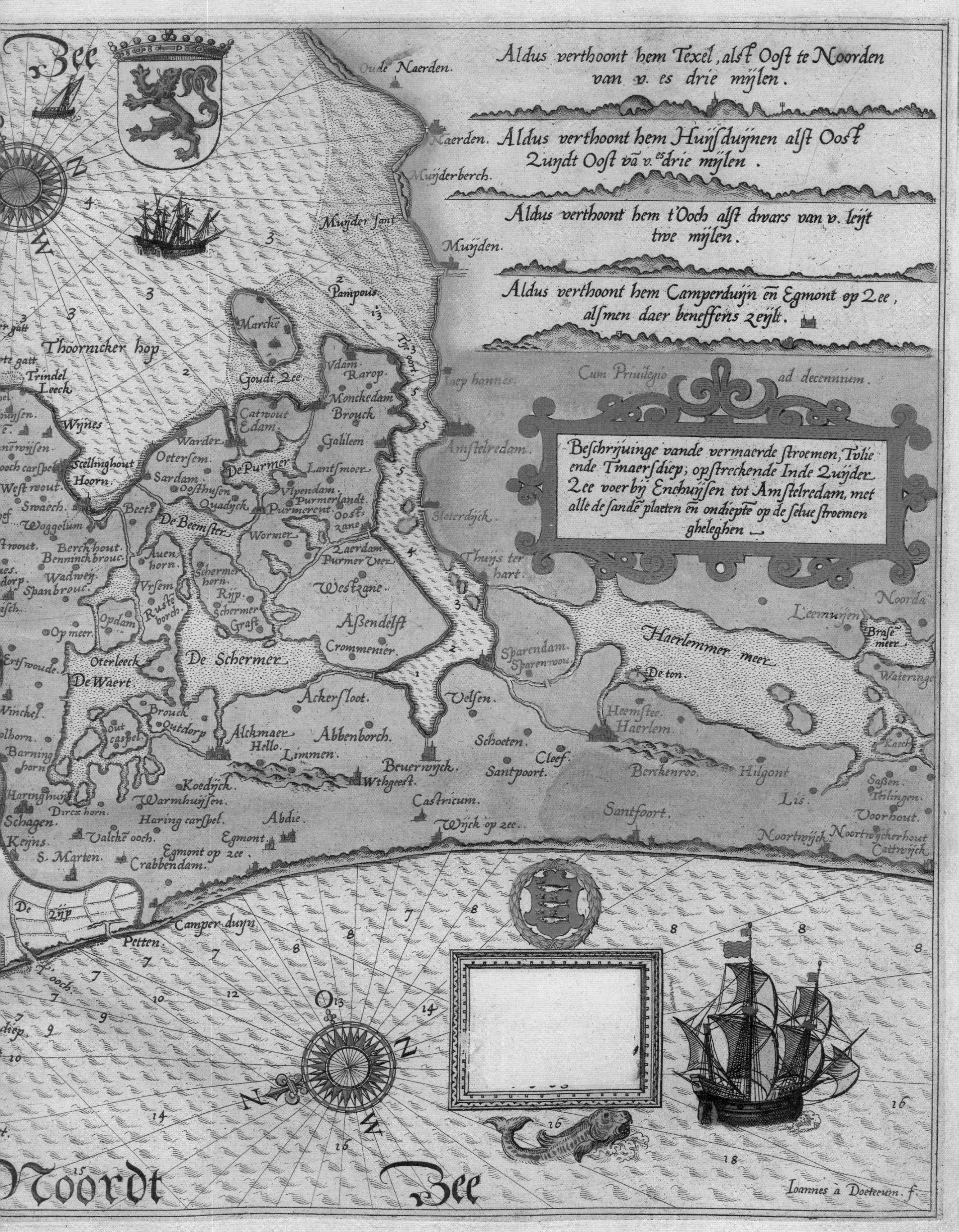
Aldus verthoont hem Texel, alst Oost te Noorden van v. es drie mijlen.
Aldus verthoont hem Huijsduijnen alst Oost Zuydt Oost vã v.es drie mijlen.
Aldus verthoont hem t'Ooch alst dwars van v. leijt twe mijlen.
Aldus verthoont hem Camperduijn ẽn Egmont op Zee, alsmen daer beneffens zeijlt.
Cum Priuilegio ad decennium.
Beschrijuinge vande vermaerde stroemen, Tolie ende Tmaersdiep; opstreckende Inde Zuijder Zee voer bij Enchuijsen tot Amstelredam, met alle de sandẽ plaeten ẽn ondieptẽ op de selue stroemen gheleghen.
Oude Naerden.
Naerden.
Muijderberch.
Muijder sant
Muijden.
Pampous
Marchẽ
Thoornicker hop
Trindel
Leeck
Goudt Zee
Vdam
Rarop
Monchedam
Brouck
Catwout
Edam
Wijnes
Warder
Galileen
Amstelredam.
Otersem.
Scellinghout
Hoorn.
De Purmer
Lantsmoer
Sardam
Oosthusen
Vlpendam
Purmerlandt
Quadyck
Purmerent
Oost zane
Beets
De Beemster
Wormer
Woggolum
Slaterdijck.
Zaerdam
Purmer Veer
Berchhout
Benninckbrouc
Auenhorn
Schermerhorn
Rijp
Schermer
Wadwey
Spanbrouc
Vrsem
Rustẽborch
Westzane
Thuys ter hart
Graft
Opdam
Aßendelft
Crommenier
Op meer.
Sparendam.
Sparenwou
Haerlemmer meer
Leemuijen
Noorda
Brase meer
De Schermer
Otterleeck
De Waert
De ton.
Wateringe
Ackersloot
Velsen.
Heemstee.
Haerlem.
Brouck
Outdorp
Out caspel
Alckmaer
Hello
Abbenborch.
Limmen
Schoeten.
Cleef.
Santpoort.
Berckenroo.
Hilgont
Beuerwijck.
Koedijck
Wtgeest.
Warmhuijsen.
Haring carspel.
Castricum.
Sassen.
Teilingen.
Lis
Abdie.
Santfoort.
Voorhout
Schagen.
Valckẽ ooch.
Egmont.
Wijck op zee.
Noortwijck
Noortwijcherhout
Cattwijck
Keijns.
S. Marten.
Egmont op zee
Crabbendam.
De Zijp
Camper duijn
Petten.
T'ooch.
Noordt Zee
Ioannes à Doetecum. f.

Ortelius WORLD MAP

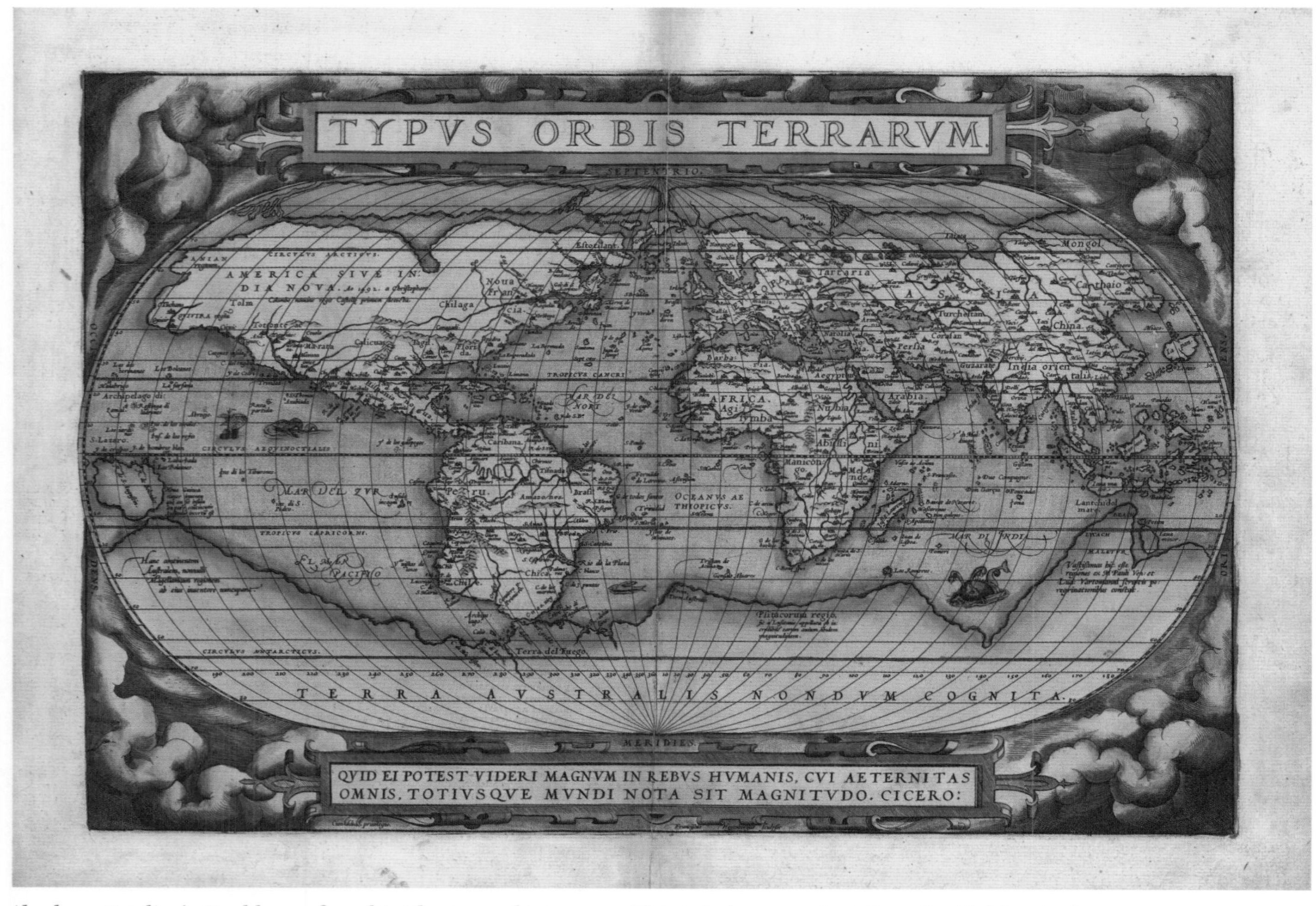

Abraham Ortelius's World map from his *Theatrum orbis terrarum.* This is just one product from the 'Golden Age' of cartography in the Low Countries. Library of Congress, Washington D.C., USA.

***Typus Orbis Terrarum* is a double-page oval planisphere designed to introduce the geographical description of the world in the *Theatrum Orbis Terrarum* – a book of maps first published in 1570 by humanist Abraham Ortelius in Antwerp and traditionally considered the first modern atlas. This influential book – published in more than thirty editions in different languages in the following forty years– established a stable template for the depiction and description of the world according to continents.**

The map displays the modern world as understood in the late sixteenth century. It shows Europe, Asia, Africa and the New World, and also a hypothetical southern content with the legend 'Terra Australis Nondvm Cognita' ('southern land, still unknown') and a smaller but equivalent counterpart in the north, at the top of the map.

In the lower margin, in a long cartouche, a quote from Cicero's *Tusculan Disputations* frames the map and situates the book within humanistic culture emphasizing the need to base even new cartographic innovations within a framework of classical learning: 'Quid ei potest videri magnum in rebus humanis, cui aeternitas omnis totiusque mundi nota sit magnitudo' ('For what can seem important in human matters to a man who knows all eternity and the vastness of the universe?'). Under that cartouche, in smaller print, the reader learns that the map was printed 'Cum Privilegio' ('with a license') by Francisco Hogenbergus.

The map, like an archive, collects a considerable amount of information, though with different levels of reliability, accuracy, and factuality. The Strait of Magellan reminds us of the circumnavigations through that passage, and the island of Nova Guinea (separated from the Southern Continent) reveals updates with data from some of the most recent explorations of the time. However, as the question of whether Asia and America were connected remained a mystery, the 'Anian regnum' in the northwestern corner of the New World was just a hypothetical geographical configuration. Some other purely fictional features complete this influential image of the world. The most prominent of those is 'Terra Australis', a continent whose existence had been inferred from some theoretical texts and from a few disconnected sightings during navigations. The idea of such an enormous southern landmass relied on the common assumption that some kind of balance or equilibrium in the distribution of lands and waters in the northern and southern hemispheres was required. Similarly, that 'Septentrionalis Terra Incognita', the unknown land in the north, mirrored this southern continent. These features shaped the geographical imagination in Europe in the early modern period, not only because the map was part of one of the best-selling atlases of its time, but also because it has been copied and published in many other books.

The long-cultivated ability of Ortelius as an illuminator of maps is evident in the details of the delightful illustrations. The oceans are full of islands, quite homogeneously distributed. Some beautiful sea creatures enrich the cartographic landscape, bringing together myth and cosmographical knowledge of the time. The interiors of the continents show pictorial representations of several rivers (highly stylized in South America), towns, kingdoms and some mountain ranges.

Ortelius's career included a broad range of activities, from colouring maps to selling coins, antiques and books. As a dealer, Ortelius was part of a large European network through which he exchanged objects and projects with many influential people, among them his friend and rival Gerard Mercator (see page 76). Before his death in 1598, Ortelius had become a well-known and respected person within the book and map trade. Doubtless his position contributed to reinforce the importance of his atlas and his world map. In the course of the seventeenth century, the types of map books multiplied and the market for them grew. Ortelius's work was a milestone and opened a new path for what afterwards would be called the 'Golden Age of Atlases' in the Low Countries. His model was followed by the most reputable firms in the cartographic market, and even his title was copied: for instance, in the 1630s the Blaeu family (see page 92) titled their atlas series with the name *Theatrum orbis terrarum, sive, Atlas Novus (Theatre of the World or New Atlas).*

Marata
Marata
Ometlan.
Cacos
Comos
Coruco.
Florida.
La Emperadada
Lucaio
Limana
Chechimicale
Cuchillo.
Culias
Tamaco.
B. de culata
Tula
Xalisko
Mechula
Hispania noua.
Panuco.
Xaques
Cuba.
Spagnola
Boriquē
S. Thomas
Anubiada
Ciguatla
Acaxcatutla
Iamaica
R. de cacatula
R. grande p. de los angelos.
Socomisco.
Trugillo.
Granata.
Mopox.
Taste
Antiochia
Benezul.
Caribes
Caribana.
yᵉ de los galopegos.
Quito
Neyua
QVINOCTIALIS
Tumbes.
Corangui
Aiauari
Aiauirizama
Guanape
Trapicari
Mapazo.
Picora.
Casma
Lima
Peru.
MAR DEL ZVR
Insulę incognitę
Maragnon fl.
Amaz
ins di S. Pedro.
Cusco.
Chichane.
Arica
Colochi
S. Ann
OPICVS CAPRICORNI.
Pisaqua
Giurumatas
Mepenes
Coquimbo
Copaiao.
Quintete.
Tarapaca
Ningata
S. Esp
EL MAR PACIFICO
yas uistas de lexos
Chili
P. de palma
Lucengo
C. de S. Maria
Chile.
Chi
R. de Salinas
Orafes P.
Caruiero p.
C. Primiero.
Archipelago.
Calis.
C. desceado
C. da Maestro.
C. di bon signal.
Terra
ARCTICVS.
210
220
230
240
250
260
270
280
290
TERRA AV

Santana
Juan de famp.º
Açores
Septa
Alger
Tunis
Bugia
Sept cites
Barbaria.
Marocco
Tripoli
Canarię insulę
TROPICVS CANCRI
Auzichi
Berdoa
Angela
Alhamara
Hair
Targa
C. blanco
Argin
Digir
Albaidi
MAR DEL NORT
Hoden
AFRICA.
y. de capo Verde
Darin
Agi
Cano
Borno
C. Verde
Tombotu
symba
Guan guara
y. de S. Bº
Melli
ripana
Solis
C. Serre liona
Cago
Mellegete
Mina
Benin
Biafar
Rio de las Amazones
C. da Verga
S. Paulo
C. de 3 puntas
Principe
R. Dangla
Gag
S. Tomas
Manicongo
S. Roque
S. Croce
S. Matheo
Nobon
Macerira
nada
Orellana
Ora
Fernãde de Loronno
Ascension
Vamba
Humos
R. de S. Domingo
Manicongo
C. Ledo
R. S. Fracesco
Brasil
G. de todos santos
OCEANVS AETHIOPICVS.
Coilla
R. S. Elena
R. estrema
P. segur
C. de arcas
Trinidad
S. Helena
C. Negro
Ascension
S. Maria
Aldea
Zimbro
C. Frio
Yslas de Miniaes
G. de S. Antonio
B. Real
G de las bueltas
S. Catelina
Cayneca
Tristan de Acuña
Rio de la Plata
Palmares
C blanco
C. Bone spei
Gonsalo Aluares
C di 3 puntas
Prom. Terre Australis
Estrecho di Magallanes
R. dolassino
Psitacorum re
sic à Lusitanis appellat
credibile earum auium
magnitudinem.
el Fuego.
300 310 320 330 340 350 360 10 20 30 40 50 60 70 80
STRALIS NON

China

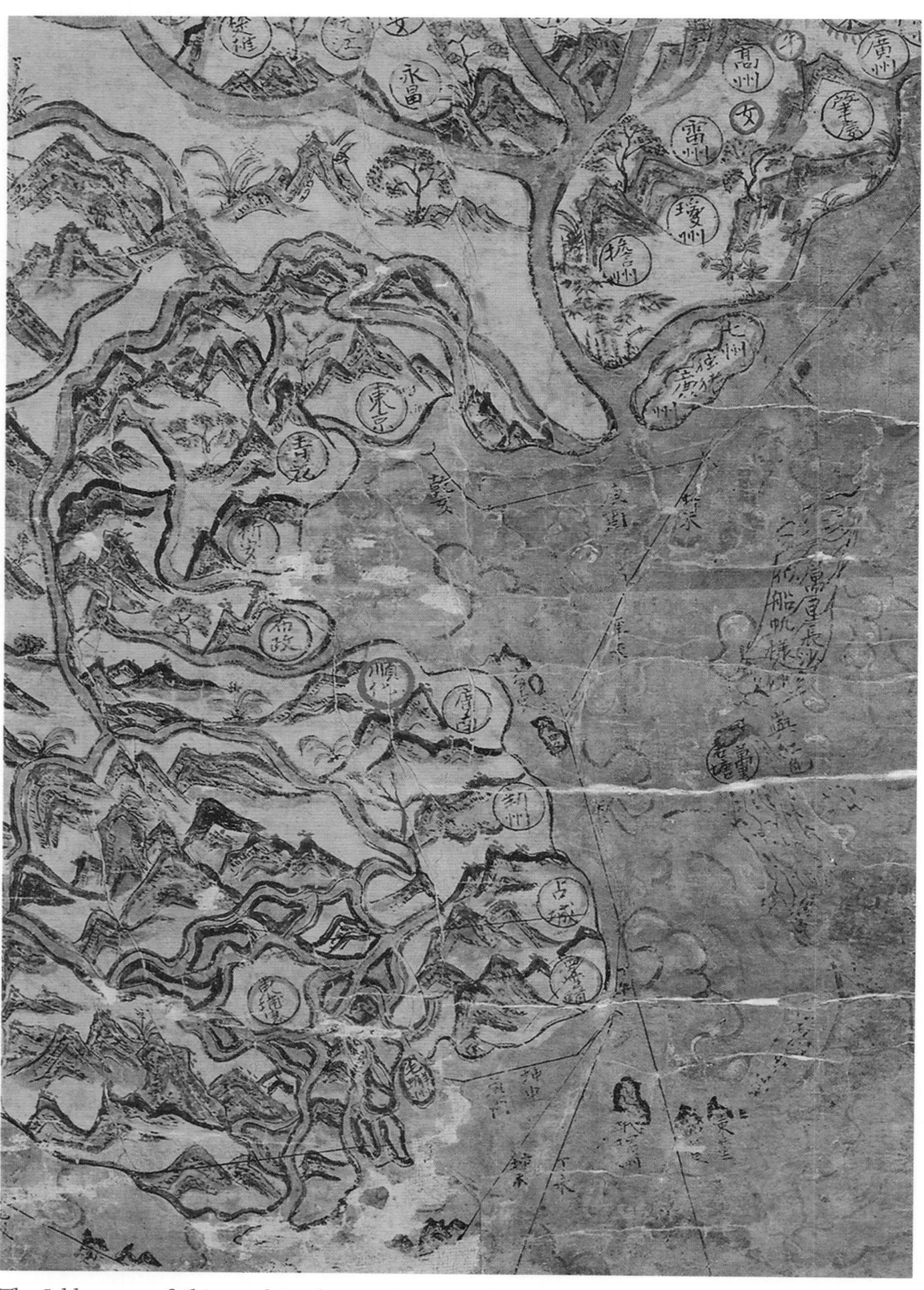

The Selden map of China and Southeast Asia – an 'independent' map which revealed for the first time the extent of Chinese interaction with the region. Bodleian Library, Oxford, UK.

THE SELDEN MAP

The Selden map is part of a small but very important collection of pre-modern Chinese maps at the Bodleian Library in Oxford, and is among the earliest Chinese maps to reach Europe. Although its ultimate provenance is unknown, it came into the possession of the London lawyer John Selden in the first half of the seventeenth century, and was bequeathed to the Bodleian in 1659.

In a codicil to his will dated 1653, Selden describes the map and an accompanying compass as having been 'taken both by an englishe comander'. It is possible that an East India Company trader took the map from another European, Japanese, or Chinese vessel in the lawless conditions of the South China Sea and brought it to London where it passed into the hands of Selden, but the records of these voyages make no mention of it.

Perhaps on account of its size and appearance, the map was kept on permanent display in the Anatomy School, and is noted in a list of the School's contents which was prepared by its Keeper, the antiquary and diarist Thomas Hearne, in 1721. It was never lost sight of, but at the same time its importance lay undiscovered. In January 2008, however, it was examined by the American scholar Robert Batchelor, who noticed two features of the map which distinguish it sharply from all Chinese maps which had been produced before it.

The first is that it is a map not just of China, but of the whole of East and Southeast Asia. Earlier and most later maps depicted China not only as the centre of the known world, but as occupying almost their entire area. China occupies less than one half of the area of the Selden map, which is centred on the South China Sea, with an equal area depicting the Philippines, Borneo, Java, Sumatra, Southeast Asia, and, notionally at least, India. The depiction of China itself is not the purpose of the map, and is copied from a standard printed map of the period.

The second distinguishing feature is the presence of shipping routes with compass bearings radiating from the port of Quanzhou on the coast of Fujian Province to all the areas covered by the map, which, to quote Timothy Brook, 'charted the commercial world as no map, East or West, had done before'. It is thus the earliest example of Chinese merchant cartography, unique in not being a product of the imperial bureaucracy. It indicates the extent of China's intercourse with the rest of the world at a time when it was generally supposed to have been isolated.

An additional feature of the map became apparent during the course of the conservation work which was undertaken following its discovery. When the old backing was removed, the main sea routes, identically drawn, were found on the reverse, showing not only that this was a first draft, but that the map was being drawn by systematic geometric techniques. It is the first Chinese map to be produced in this way, and its use of voyage data obtained from a magnetic compass, and distances calculated from the number of watches, is a technique with no western parallel.

Unfortunately, it is not known exactly when the map was drawn, or who drew it and for what purpose. It is elaborately decorated with landscapes and plants, and was almost certainly produced for reference in the house of a rich merchant rather than for use at sea. The most we can derive from recent scholarship is that it was probably produced in the early seventeenth century by a Chinese, as Chinese sources are used for the place-names on the map and also the shipping routes, and the compiler was probably based in Southeast Asia, as the map's depiction of that area was to remain the most accurate for another two centuries.

Blaeu

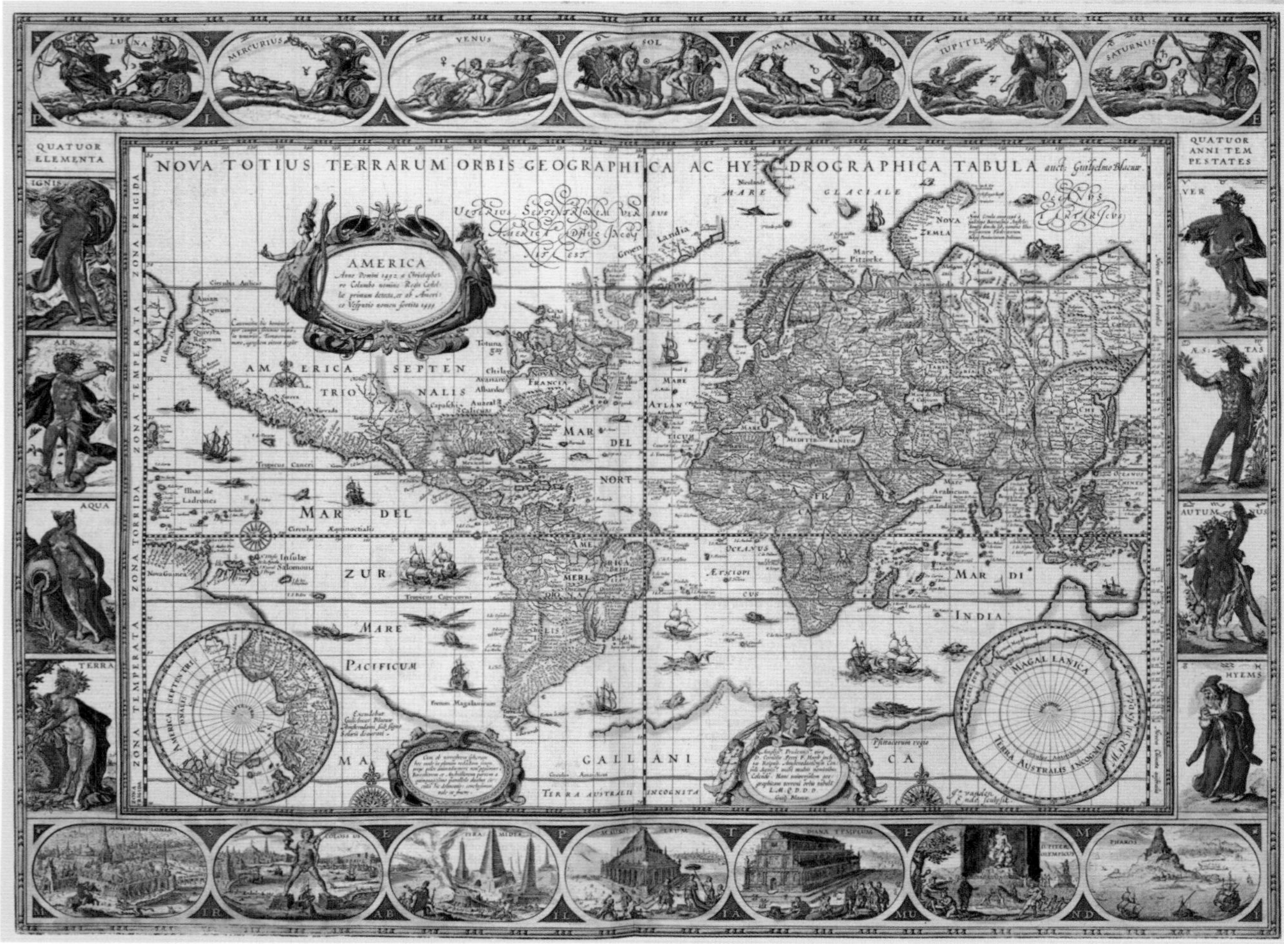

A beautifully illustrated map of the World from the Blaeu publishing house – the leading European map publishers of the time. British Library, London, UK.

MAPPING DOMINANCE

In the sixteenth century, the Dutch embraced the Reformation and Protestantism and rebelled against the catholic Spanish empire, establishing the independent Dutch Republic in 1588. The small republic grew to be a world power based on ships, commerce, and cartography. By the mid-seventeenth century, the Blaeu family had become one of the largest mapmakers in Europe.

In the seventeenth century, the Netherlands ruled an empire stretching from New Amsterdam (today's New York City) to Batavia (today's Jakarta, Indonesia). In Amsterdam, at the centre of this empire, the Blaeu publishing house started selling maps in 1599. Blaeu's globes and maps were popular with Dutch merchants and sailors. Willem Janszoon Blaeu's first world map, issued in 1606, highlighted Dutch knowledge and discoveries.

The principal rivals of the Dutch in the East Indies were the Portuguese, and a bitter battle for supremacy took place between the two countries during the seventeenth century. The Dutch East India Company (Verenigde Oost-Indische Compagnie or VOC), formed in 1602, became one of the world's most powerful trading companies, usurping Portugal in the Asian spice trade by 1641 in places like Ceylon (Sri Lanka) and the Spice Islands (Indonesia). The VOC made Batavia their Asian headquarters in 1619. The Dutch merchant fleet, Europe's largest, traded throughout Asia and brought commodities such as nutmeg and cinnamon back to Amsterdam; but the ships needed accurate maps for worldwide trade. The Blaeus, by being appointed as cartographers to the VOC, gained access to secret map information denied to mapmaking rivals. The political and business connections of the Blaeu family increased the demand for their maps.

As the Dutch prospered, the demand for maps and atlases increased for businesses (merchants and sailors), and also among the general public. The affluent Dutch population were curious about other countries and Blaeu wall maps came to express status in Dutch society, often taking the place of paintings in Dutch homes and businesses. Blaeu maps and atlases satisfied customers by offering the best in Dutch cartography, illustrated with the finest artwork. For example, border art for the world map, labelled in Latin, depicts the sun, moon, and the five known planets along the top. The bottom panels reveal the Seven Wonders of the World. Panels on the right feature the four seasons, and those on the left illustrate the four elements of (from top) fire, air, water, and earth. On the map, the continents of Antarctica and Australia remain largely unknown ('Terra Australis incognita'). The Arctic regions of North America and Greenland are not drawn, and Korea appears as an island.

After Willem Blaeu died in 1638, his son Joan Blaeu took over the company, and he produced larger and larger atlases culminating in the *Atlas Maior* (Great Atlas) in 1662, which consisted of eleven volumes holding some 600 maps. This atlas was the most expensive publication of the seventeenth century and was also considered the best. It totalled almost 4,000 pages of maps and text in the Dutch language and 5,300 pages in the French language edition.

The fortunes of the Blaeu business followed the rise and fall of the Dutch Republic. The waning of the Dutch empire saw the loss of Chinese trade in 1662 and of New Amsterdam in 1664 (the victorious English renamed it New York). In 1672, France declared war on the Dutch Republic and invading French forces were only stopped when the Dutch opened the dykes to flood the borderlands of Holland. The Blaeu family had competitors in the mapmaking field and this war hurt trade as it reduced the empire's wealth. Much of the money to buy maps was gone by the late seventeenth century. Also in 1672, the Blaeu mapping studio was destroyed during the great fire in Amsterdam, and many of the copperplates for maps were lost to the flames. Joan Blaeu died the following year. The Blaeu publishing house continued until 1712, but its golden age had passed.

MARE
LAN
TICUM
Asores Ins.
Brazil
Anglia
Londen
Germania
Gallia
Polonia
Hungaria
Hispania
Sardinia
Cicilia
MARE
MEDITER
RANEUM
Pontus Euxinus Mare Maior
Natolia
Cyprus
Barbaria
Biledulgerid
Numidia
Libia Interior
Sarra
A
FR
I
CA
Guinea
Zanfara
C. Verde
C. Blanco
I. S. Thome
OCEANUS
ÆTHIOPI
CUS
S. Helena
I. Atention
I. Loando
Monomota
C de Bona Esperaca
I. Tristan da cunha
I. Gonçalo Alvare
Madagascar

TARTA
RIA
Karakythay
Murus 400 leucarum
Mare de Sala ol Caspium
Salmarchand
Hircania
Corasan
Bramas
India Extra
Tipura
Chesimur
Multan
India Intra
Guzarata
Ormus
Quinzay
CHI
NA
Nanquin
Quicheu
Foquiam
Cantam
Corea
Lipan
OCEANU
CHINE
Philippinæ Ins.
Luconia
Mare Arabicum et Indicum
Golfo de Bengala ol Gangeticus Sinus
Ceilan
Sian
Patana
Borneo
Celebes
Gilolo
Iava
I Timor
Terre alta
MAR DI INDIA
Beach
Fretum Magallanicu
MAGALLANICA
del Fuego

van Keulen

Reliable charts were essential for the safe passage of mariners in the Age of Discovery, when merchant companies of seafaring nations sailed far from home, in search of commerce and territories. In that era, the *Vereenigde Geoctroieerde Oostindische Compagnie* or VOC, the Dutch East India Company, became wealthy and influential, conducting extremely profitable trade with the Spice Islands of the East Indies (Indonesia). It also established bases in these islands, where the Dutch settled and later colonized.

The VOC had its own cartographers and for a century and a half, from its inception in 1602, maintained a secret atlas of ocean charts. *Nieuwe Pascaert van Oost Indien* – shown here – was one such early secret chart, which was copied in manuscript to restrict distribution and to ensure that only Dutch mariners – and not their rivals for East Indian trade, principally the Portuguese – benefited from it. The chart extends from the Cape of Good Hope to Japan and also includes an accurate outline of western Australia. It was printed in Amsterdam in 1680 by publisher Johannes van Keulen.

By the mid eighteenth century, East India Companies of Portugal, France and Britain had made their own marine charts of the Indian Ocean routes. In 1753, the sixth and final volume of the flagship Keulen family atlas, the *Nieuwe Lichtende Zee-Fakkel*, or *'New Shining Sea Torch'*, was published. It included this chart from the secret atlas, for by then it was no longer necessary to protect it from competing East India Companies.

This is a map rich in several mapping traditions, seamlessly integrated. Following Ptolemaic geography (see page 16), longitudinal positions are marked from the Isle of Ferro. Using portolan charting conventions (see page 40), compass roses indicate thirty-two directions and are marked as radiating lines. Mercator's invention of rectilinear graticules on navigation maps, to show constant bearing between meridians (see page 76), has also been introduced. Every one degree of latitude and fifteen minutes of time are marked on a vertical line for easy reference of ships' captains.

Nieuwe Pascaert van Oost Indien is a fine example of a sea chart of this period. It was compiled from information gathered from mariners who sailed on this route. Numerous coastal ports and archipelagos, which mariners anchored at routinely, are marked. These include the Seychelles, the Maldives, the Andaman Islands, Indonesia, the Philippines and Japan.

The chart gives a fairly realistic depiction of the coastlines, although the shape of India is far from accurate. At this time the Dutch ships touched only the southern tip of India and Sri Lanka and sailed on towards the Spice Islands further east. The Spice Islands themselves – Sumatra, Java, Bali, Lombok, Flores, Timor, and the straits and seas around them, are easily recognizable by their correct shapes, names and relative positions.

The Dutch had already landed on the west coast of Australia, where the earliest date marked is 1616. Inland areas and open oceans have no place in this map, for it is a classic portolan. The ports are named and labels placed on land, leaving the waters for drawing islands, rocks, shallows and rhumb lines to aid navigation.

The cartouche enclosing the title *Nieuwe Pascaert van Oost Indien* shows knights and young Duke Henry of Silesia, beheaded in battle against the Tartars in 1241. This is an allusion to the Mongol warriors who reached Europe on horseback and plundered and occupied many of its eastern habitations. Perhaps it was a suggestion to merchants of the dangers lurking in remote eastern kingdoms.

At the bottom of the map is an illustration of a name plate held by two dolphin-like sea creatures. On it is inscribed the name of Johannes van Keulen's firm 'In de Gekroonde Lootsman' or 'In the Crowned Pilot'.

This portolan chart was made in the golden age of the VOC. Hundreds of ships followed the 'spice route' to the Dutch East Indies factories in and around Batavia (present-day Jakarta), and then on to Japan. This secret chart enabled Dutch sailors to keep ahead of their rivals in navigating to these remote destinations seeking spices and silk, and then to return safely to Europe.

EAST INDIES SPICE TRADE

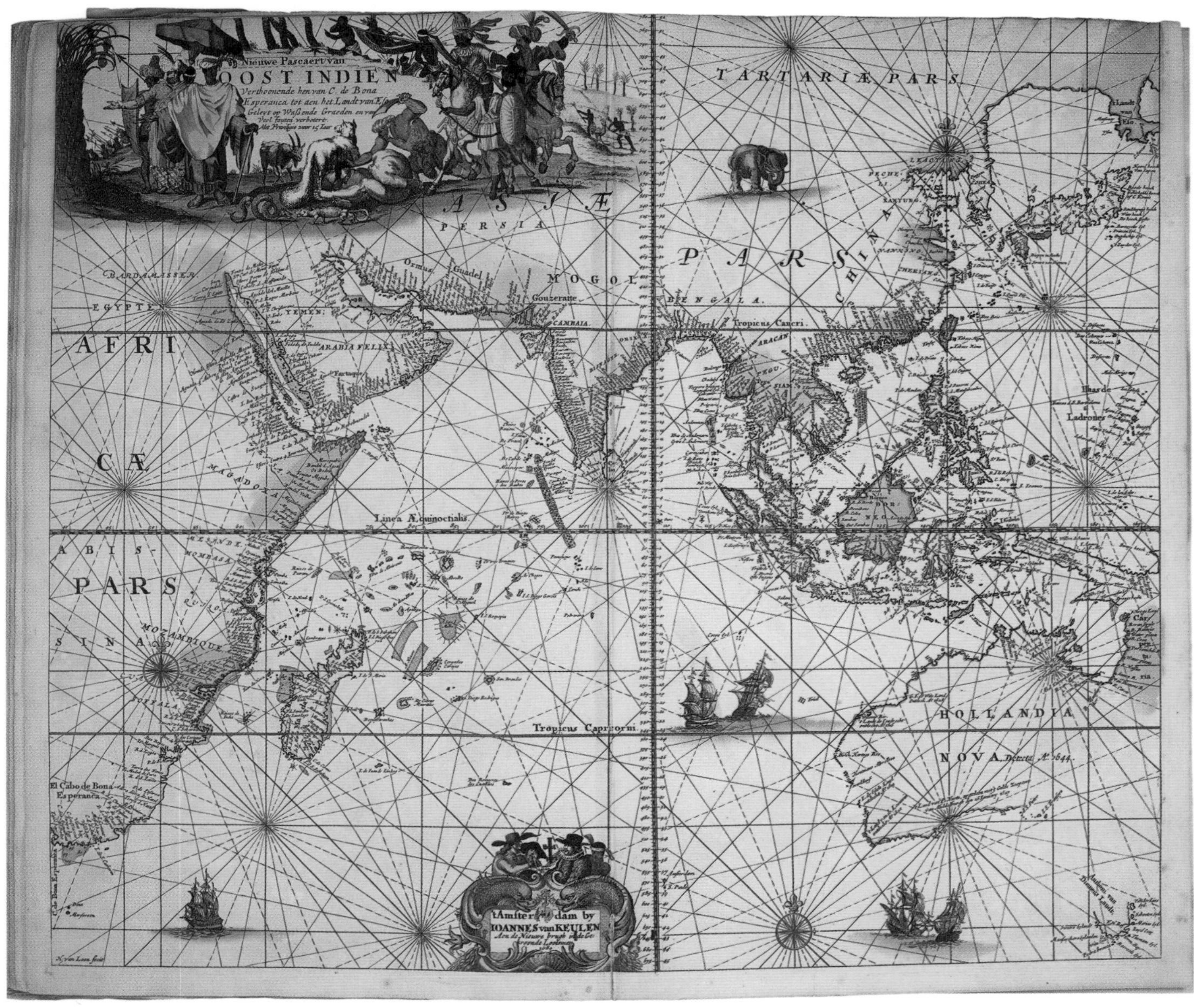

Nieuwe Pascaert van Oost Indien maritime chart by Johannes van Keulen, 1680 from the 'Secret Atlas'. Later published by Johannes van Keulen II in 1753. National Maritime Museum, Greenwich, London, UK.

CAMBAIA.
B E N G A L
Bengala
ORIXA
BISNAGER
Decan
Canara
Coromandel
Goa
Palicatte
Ceylon
Negombo
Colombo
C. Comorin
Panova
Baixos de Padua
Cheiro Baniani
I. Tatimo
De Pracel
Andaro
Malique
De Cubile
Manacopor
Paraque
Baixos de Peros dos Banhos
Maldiva
Ouro
Æquinoctialis.
80
85
90
95
100
105
110
115
P. Nias
As Sete Irmanos
Os tres Irmanos
Abrolho
S. Francisco
As Chagas
Pomoluque
I. de Gamo
Adu
Candu
I.d. Diego Garcia
Baixos de S. Miquel
Agalega
I.d. Roquepiz
Polvrera
Corgados Carajos
San Brandao
Diego Rodrigus
Do Cierne Mauritius
Bulango
Chedube
Blau Water
Priparis
Ilhas Cocus
Hoog Eyl.
Andomaon
Ilhas de Andomaon
Carnicubar
I. de Raze
Nickubares
Palo Way
P. Ronde
Cocos Eyl.
Verckens Eyl.
Cocos Eyl.
Tropicus Capricorni.

s Cancri
QUANTUNG
FOKIEN
I. Formosa
Dos Reis
Amsterdam
Malo Abrigo
I. Disierta
Das Columnas
Una Columna
Desierta
Tabaco Miguel
Tabaco Xima
Tansay
I. de Pratt
5 Eylanden
Aynam
SIAM
Cambodia
Con Chin china
C. del Engano
C. Baiador
P. d. Paxoras
G. d. Matalahombre
P. de Mandato
Lu
coma
Strecho d. Mindoro
Mindoro
Strecho d. Manilha
B. de la baja
Ilhas de
Baixos d. S. Bartholome
Ladrones
C. del Spirito Santo
Tandaia
Paßage de S. Clara
Saya Vedra
Sahavedra
St Iuan
I. dos Matelotes
I. dos Arecifes
Minda
nao
B. d. la Rosurecion
C. Bicay
S. Ioannos
I. das Palmeiras
Carangano
I. d. Tulaon
Sangust
Sian
Moratay
I. Gilolo
Bohos
Ternate
Tidor
de los Mart
I. d. Aves
Bocasa
Borneo
BOR
NEO
R. de Borlo
Domibono
R. Colca
Sambas
Out Sambas
Aert Grisels hoeck
P. Condor
P. Cofyn
P. Ridang
P. Capas
P. Timon
P. Parcelar
Malacca
125
135
140
145
150
155
160
Celebes
Macassar
Bouro
Boutón
Bachan
Ceram
Banda
7 Eylanden
Kay
Amboina
Gebroken hoeck
Terra dos Papous
Willem Schouten
I. Armoa
Sneeuberg
Nova
Keer
Valsche Caap
Salayer
Pater Noster
Iava
Balanbangan
Baly
Lambocs
Cumbava
Flores I.
Terra Alta
I. Timor
Timor Landt
I. Schilpads
Arauw
Selam
Mony
Straet van Sunda
Iava
Banca
Vuyle hoec
Cocodrils Eyl.
Aernhems Landt
Van Diemens Landt
Baij van Diemen
Lammen bocht
C. Maire
C. vanden Lim
Abel Tasmans Riv.
R. van Alphen
C. van Diemen
R. Maetsuycker
G. F. de Wits Land.
Ontdeckt Ao. 1628
Trial
HOLLANDIA
Willems Riv.
't Landt de Eendracht
ontdeckt Ao. 1616
Dirck Hartogs Ree
NOVA Detecta Ao. 1644.

IMPERIAL RUSSIA

Map of Russia from the first official geographic atlas of the Russian Empire compiled and published by the Russian Academy of Sciences in 1745. Overleaf: comparative extracts of this map (left) and *Asia (General Map)* from *The Times Atlas of the World*, 1895 (right).

Russia was the European country whose boundaries expanded most dramatically between the seventeenth and nineteenth centuries This expansion created a vast hinterland for the Tsars (the rulers of Russia) which needed mapping, both for practical reasons and as part of a new self-image of an empire which spanned both Europe and Asia. The map of Russia in 1745 already shows its boundaries encompassing most of Siberia, but by 1895 it had pushed forward dramatically in Central Asia, the Caucasus, the Crimea, Finland and Poland.

The earliest evidence of cartography in Russia is the Tmutarakan Stone, which bears an inscription relating to land measurements on the Kerch Strait (between the Black Sea and the Sea of Azov) in 1068, but it was the wholesale expansion into Siberia between 1580 and 1639 which provided the real impetus to centrally-directed mapping. The most important early map of Siberia was Godunov's in 1667, which appeared just after the first Russian printed map in 1658. This provided a model which was built upon by Semyon Remezov in his great *Siberian Drawing Book* (1697-1711) ,which included town maps of settlements such as Tobolsk and Irkutsk as well as a wealth of topographic information about the Siberian interior.

Scientific cartography gained a foothold in Russia under the reformist Tsar Peter the Great (1682-1725). A class in geodesy was established at the state Naval Academy in 1716, which began to produce naval charts in 1721. The Academy supplied the expertise for a nationwide geodetic survey which was set up in 1720 and which worked for twenty-four years to map 200 of Russia's 285 districts. From 1739 it operated under the Geographic Department of the Academy of Sciences and the end result was the 1745 Atlas of Russia, and a general map of the whole country which provided the first accurate cartographic representation of its territory. Framed by the Romanov coat of arms with its double-headed eagle, the great expanse of Siberia, tinted green, shows how far the empire had spread its wings east of Moscow. Additional information in the East was provided by the Great Northern (or Second Kamchatka) Expedition led by Vitus Bering from 1733–43 which accurately mapped the northeast tip of Siberia and discovered Alaska (which was soon to become the easternmost extension of the Tsarist empire).

As Russia's lands expanded further, with the aggressive policies under Catherine the Great (1762–96) seeing Poland partitioned and partly occupied by Russia from 1772, and the conquest of the Crimea in 1786, so the practical need for maps of the conquered territories grew, alongside the more ideological demand for representations of the extent of the new empire. A new civilian land survey had been established in 1763, which would labour for 122 years before its task was complete, its work supported by the creation of a State Geographic Department in 1786. Meanwhile the army, which had become more conscious of the need for accurate maps during the Napoleonic Wars, began active topographical work in 1815, particularly in the newly acquired territories of Finland and Poland, and established a Military Topographic Corps in 1822.

The new shape of Russia was depicted in maps such as V.P. Pyadhyshev's 1834 *Geographical Atlas of the Russian Empire, the Kingdom of Poland and the Grand Duchy of Finland* and its growth celebrated by Ivan Akhmatov's 1821 *Historical, Chronological, Genealogical and Geographic Atlas of the Russia State* which included maps tracing the expansion of Russia from the early days of Kievan Russ in the ninth century to its new imperial grandeur in the nineteenth. Yet more was to come. Even before the military General Staff decided to conduct a new survey in 1863, Russia had projected its power into Turkestan, while the next year its conquest of the Caucasus was completed with the final defeat of the Circassians. Shortly after the survey completed its work in 1872 Russia absorbed even more of Central Asia, reducing Bukhara and Khiva to the status of protectorates and annexing Kokand in 1876. The map of Russia in 1895 shows a country which had expanded significantly along all its frontiers since 1745 (although it had sold Alaska to the United States in 1867) The ever-expanding boundaries had provided a moving target against which Russian cartographers honed their skills and growing technical expertise and yielded a series of maps which bolstered Russia's growing self-image as a powerful Eurasian empire.

ЕРНОЙ ОКIА
НОВАЯ ЗЕМЛЯ
БѢЛОЕ МОРЕ
АРХАНГЕЛО ГОРОДСКАЯ ГУБЕРНIЯ
ОБСКАЯ ГУБА
ТАЗОВСКАЯ ГУБА
ТОБОЛЬСКАЯ ПРОВИНЦIЯ
ЕНИСЕИСКАЯ ПРОВИНЦIЯ
ИРКУТСКАЯ ПРОВИНЦIЯ
АСТРАХАНСКАЯ ГУБЕРНIЯ
КАЗАЧЬЕЙ ОРДЫ СТЕПЬ
СТЕПЬ КЪ РѢКѢ ИШИМУ
АРАЛЬСКОЕ МОРЕ
ВЛАДѢНIЕ КАРАКАЛПАЦКОЕ
ХВАЛЫНСКОЕ МОРЕ
АСТРАХАНЬ

NORWAY
SWEDEN
BARENTS SEA
KARA SEA
Novaya Zemlya
RUSSIAN EMPIRE
EAST SIBERIA
WEST SIBERIA
MONGOLIA
Gobi (Shamo)
CHINESE EMPIRE
TIBET
CHINA
MANCHURIA
PERSIA
AFGHANISTAN
BALUCHISTAN
ARABIA
BLACK SEA
CASPIAN SEA
85 ft. below the Mediterranean
BALTIC SEA
PERSIAN GULF
Gulf of Bothnia
Mesopotamia
Transcaspian
Semipalatinsk
Semirechinsk
Akmolinsk
Tobolsk
Yeniseisk
Irkutsk
Moscow
Kazan
Tomsk
Omsk
Yakutsk
Bokhara
Samarkand
Tashkent
Kashgar
Yarkand
Khotan
Peking
Delhi
Constantinople

Halley
GLOBAL MAGNETISM

The coming of the Enlightenment in seventeenth-century Europe heralded a practical approach to the gathering and analysis of scientific data and with it the first thematic maps, which displayed scientific phenomena in cartographic form. Edmond Halley's maps of magnetic variation were among the earliest and most influential of the genre.

Edmond Halley, a distinguished astronomer best known for the comet which was later named for him (after he predicted in 1682 that it would return approximately seventy-five years later), was deeply involved in the late seventeenth-century controversies over the Earth's magnetic field. As early as the sixteenth century it had been observed that magnetic compasses displayed both declination, pointing slightly away from true north and inclination, dipping below the horizontal, as though pointing to a place below the Earth's surface. By the 1660s it was believed that careful measurement of these variations might assist in determining longitude at sea, a problem which had long vexed sailors and scholars, and the newly founded Royal Society sponsored voyages to assist such surveys.

In 1698 Edmond Halley set out on the *Paramour* to 'compass the Globe to make observations on the Magneticall Needle', a voyage he repeated the next year. The Royal Society also hoped he would find the 'Terra Incognita' or Great Southern Continent which was supposed to lie to the south of Africa. Halley failed in the second objective – the ship was turned back by icebergs at 52 degrees South – and his observations at sea were not precise enough to help with the first. He also found no confirmation of his own view that the Earth had four magnetic poles, two in each hemisphere.

He did, though, use the data he had gathered to create a map of magnetic variation, showing this by what he called 'Curve Lines' or lines along which the variation was equal. It was the first successful cartographic use of isoclinic curves (lines which show equal occurrence of a particular phenomenon), later to become a mainstay of thematic cartography. Halley's first map in 1700 of the Atlantic (which he dedicated to King William III) was supplanted in 1702 by his global map of magnetic variance. These two maps, together with his earlier chart of winds in the Atlantic, and his map of the path of the 1715 total eclipse of the Sun across England, make Halley one of the true pioneers of thematic cartography.

OPPOSITE PAGE – Africa and the Indian Ocean from *A new and correct sea chart of the whole world shewing the variations of the compass as they were found in the year M.D.CC* by Edmond Halley. National Maritime Museum, Greenwich, London, UK.

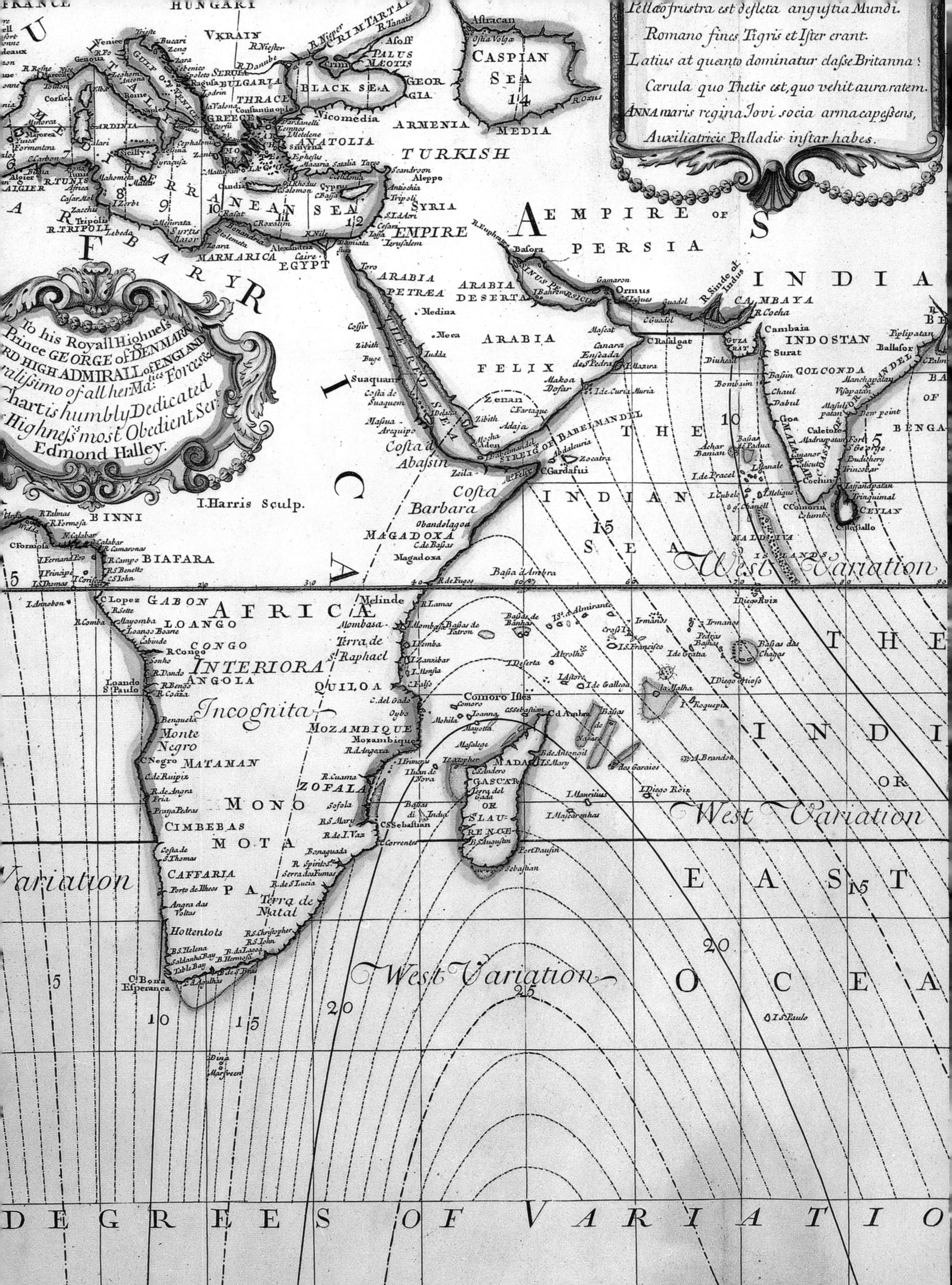

Pellæo frustra est defleta angustia Mundi.
Romano fines Tigris et Ister erant.
Latius at quanto dominatur classe Britanna:
Cærula quo Thetis est, quo vehit aura ratem.
ANNA maris regina Jovi socia arma capessens,
Auxiliatricis Palladis instar habes.
To his Royall Highness Prince GEORGE of DENMARK
LORD HIGH ADMIRALL of ENGLAND
Generalissimo of all her Ma.ties Forces &c.
This Chart is humbly Dedicated
by his Highness's most Obedient Serv.t
Edmond Halley.
I. Harris Sculp.
HUNGARY
VKRAIN
CRIM TARTARY
PALUS MEOTIS
CASPIAN SEA
BLACK SEA
GEORGIA
THRACE
GREECE
ARMENIA
MEDIA
ANATOLIA
TURKISH EMPIRE
SYRIA
Aleppo
MEDITERRANEAN SEA
BARBARY
AFRICA
MARMARICA
EGYPT
EMPIRE OF PERSIA
SINUS PERSICUS
INDIA
ARABIA PETRÆA
ARABIA DESERTA
ARABIA FELIX
Medina
Mecca
THE RED SEA
STREIG OF BABELMANDEL
CAMBAYA
INDOSTAN
GOLCONDA
COAST OF MALABAR
COAST OF COROMANDEL
CEYLAN
THE INDIAN SEA
MALDIVA ISLANDS
Costa Barbara
MAGADOXA
Melinde
ZANGUEBAR
AFRICÆ INTERIORA Incognita
LOANGO
CONGO
ANGOLA
GABON
BIAFARA
BINNI
QUILOA
MOZAMBIQUE
ZOFALA
MONOMOTAPA
CIMBEBAS
MATAMAN
CAFFARIA
Terra de Natal
Hottentots
C. Bona Esperanca
Comoro Isles
MADAGASCAR OR S.T LAURENCE
I. Mauritius
I. Mascarenhas
THE INDIAN OR EAST OCEAN
West Variation
DEGREES OF VARIATION

Smith GEOLOGICAL MAP

The late eighteenth century saw an upsurge of interest in geology, as scientific explanations of how geological strata had been laid down took primacy over religious ones. This saw a bitter debate between Neptunists, who considered they had been formed by heat from the Earth's core and Plutonists who argued that they had been produced by precipitation from the oceans. The Industrial Revolution, and the increased need to identify productive strata of coal and iron, inspired a more practical interest, resulting in the first geological maps, notably the one of Britain produced in 1815 by William Smith.

William Smith's comparatively humble background (born in 1769 to a village blacksmith) proved no obstacle to his establishing himself as the founder of modern British geology. Smith spent much of his early career working on the canals which were proving themselves vital arteries along which coal, essential fuel for Britain's burgeoning Industrial Revolution, could travel. As chief engineer to the Somerset Coal Canal from 1794 to 1799, Smith noted, from the rocks revealed in making cuttings for the canal, the layers, or strata, in which rocks were formed and that in coal-bearing formations the same sequence of these layers was repeated: sandstone, siltstone, mudstone, non-marine, marine and coal. He also observed the fossils to be found in any particular stratum were similar and by 1799 had devised a way to identify strata from the type of fossils found within them (which he finally published in 1816 as *Strata Identified by Organized Fossils*). Smith began to travel throughout England and Wales, taking note of the rock formations and layers he found on his journeys. He produced his first small scale geological map, of the area around Bath, in 1799. He used hand-colouring to indicate the different rock types – for example, yellow for Oolite and blue for Lias. Smith next produced a small scale map of England and Wales in 1801, using a base provided by the distinguished map publisher John Cary. Yet this was not nearly the full extent of his ambitions and for the next fourteen years he laboured on a far grander project. His *A Delineation of the Strata of England and Wales with part of Scotland* – a map measuring 1.8 m x 2.7 m (6 ft x 9 ft) in fifteen sheets, each of which took him over a week to hand-colour – was completed in 1815. The map showed in detail the geological composition of a large part of Britain, covering an area of 129,000 sq km (50,000 sq miles).

Although not quite the first geological map – Guettard's 1780 *Atlas et description minéralogiques de la France* had preceded it – Smith's achievement was prodigious in translating the three dimensional structure of rock formations (by deepening the hue to show distance below the ground) onto a two-dimensional map and in his comprehensive survey which revealed the geological face of an entire nation.

OPPOSITE PAGE – The first geological map of Britain, published in 1815 by British geologist William Smith, often described as the father of English geology. The map is dedicated to British naturalist Sir Joseph Banks. British Library, London, UK.

A DELINEATION OF THE STRATA OF ENGLAND AND WALES, WITH PART OF SCOTLAND;
EXHIBITING THE COLLIERIES AND MINES,
THE MARSHES AND FEN LANDS ORIGINALLY OVERFLOWED BY THE SEA,
AND THE VARIETIES OF SOIL
ACCORDING TO THE VARIATIONS IN THE SUBSTRATA,
ILLUSTRATED by the MOST DESCRIPTIVE NAMES
BY W. SMITH.
To the Right Hon.ble Sir Joseph Banks Bar.t P.R.S.
This MAP is by Permission most respectfully dedicated
by his much obliged Servant
W. Smith
THE GERMAN OCEAN
IRISH SEA
THE
FIRTH OF FORTH
CAERNARVON BAY
CARDIGAN BAY
St. GEORGE'S CHANNEL
BRISTOL CHANNEL
THE WASH
THAMES
EXPLANATION
THE ENGLISH CHANNEL

Cassini Carte de France

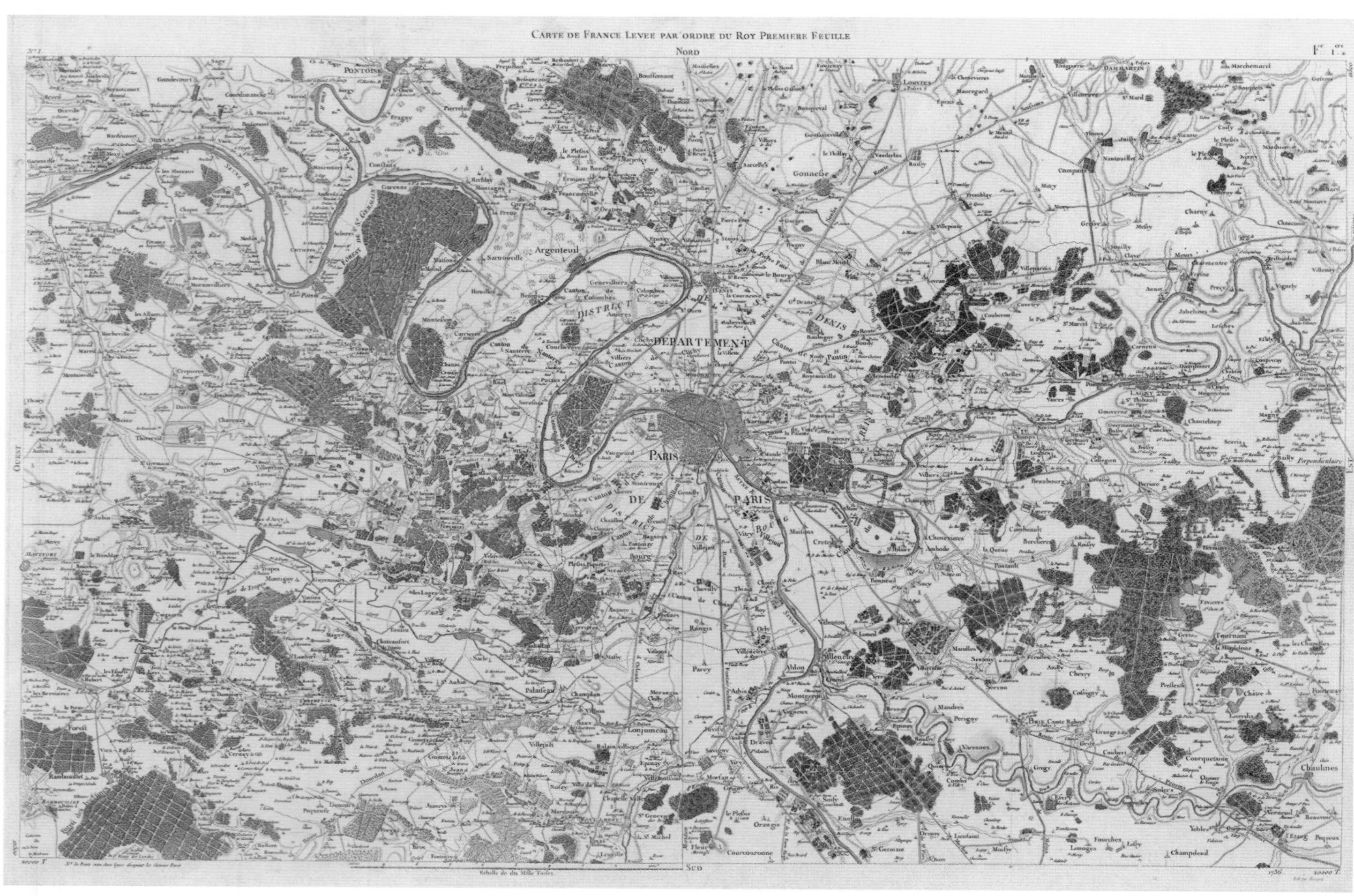

Map of Paris, published in 1788, by César-François Cassini de Thury. This map was part of a second survey of France which took nearly seventy years to complete. Library of Congress, Washington D.C., USA.

MAPPING A NATION

In 1744 César-François Cassini de Thury completed work on a survey of the whole of France, culminating in the publication of eighteen maps of the country – the first systematic cartographic survey of a nation state. Subsequently Cassini was commissioned to create a more detailed national map series, which consisted of 180 map sheets, including this map of Paris and its environs. The maps represented the culmination of nearly a century's work by three generations of the Cassini family, which put France at the forefront of European cartography.

France had not been at the forefront of innovations in mapping in the sixteenth century and the country's comparatively modest overseas empire did little to promote it. Two developments, one political and one scientific transformed this conservatism. The emergence of a strong centralized state in France, directed by an absolutist monarch meant that the government became able to undertake large-scale enterprise that other, less dirigiste, countries could not. Louis XIV required maps to delineate his borders and demarcate his realm and the *Cartes générales de toutes les provinces* (1658–9) of Nicolas Sanson answered this need.

Yet Sanson's atlas was not produced on the basis of a scientific survey. It needed the application of the technique of triangulation, devised by the Dutch mathematician Gemma Frisius around 1533, to produce one. By laying out a measure base line of known length, and determining the distance between two points on it, the distance to a third point could be calculated by creating a triangle linking all three, measuring the angles between them and then using trigonometry to calculate the length of the triangle's sides.

Louis XIV's talented finance minister Jean-Baptiste Colbert was keen to improve on Sanson's map and asked provincial officials to submit every map of their region that could be found to be corrected and form the basis of a new national map. But few provinces did so and it was clear a new survey was needed. Colbert turned to luminaries of the newly founded Académie des Sciences and the Paris Observatory and so a partnership was formed between Abbé Jean Picard and the Italian astronomer Giovanni Cassini to conduct a new survey. In 1669, Picard established a meridian line between Malvoisine south of Paris and Sordon near Amiens and nine years of triangulation work yielded a map of the Paris region in 1678 on a scale of 1: 86,400. A similar survey of France's coastline between 1679 and 1682 led to the publication of a *Carte de France corrigée*, which caused some consternation when it revealed that France's area was a fifth smaller than had been supposed.

Louis XIV's campaigns in the Spanish Netherlands in the 1680s and the Spanish War of Succession (1701–13) led to a suspension of the surveying and it was only in 1733 that Cassini's grandson César-François (known as Cassini III) resumed the process on the orders of Louis XV's Controller-General Philibert Orry. For eleven years the surveyors toiled in often adverse conditions; one was even hacked to death by villagers who thought his surveying tools were instruments of witchcraft. In 1744, however, Cassini III published a map of France in eighteen sheets. It lacked any indication of settlements or other topographic features, showing the country instead as a dense network of triangles.

In 1746 Louis XV asked Cassini III to begin yet another new survey which would fill in the topographical detail missing from the previous map. Cassini planned an eighteen-year programme of surveying and publication, with ten maps to appear each year, making a total of 180 sheets, also on a scale of 1:86,400. Teams of engineers were to resurvey triangulation points (setting up two-metre high pyramids where there was no suitable landmark) but also to keep logs of topographical date. Once the maps had been drawn these were to be signed off by local notables (on pain of a hefty fine for the engineer if he did not do so). The work began in 1748, but it took eight years for the first two sheets (of Beauvais and Paris) to be produced. With their unparalleled level of accurate detail of rivers, mountains, villages and towns, they were the first truly modern maps of a nation.

No sooner had Cassini III presented the Beauvais map to Louis XV, than state funding was cut, as costs had so vastly overrun. A private company, the Société de la Carte de France, was set up with fifty shareholders who each subscribed 2,400 livres to fund the map's continuance, while copies of each sheet were sold for four livres as they were published. Progress was swifter now, and by 1760 fifty maps had been produced, but only thirty-eight more came out between 1760 and 1770. By the time the company was nationalized by the French Revolutionaries in 1793, the project, now directed by Cassini III's son Jean-Dominique (Cassini IV), was tantalizingly close to completion, with just eleven maps yet to be engraved and part of Brittany surveyed but not yet drawn.

So valuable were the plates of Cassini's maps deemed to be by the French state – Napoleon's army had made ample use of them during the wars – that they were never returned to him, despite a lawsuit in 1818. By then, however, they were nearly redundant. A new survey of France had already begun to produce an improved map of the country, a process that was only finally complete with the publication of the last sheet (from a total of 273) in 1880.

SEYNE R.
Triel
Chanteloup
Verneuil
Vernouillet
Maurecourt
Conflans
Andrefis
Garenne
Herblay
Montagny
le Plefsis Bouchart
Andilly
Marjency
Eau Bonne
Ermont
Franconville
Soify
Mont morency
Deuil
Cormeil en Parisis
la Frette
St. Gratien
Sannois
Epinay
Medan
Acheres
Carrieres
FOREST DE ST. GERMAIN
Maisons
le Mesnil
Sartrouville
Argenteuil
Houilles
Befons
Genevilliers
Colombes
Canton de Colombes
DISTRICT
Anieres
POISSY
Mantes
Orgeval
Chambourcy
Egremont
St. GERMAIN
Montefson
Carrieres St. Denis
Canton de Nanterre
Nanterre
Courbevoye
Clichy
Forest de Marly
Fourqueux
Mareil
Chatou
Croissy sur Seine
Ruel
Putaux
Neuilly
Villiers
Feucherolles
Lanluet
l'Etang
Marly
Louvecienne
Bougival
Mont Valerien
Surefnes
Boulogne
Chaillot
Chavenay
St. Nom
Noisy
Bailly
Garches
St. Cloud
Canton de Passy
Passy
Auteuil
Invalides
Ecole Militaire
Vaugirard
PARI
Dreux
Villepreux
Rennemoulin
Rocquencourt
Marne
Villedavre
Sevre
Bellevue
Billancourt
Ifsy
Montrouge
les Clayes
Fontenay
St. Antoine
Canton Vanves
DE
DISTRICT
Meudon
Chatillon
Bois d'Arcy
Montreuil
VERSAILLES
Viroflay
Chaville
Clamart sous Meudon
Canton de Arcueil
Bagneux
Fontenay aux Roses
Flancourt
Velify
Villacoublay
Bourg la Reine
Plefsis Piquet
Sceaux
Trapes
Montigny
Guyencourt
Buc
Cour Roland
Bois de Trapes
Voifins le Bretoneux
les Loges
Jouy
Chatenay
Antony
Bievres
Verrieres
Toufsu
Chateaufort
Magny
Port Royal
Igny
Vauhallan
Sacle
St. Lambert
Milon
Villiers le Bacle
Vifsous
Mafsy
St. Aubin
Palaiseau
Champlan
Dampiere
St. Forget
CHEVREUSE
St. Remy
Gif
Orfay
Bures
Vilbon
Saux
Lonjumeau
Choifel
les Trous
Gometz le Chatel ou St. Claire
Villejuft
Cernay la Ville
les Molieres
Gometz la Ville
Balainvilliers
Route d'Orleans
Chilly

le Thillay
Vauderlan
Roisy
Nantouillet
Sarcelles
Gonnesse
Route
Compans
Groslay
la Madeleine
Arnouville
Pt. Tremblay
Mitry
Tremblay
St. Mesme
Bonneuil
Senlis
Mory
Pierre Fitte
Garges
Villepinte
Gressy
Canton
Stains
Messy
Dugny
Souilly
de Pierre Fitte
Blanc Mesnil
Aunay
Claye
le Bourget
Villeparisis
St. Remy
Sevran
S.t DENIS
la Courneuve
Gd. Drancy
Courtry
Aubervilliers
DENIS
Livry
Couberon
le Pin
St. Marcel
Route
Baubigny
Bondy
TEMENT
la Villette
Canton
de Noisy
Pantin
Montfermeil
la Chapelle
Romainville
Chelles
Brou
Pompone
Belleville
Villemomble
Rosny
Montreuil
Gagny
Vaires
LAGNY
Bagnolet
Canton
Charonne
Montreuil
Neuilly
Gournay
Gouverne
Noisiel
Guermantes
Bussy
Canton
la Pissotte
de Vincennes
Fontenay
REINE
Noisy le Grand
Champs
Torcy
St. Germain
Vincennes
St. Mandé
Nogent
Logne
Collegien
Bercy
Conflans
Charenton
Bry sur Marne
Villiers
Beaubourg
Croissy
PARIS
Ivry
BOURG
Champigny
Emery
Vitry
Villejuif
Maisons
Creteil
Canton
St. Hilaire
Chenevieres
Amboile
Combeault
Berchieres
la Queue
Roissy
Pontault
Choisy le Roy
Thiais
Bonneuil
Sucy
Noiseau
Chevilly
de Choisy
le Roy
Ozouer
Rungis
Orly
Valenton
Limeil
Boissy
Lesigny
Villeneuve le Roy
Villeneuve
Marolles
Senteny
Ferolles
Attilly
Chevry
Parey
Ablon
Villecresne
Yeres
Crosne
Servon
Athis
Montgeron
Cossigny
Vigneux
Brunoy
Mandres
Perigny
BRIE Comte Robert
Epinay
Grange
Juvisy
Draveil
Boussy
Varennes
Grisy
Coubert
Savigny

Cook
CHART OF NEW ZEALAND

In 1642, Abel Tasman, whilst searching for a mythical southern continent, discovered what is today known as New Zealand. Tasman's voyage resulted in only a small but nevertheless highly significant cartographic record of what was named by Dutch cartographers as 'Nieuw Zeeland'. However, it would take the skill of a British naval navigator and hydrographer to locate and 're-discover' the islands for Western civilization well over a century later.

In 1768 arrangements were made between the Royal Society in London, King George III and the Admiralty to send Lieutenant James Cook R.N. (1728–1779), who had undertaken extensive hydrographic surveys in North America, to search for a great southern continent. Combined with this objective was the Royal Society's desire to observe a transit of Venus, which had been calculated to occur in 1769. Thus Cook was given command of H.M.S. *Endeavour*, which transported himself, his crew and a scientific party of gentlemen, including naturalists Sir Joseph Banks (1743–1820) and Dr Daniel Solander (1733–1782), to the ends of the known world and beyond.

An essential part of Cook's voyage involved the detailed surveying and mapping of the great uncharted parts of the southern hemisphere. Subsequently hundreds of observations were made recording the ship's position and the numerous new discoveries, many of which had not been seen by Western civilisation. Thus on 7 October 1769 New Zealand was sighted by Nick Young, one of the ship's boys, and Cook oversaw the monumental task of surveying New Zealand's coastline for the first time.

The re-discovery of New Zealand took place during an era when the super powers of the day were actively involved in colonial expansion and scientifically based voyages of exploration. It was a significant moment in world history when Britain claimed a territory of just over 259,000 sq km (100,000 sq miles), which not only added to its empire but gave it a strategic foothold in the Pacific. Under Cook's leadership a detailed geographic picture of New Zealand was recorded for the first time, showing it as two main islands littered with a myriad of smaller ones. For New Zealander's this *Chart of New Zealand explored in 1769 and 1770* represents the first depiction of their country. Many of the place names on the map remind us of those who made the voyage and their experiences, and also of their sending nation – a lasting image of this age of exploration and colonial expansion, and of the brave group of men aboard H.M.S. *Endeavour*.

OPPOSITE PAGE – Chart of New Zealand from Captain Cook's explorations in 1769-70 aboard HMS *Endeavour*. British Library, London, UK.

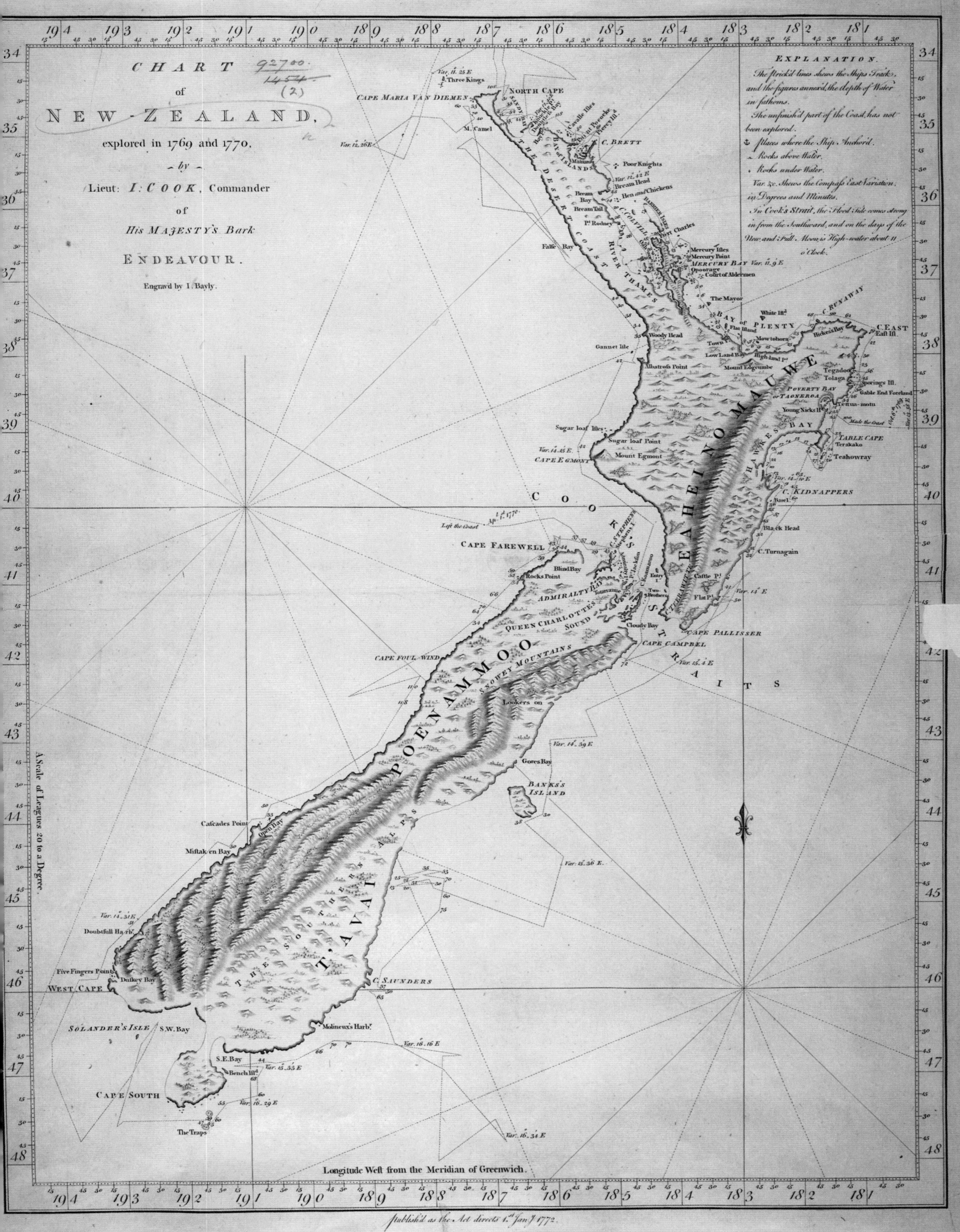

CHART of NEW-ZEALAND, explored in 1769 and 1770, by Lieut: I: COOK, Commander of His MAJESTY's Bark ENDEAVOUR.
Engrav'd by I. Bayly.
EXPLANATION.
The strick'd lines shews the Ships Track, and the figures annex'd, the depth of Water in fathoms.
The unfinish'd part of the Coast, has not been explored.
Places where the Ship Anchor'd.
Rocks above Water.
Rocks under Water.
Var. &c. Shews the Compass East Variation, in Degrees and Minutes.
In Cook's Strait, the Flood Side comes strong in from the Southward, and on the days of the New, and Full Moon, is High-water about 11 o'Clock.
EAHEINOMAUWE
POENAMMOO
T'AVAI POENAMMOO
COOK'S STRAITS
CAPE MARIA VAN DIEMEN
NORTH CAPE
Three Kings
C. BRETT
Poor Knights
Bream Head
Hen and Chickens
Mercury Isles
MERCURY BAY
Court of Aldermen
The Mayor
BAY of PLENTY
White Isl.
C. RUNAWAY
C. EAST
East Isl.
Mount Edgcumbe
Tegadoo
Tolaga
POVERTY BAY
Young Nicks Hd.
TABLE CAPE
Teahowray
C. KIDNAPPERS
HAWKES BAY
Black Head
C. Turnagain
Castle Pt.
Flat Pt.
CAPE PALLISSER
CAPE EGMONT
Mount Egmont
Sugar loaf Isles
Albatross Point
Woody Head
Gannet Isle
False Bay
RIVER THAMES
THE DESERT COAST
CAPE FAREWELL
Blind Bay
Rocks Point
ADMIRALTY BAY
QUEEN CHARLOTTES SOUND
Cloudy Bay
CAPE CAMPBEL
CAPE FOUL WIND
SNOWEY MOUNTAINS
Lookers on
Gores Bay
BANKS'S ISLAND
Cascades Point
Open Bay
Mistaken Bay
THE SOUTHERN ALPS
Doubtfull Harb.
Five Fingers Point
Duskey Bay
WEST CAPE
C. SAUNDERS
Molineux's Harb.
SOLANDER'S ISLE
S.W. Bay
S.E. Bay
Bench Isl.
CAPE SOUTH
The Traps
A Scale of Leagues 20 to a Degree.
Longitude West from the Meridian of Greenwich.
Publish'd as the Act directs 1st Jan. 1772.

Flinders
CHART OF AUSTRALIA

Since the discovery in about 1606 by a Dutch ship of *Terra Australis Incognita*, or the unknown southern land, there was a desire to survey and establish the true extent of this new found land. After Lieutenant James Cook surveyed and charted the east coast of Australia, in 1770, this great land mass was claimed for Great Britain and King George III. Subsequently British, French and Spanish governments sent naval hydrographers to survey the vast areas of uncharted waters.

A young midshipman, Matthew Flinders R.N. (1774–1814), undertook a voyage from Port Jackson (Sydney, Australia) in 1795, during which he proved his talents as a cartographer and navigator. Following a further voyage, as a lieutenant, in 1798 during which he established that Tasmania was not part of the mainland, the Admiralty once again sent Flinders back to Australia. Thus in 1801, in command of H.M.S. *Investigator* he set out to prove that the land mass we know today as Australia was a single continent. He recorded hundreds of dangers and miles of uncharted coastline on the other side of the world, becoming the first European to circumnavigate the continent.

After two years of surveying, during which he survived both shipwreck and a hazardous open boat voyage, the schooner he commanded was forced to put in at Mauritius, which was under the control of the French, who at that time were at war with Britain. Due to an unfortunate discrepancy with his papers he ended up spending six years as a prisoner of war there, allowing him plenty of time to draw up his charts and prepare his journal for publication. He returned to England where his charts and account of his voyage, *A Voyage to Terra Australis*, were published the day before his death in 1814. In that publication Flinders wrote of the naming of this great continent, 'Had I permitted myself any innovation upon the original term, it would have been to convert it into Australia; as being more agreeable to the ear, and as an assimilation to the names of the other great portions of the earth.'

His chart recording the word 'Australia' for the first time is viewed by many people as Australia's birth certificate.

OPPOSITE PAGE – Chart of Australia by captain Matthew Flinders, produced in 1804 from his surveys during the first circumnavigation of the continent. UK Hydrographic Office, Taunton, UK.

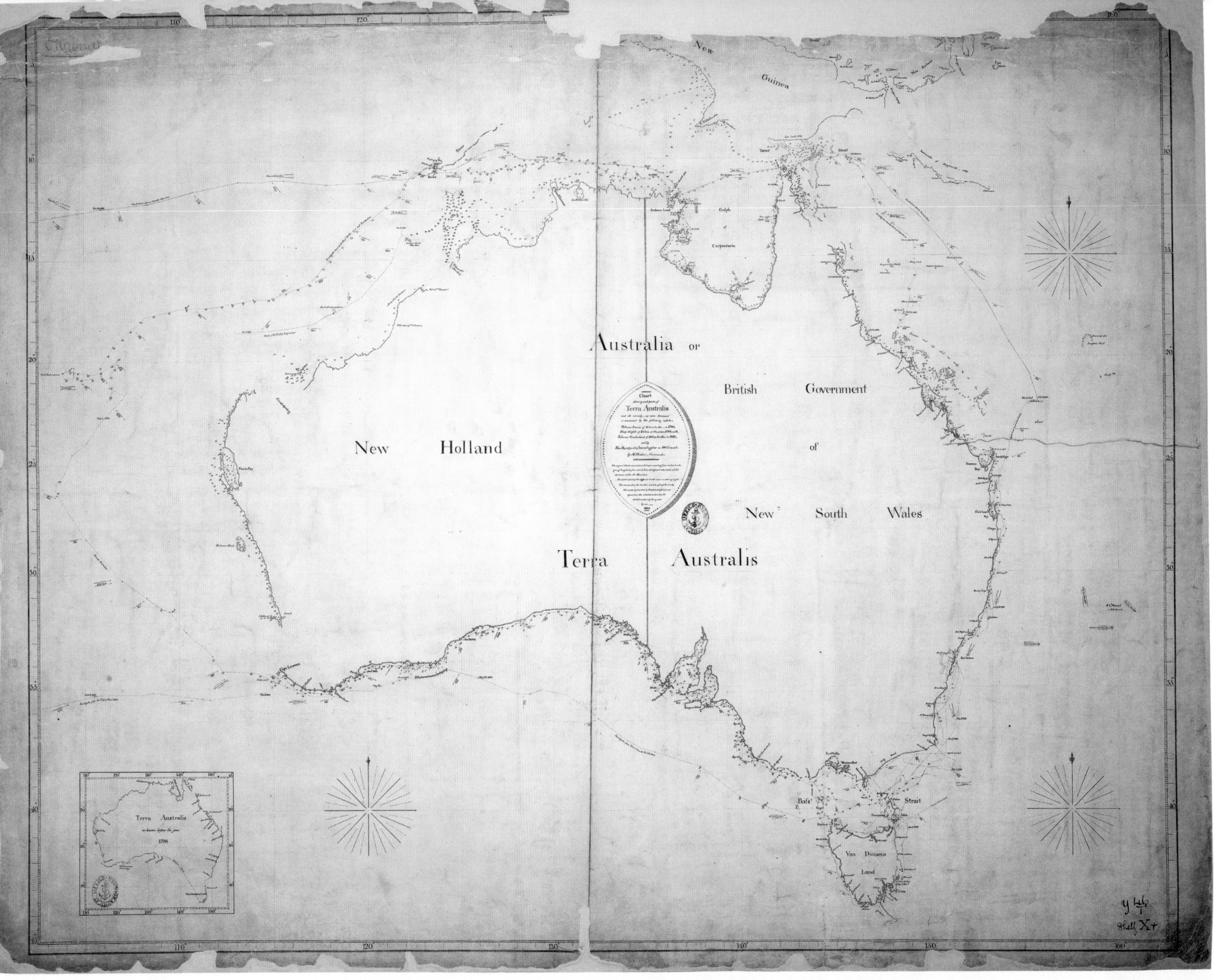

Australia or
British Government
of
New South Wales
New Holland
Terra Australis
Chart
Terra Australis
New
Guinea
Gulph
Carpentaria
Strait
Bass' Strait
Van Diemens
Land
Terra Australis
1798

James Rennell

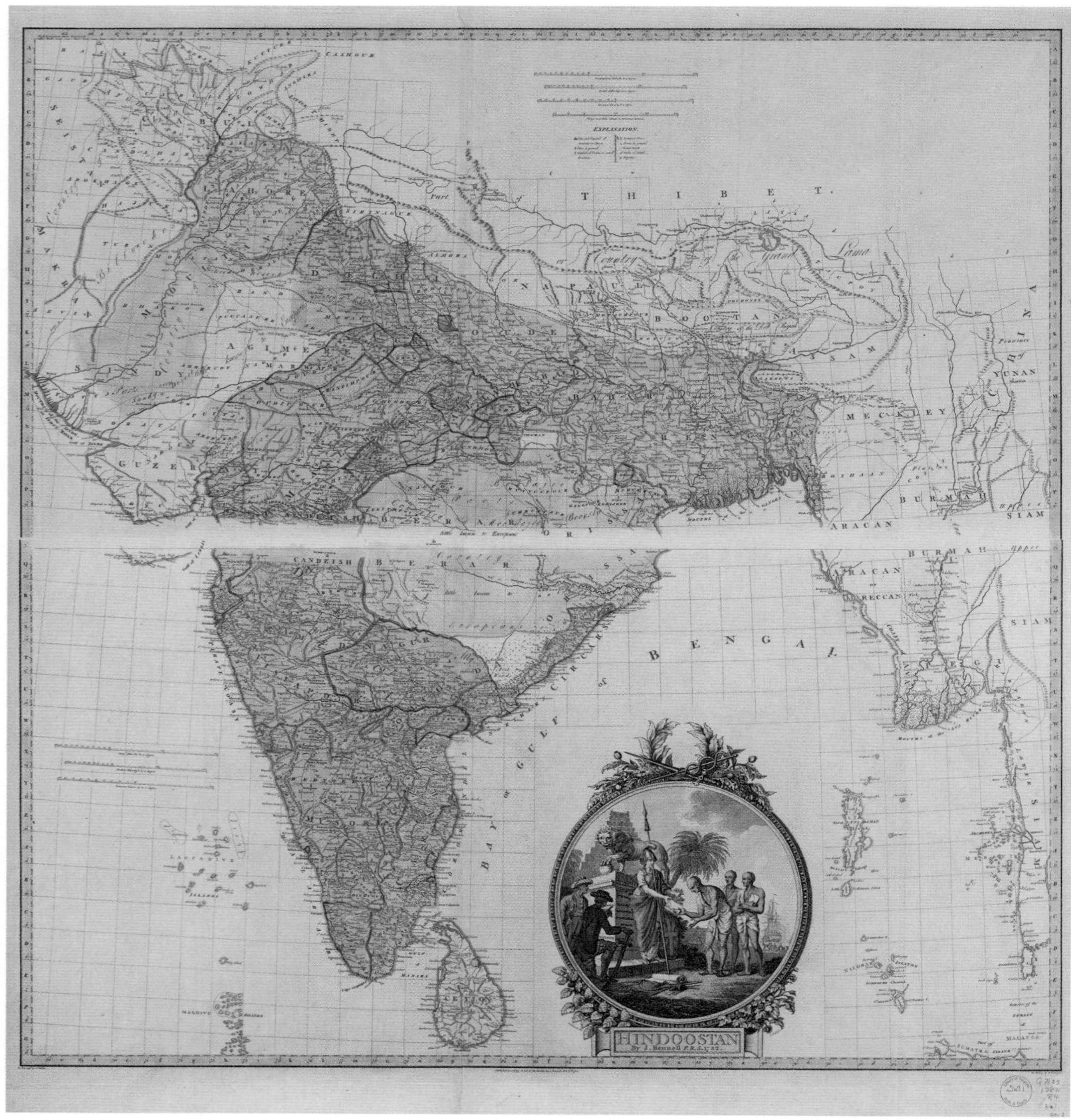

Hindoostan, map by James Rennell from 1782. Library of Congress, Washington D.C., USA.

MAP OF INDIA

***Hindoostan* was known to the world from early times, but reliable maps depicting its coasts, lands, rivers and mountains were few. James Rennell's map *Hindoostan*, engraved and published in London in December 1782, is undoubtedly the first map to show the near correct outline of the coast and detailed internal information. The map marks, in pink, areas occupied at that time by the British East India Company (EIC), including the 'theatre of war and commerce' – the extensive territories in Bihar and Bengal.**

As the EIC's Surveyor General in Bengal from 1764 to 1777, James Rennell had personal knowledge of its geography. For regions beyond, he relied on earlier maps made by military surveyors on their marches, particularly in the south. He referred to the authoritative account of Emperor Akbar's administration, the *Aiyin-e-Acbari*, to mark the provinces. Yet his map has scanty information in the north and west, places then unfamiliar to Europeans. Astronomical observations taken at several cities enabled Rennell to draw the shape of the country more accurately than shown in earlier maps.

The Map *Hindoostan*, 1782, covers the area between 65°E to 102°E and 5°N to 36°N and was hand coloured. It included Ceylon (Sri Lanka), coastal Burma (Myanmar) and Thailand. A large map, made up of two printed sheets joined together horizontally, it measured approximately 81 cm (32 in.) square. Drawn on a scale of 1 inch to 1 equatorial degree, comparative scale lines show Geographical miles, British miles and Common Coss, the local distance measure. Latitudes and longitudes were marked on the subcontinent.

The unusual cartouche is an allusion to the 1781 Act of Parliament which assured the natives in Bengal the freedom to continue with their cultural practices. It shows a lion resting his paw on a globe, projecting Britain as a dominant world power. A three mast Indiaman points to the importance of seafaring in trade. In the foreground three *pundits*, or learned men, offer the scroll containing ancient Hindu laws of religion, the *shastras*, to Britannia, for safekeeping. Two *sepoys*, native soldiers, point to the inscription on the pedestal recording British victories in battles. Various instruments lie on the ground, indicating their usefulness in survey and mapping.

This intricate picture is framed in a wreath of flowers and leaves of poppy and pepper, representing opium and spice. A caduceus, symbol of commerce, and a sword, symbol of military power, rest on top. This cartouche represents the EIC's interests in India in the late eighteenth century: commerce in opium, spices and textiles; armed occupation of territories; dependence on seafaring and non-interference in matters of native cultures. This cartouche was withdrawn from later editions of the map, perhaps because of changes in attitude towards the natives among British opinion makers and a growing influence of missionaries.

Hindoostan was an exceptional achievement, not just for Rennell personally, but also for the EIC – a merchant company which had surveyed and mapped the provinces of Bengal and Bihar, where they had negotiated with the Mughal rulers the rights to collect taxes. They mapped it to know the best sources of its resources: indigo, opium, saltpetre, salt, minerals, textiles, river transport routes, and more.

Rennell remarked in the *Memoir of a Map of Hindoostan; or the Mogol's Empire*, published in 1783, '*... it must not be forgotten, that the EIC have caused a mathematical survey to be made, at their own expense, of a tract in extent to France and England taken together; besides tracing the outline of near 2000 miles of sea coast, and a chain of islands in extent 500 miles more.*'

James Rennell, widely acknowledged as the 'Father of Indian Geography', later added to and improved his map of Hindoostan as the EIC occupied new territories and collected information across the country. Significantly, one purpose of this map was to broadcast to the world the extensive territories that it controlled in Hindoostan. It not only shows the provinces of Bihar and Bengal as settled EIC territories, but also coastal settlements north of Mumbai, around Madras and present day Andhra Pradesh, along the Bay of Bengal. In another couple of decades, the British EIC would dominate the subcontinent.

AURUNGABAD
Dowlatabad
BOMBAY
Salsette
Bandora
Basseen
Choul
POONAH
Sevendrook
Dabul
Sattarah
PIRATE COAST
BEJAPOUR or VISIAPOUR
VISIAPOUR
HYDRABAD
Golconda
Beder
Calberga
Callian
Sollapour
Indour
Europe
Godavery River
Nizam Ally
Warangole
Rajamundry
Condavir
Masulipatam
Kistna River
Kistnah R.
Raichore
Adoni
Condanore
Gutti
Ongole
GOA
St. George's I.
Cape Ramas
Oister Rock
Angediva
Onore
Pideon I.
Barcelore
St. Marys I.
Palmera Rocks
Mangalore
Bednore or Hyder Nuggur
BEDNORE
Harpanelly
Sanore
Bisnagur
Chittledroog
ANANTPOUR
Anantpour
Cudapa
Nellore
Pullicate
MADRAS or Ft. St. George
ARCOT
Arcot
Chinglepet
Sadras
Gingee
Pondicherry
Ft. St. David
Cuddalore
Portonovo
Devicotta
Tranquebar
Karical
Negapatam
Point Calymere
Point Pedro
Jafnapatam
MYSORE
SERINGAPATAM
Mysore
Bangalore
Tellicherry
Mahe
Calicut
Tanore
Paniani
Cranganore
COCHIN
St. John
St. Andrew
Quilon
Anjenga
Travancore
Cape COMORIN
DINDIGUL
Dindigul
Tritchinopoly
Madura
Tutacorin
Manapar
Adam's Bridge
Manar
GULF of MANARA
CEYLON
Trinkamaly
Pigeon I.
Molodive
Candy
Negombo
COLUMBO
Callitoor
Barbereen I.
Point de Galle
Red Bay
BAY
LACCADIVE ISLANDS
Bank of Charbaniang
Bank 20 to 35 fathm
Kelay Island
MALDIVE ISLANDS

GULF of BENGAL
NORTHERN CIRCARS
CICACOLE
White Pagoda
Black Pagoda
Jagarnaut Pagoda
Narsingapatam
Manickpatam
Maloud
Ganjam
Itchapour
Barra
Poondy
Nowapara
Calingapatam
Cicacole
Bimlapatam
Singapilly
Visagapatam
Settiaveram
Penticotta
Jagrenatpour or Cock Nanara
Yanam
Point
Capt. Clough's
Great
Great Centinel
DUNCAN'S
Little Centinel
Little
PLASSEY
PONDICHERRY
BUXAR
HOLINGUR
E. Edwards del.
J. Hall sc.
HINDOOSTAN
By J. Rennell F.R.S. 1782.

Abel Buell

The Treaty of Paris (1783) formally ended the Revolutionary War (1775–1783) between Great Britain and the new United States of America. The first American-made map to show the territorial extent of the United States was published by Abel Buell in 1784, with the title, *A New and Correct Map of the United States of North America Layd Down from the Latest Observations and Best Authorities Agreeable to the Peace of 1783*.

Peace talks recognizing US independence started in Paris in 1782 shortly after allied American and French forces defeated the British army at Yorktown, Virginia. Benjamin Franklin, one of the American peace negotiators, wanted the British to cede Canada to an independent America. In the end, the US did not get Canada, which remained British; but the Treaty of Paris confirmed Britain's former thirteen colonies as independent, with territorial boundaries stretching from the Atlantic coast westward to the Mississippi River, as shown on the Buell map.

The mapmaker, Abel Buell, born in Connecticut in 1743, was a silversmith, type founder, and engraver. In 1769, he established the first English-language type foundry in America. His 1784 map is based on a variety of sources, including John Mitchell's 1755 map of North America, which was used to establish boundaries for the Treaty of Paris. As an engraver, Buell was able to beat a rival cartographer, William McMurray, by printing his own map.

The map, a masterpiece of cartography, measures 109.4 cm x 122.8 cm (about 43 x 48 inches), comes in four sections, and is beautifully hand coloured. The map displays the patriotic pride of the new republic by prominently featuring its flag in the large cartouche (lower right). The flag's thirteen stars and stripes, representing the original thirteen states, is lit by the sun, and the flag's staff is carried by an angel-like goddess.

States get large labels on this map except for New York, which was in a boundary dispute with Connecticut over the westward extension of state lands. Buell, a Connecticut citizen, favoured the territorial claims of his home state, and so the name for New York is missing from the map. In fact, competing state claims to western lands, as shown on Buell's map, threatened to pull the new nation apart. However, in 1784 Virginia ceded its land claims north of the Ohio River to the new federal government, and other states soon relinquished their territorial claims too.

Abel Buell's map reflects a key moment in the history of the United States and claims a number of firsts: 'The first map of the newly independent United States that was compiled, printed, and published in America by an American,' according to the US Library of Congress. Also, it was the first American-made map to be copyrighted and to feature the new US flag. Finally, it is a rare map, with only seven known copies in existence. The copy currently on display at the Library of Congress was purchased for a record price of $2.1 million (£1.2 million) in 2010.

UNITED STATES

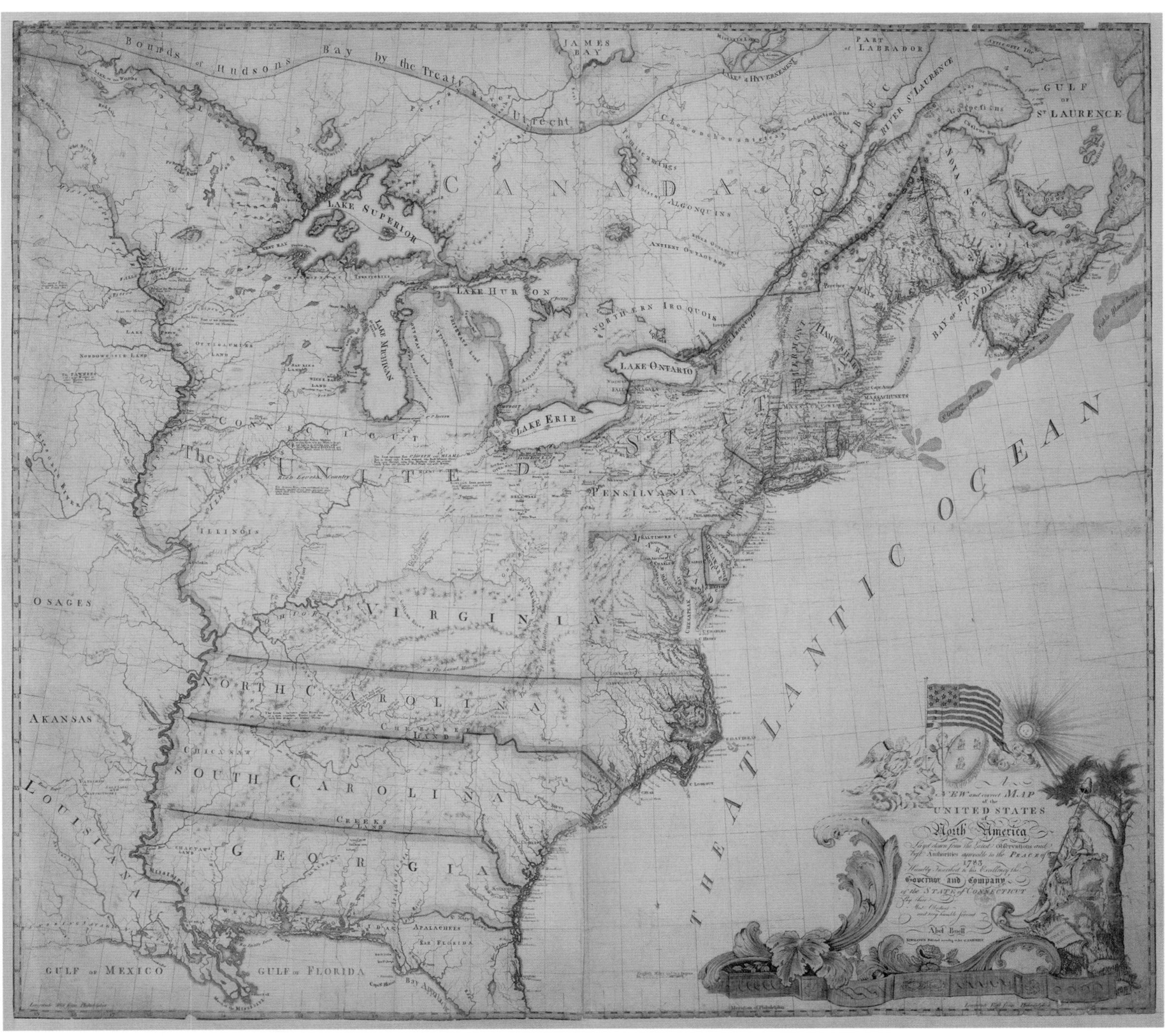

Abel Buell's 1784 map of the United States. This was the first map of the US produced in America, by an American. British Library, London, UK.

LAKE HURON
LAKE MICHIGAN
LAKE ERI
DETROIT
WEST BAY
THE CHIPEWAY TERRITORIES
FALLS of St. ANTHONY
LAKE PEPIN
Chipeway River
NODDOWESSIE LAND
OTTIGAUMIES LAND
SAUKIES LAND
WINNE BAGO LAND
The PAWNEES
CONNECTICUT
The UNITED
Rich Level Country
ILLINOIS RIVER
ILLINOIS
MISSISSIPPI
MISSOURI RIVER
Missouri River
Cabokin
Wabash River
OHIO RIVER
DELAWARE
OSAGES
VIRGINI
The Laurel Mountains
The Allegheny Mountains
NORTH CAROLINA
CHERAKEES LAND
AKANSAS
CHICASAW
SOUTH CAROLINA
Santee River
CREEKS LAND
CHACTAWS LAND
GEORGIA
SAVANNAH
MISSISIPI R
OUISIANA
WEST FLORIDA
APALACHEES
East FLORIDA
St. Mary River

A NEW and correct MAP
of the
UNITED STATES
of
North America
Layd down from the Latest Observations and
best Authorities agreeable to the PEACE of
1783
Humbly Inscribed to his Excellency the
Governor and Company
of the STATE of CONNECTICUT
By their
Most Obedient
and very humble Servant
Abel Buell
NEWHAVEN Published according to Act of ASSEMBLY
THE ATLANTIC OCEAN
ANTIENT OUTAOUAOS
RIVER OUTAOUAIS
NORTHERN IROQUOIS
LAKE ONTARIO
FALLS of NIAGARA
Niagara
VERMONT
HAMPSHIRE
MASSACHUSETS
Province of MAIN
BAY of FUNDY
Sable Island Bank
St Georges Banks
Cape Anne
BOSTON
Cape Cod
LONG ISLAND
JERSEY
PHILADELPHIA
ENSILVANIA
BALTIMORE C
DELAWARE
CHESAPEAK
C CHARLES
C HENRY
C MAY
C HINLOPEN
CHATTERAS
C LOOKOUT
C FEAR
PORTSMOUTH

Lewis and Clark

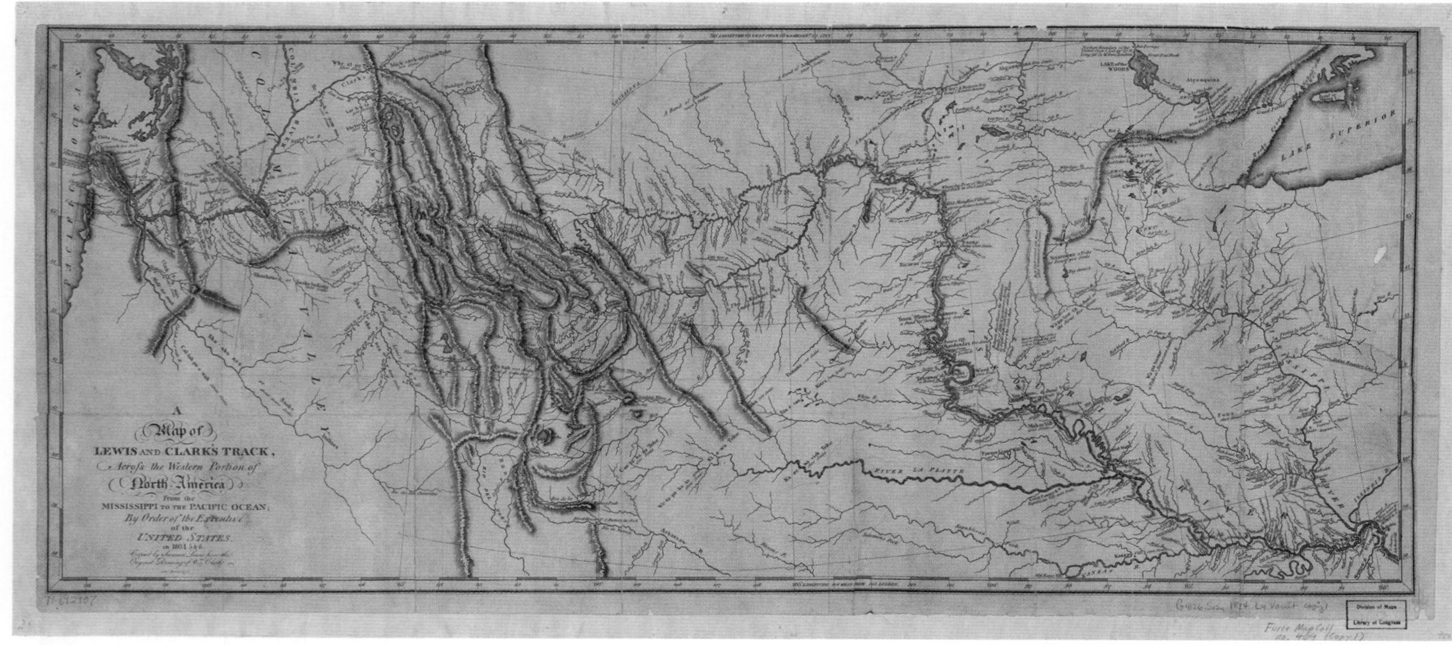

The first detailed map of the river systems of western North America, produced following Lewis and Clark's expedition of 1804–06. Library of Congress, Washington D.C., USA.

ACROSS NORTH AMERICA

The Lewis and Clark expedition, despatched in 1804 by President Thomas Jefferson to map the vast territories in the West, which had been added to the United States by the Louisiana Purchase the year before, spent two years exploring and surveying from St Louis to the Pacific Ocean. The expedition provided a wealth of new information for cartographers, particularly in the area of the Columbia river system shown by this map.

The land which the United States acquired in 1803 under the terms of the Louisiana Purchase covered an area roughly a quarter the size of Europe and was very poorly known. The government had no official programme of mapping and the best available source, an 1802 map by Arrowsmith, included large blank areas in the West. It was with this in mind, and the need in particular to counter British claims in the extreme northwest, that President Jefferson commissioned his former private secretary Meriwether Lewis and William Clark to explore the area from the Missouri River, across the Rocky Mountains and, if possible, to reach the Pacific coast.

The expedition, consisting of thirty-three men, set off from St Louis on 14 May 1804, just two months after the formal transfer of the Louisiana territories to the United States. The Corps of Discovery, as they became known, travelled mainly along navigable waterways, at first by barge down the Missouri and then by canoe to cross the Continental Divide and enter the Clearwater, Snake and Columbia rivers. The expedition reached the Pacific coast in November 1805, its way smoothed by the generally cordial relations Lewis and Clark established along the way with Native American tribes, such as the Mandan, and the information which these groups were able to provide them with. The help provided by Sacagawea, a Shoshone woman married to a French-Canadian fur trader, proved invaluable, both as an interpreter, and as a reassurance to the Native Americans whose lands they crossed, who saw a party travelling with a woman as less threatening.

Having overwintered at the estuary of the Columbia River, the Corps of Discovery made far better time on the return journey, reaching St Louis in September 1806 after just five months' travel. Jefferson had tasked them to explore the Missouri and adjoining rivers which might offer 'direct and practicable communication across the continent or for the purpose of commerce' – a rather vague brief. This was not made any easier by the prevailing notion that the headwaters of all the main rivers in the regions (including the Missouri and Columbia) were in roughly the same area and the view that the Rockies were a fairly gentle, low mountain range which would be easy to cross. The Lewis and Clark expedition dispelled both these misconceptions and provided the first reasonably accurate maps of the Missouri and Columbia river systems. Clark was the Corps of Discovery's main cartographer, compiling some 137 of the 140 maps which the expeditions made. Each day he made sketch maps of the route the Corps took and when they were in winter quarters he used these to create regional surveys, combining them with information he gathered from Native Americans.

After their return, Clark transferred all of his maps onto a single large map measuring 91 cm x 151 cm (36 in. x 60 in.) showing the whole of the West. The manuscript was later reduced and redrafted by the Philadelphia cartographer Samuel Lewis, and, after this version was engraved by Samuel Harrison, it was finally printed and published in 1814. Where before there had been blanks in the map now the relief of the Rockies and the course of the Columbia and its tributaries were shown accurately. The mapping of the Columbia estuary around modern Seattle also provided precious information about a region disputed between the United States and British-controlled Canada until the Oregon Treaty of 1846 settled the frontier. The expedition also gave a boost to American cartography in general, leading to the establishment of an army Corps of Topographical Engineers in 1818 which carried out a regular programme of surveys with the aim of improving military and government mapping. Further expeditions were sent out, such as those led by John Charles Fremont between 1842 and 1845 to survey the Rocky Mountains, Oregon and Upper California. The United States became more densely and more accurately mapped than ever before, but in terms of matching practical exploration to cartographic expertise, the Lewis and Clark map was truly ground-breaking.

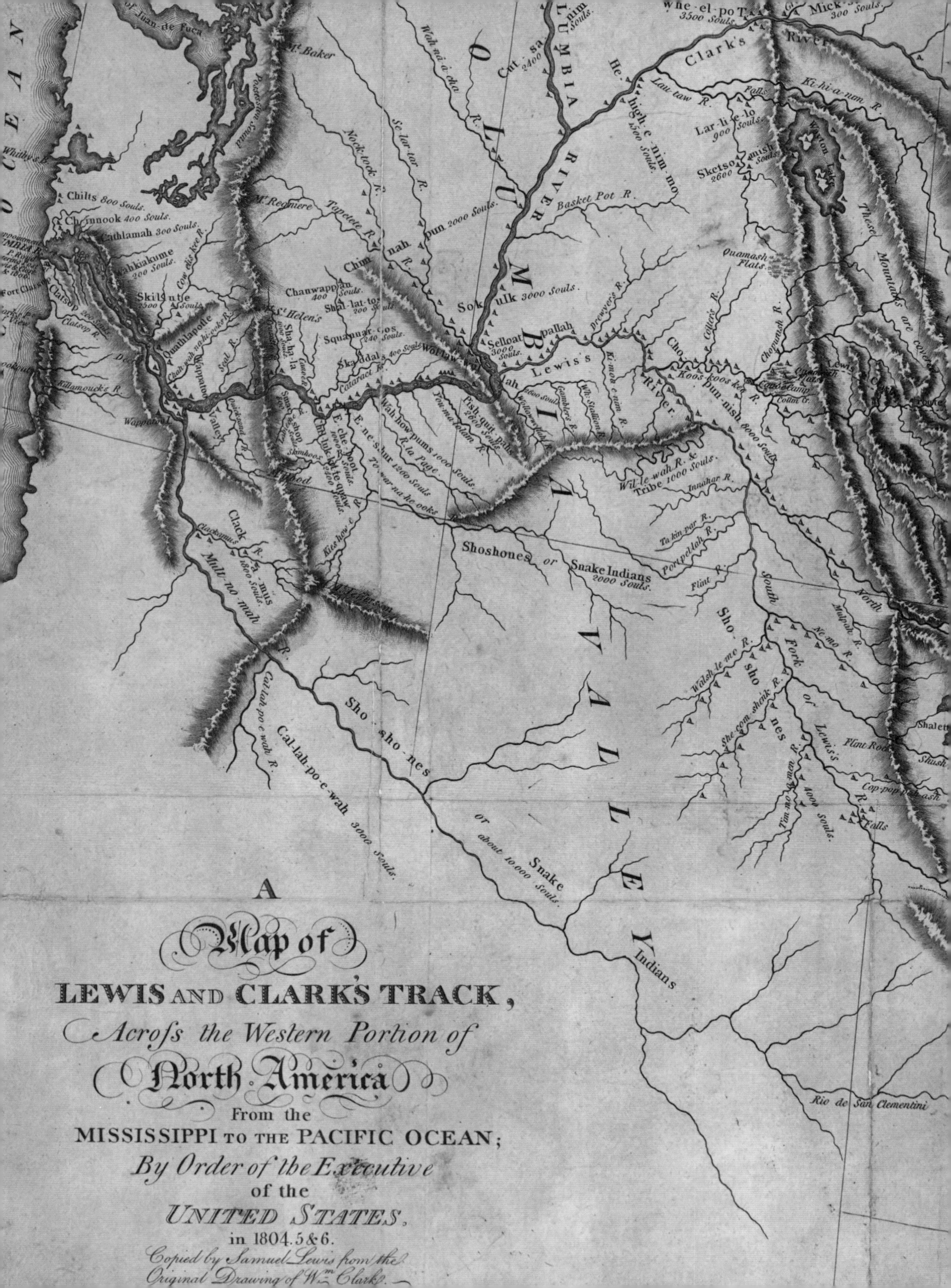
A
Map of
LEWIS AND CLARK'S TRACK,
Across the Western Portion of
North America
From the
MISSISSIPPI TO THE PACIFIC OCEAN;
By Order of the Executive
of the
UNITED STATES,
in 1804.5&6.
Copied by Samuel Lewis from the
Original Drawing of Wm. Clark.
COLUMBIA
COLUMBIA RIVER
VALLEY
Clark's River
Lewis's River
Chilts 800 Souls.
Chinnook 400 Souls.
Cathlamah 300 Souls.
Wahkiakume 200 Souls.
Skilloote 2500
Clatsop 200 Souls.
Clatsop R.
Killamouck's R.
Wappatoo I.
Wappatoo Valley
Seal R.
Quicksand R.
Cow elis kee R.
St. Helen's
Mt. Baker
Mt. Regniere
Mt. Jefferson
Possession Sound
Whitby's B.
Chanwappan 400
Shallattoos 200 Souls
Squannaroos 240 Souls
Skaddals 400 Souls
Sokulk 3000 Souls.
Selloat pallah 3000 Souls.
Chimnahpun 2000 Souls.
Cataract R.
Tapetete R.
Nocktock R.
Selartar R.
Wahnaacha R.
Cutsahnim 2400
Basket Pot R.
Quamash Flats
Clarkmus
Clackamus
Multnomah
Mull no mah R.
Callahpoewah R.
Cal-lah-po-e-wah 3000 Souls.
Sho sho nes
Shoshones or Snake Indians 2000 Souls.
Snake or about 10,000 Souls.
Indians
Wil le wah R. & Tribe 1000 Souls
Innahar R.
Ta kin par R.
Port pel lah R.
Flint R.
Koos koos kee R.
Chopunnish 8000 Souls.
Colter's R.
Drewyer's R.
Kimoh e nim R.
Lau taw R.
Ki-hi-a-nan R.
Lar li e lo 900 Souls.
Sketsomish 2600
Whe-el-po 3500 Souls.
Micksucksealton 300 Souls
These Mountains are covered
South Fork of Lewis's R.
Walsh le me R.
She com sheink R.
Tim no a men R.
Flint Rock
North
Falls
Rio de San Clementini
Tower Creek
Chopunnish R.
Canoe Camp

Northern Boundary of LOUISIANA.
A Band of Assiniboins 2000 Souls.
Black Foot Indians 3500 Souls.
Point of Observation
Maria's R.
Battle R.
Tansey R.
Medicine R.
Dearborn's R.
FALLS.
Stone Walls
MISSOURI RIVER
Milk R.
Stonewall Cr.
Thomson's Cr.
Windsor Cr.
North Mountain Cr.
Wiser's Cr.
Bratton's R.
Porcupine R.
Martha's R.
Little Dry R.
Big Dry Cr.
Big Dry R.
York's Dry R.
Samuel R.
Slaughter R.
Judith R.
Muscle Shell R.
Little Wolf R.
Manuel's Fort in 1807
Pompey's Tower
Here Capt. Clark made Canoes to descend the R.
The Keeheet sa 4000 Souls.
Shannon's Cr.
Little Big Horn R.
Little Horn
Tongue
Rosebud R.
Pryor's Fork
Clark's Fork
Rocky Mountains
Oat la shoot 400 Souls
Lewis's R.
Wisdom R.
Jefferson's R.
Philanthropy R.
Madison's R.
Gallatin's R.
Sho-sho-nes
Po-hah 1000 Souls Band of Snake Ind.
Colter's route in 1807.
Yep pe Band of Snake Ind. 1000 Souls.
Stinking Water R.
Boiling Spring
Salt Fork
Big Horn R.
Southern
Wiser's R.
Henry's R.
Lake Riddle
Cas ta ha na Tribe 1500 Souls.
Ki a wa Tribe
We ta pa ha to and 1800 Souls.
Ka ne na vish Tribe 1800 Souls.
Dotame 120 Souls.
Cataka 300 Souls.
Sta e tan or Kites
Rio de la Plata
Highest Peak
Block House U.S. Factory in 1806.
Rio de Nanesi
Arkansaw R.
Third Fork
Second Fork
Kanzas R.
RIO DEL NORTE.
Colter's R.
Bands

1852 THE VICTORIAN ERA

The Industrial Revolution which began in the United Kingdom in the mid-eighteenth century depended critically on iron and coal to build and fuel its factories. The problem of transporting these raw materials was solved by the development of a dense network of canals and railways, and their spread was mapped by men such as George Bradshaw, whose 1852 railway map of Britain shows the system just after all the main English railway lines had been built.

The very earliest railways in Britain were built to serve coal and iron ore mining areas, enabling the quarries to transport their raw materials faster (and so more cheaply) to towns, factories and ports. Most of the earliest experiments used wooden rails and carts hauled by horses, but in 1804 Richard Trevithick, a Cornish mining engineer, constructed a steam locomotive for the Penydarren Iron Works near Merthyr Tydfil, which hauled 10 tons over 15.3 km (9½ miles). Although the engine was then put to use driving a steam hammer, the idea spread and, combined with the perfection of iron rails by the 1820s, drove the building of a network of railways serving all the main mining areas.

The first public railway in England was the Surrey Iron Railway from Wandsworth to Croydon which opened in July 1803 to service oil-cake mills and move iron from barges on the Thames. By 1807 a prototype passenger service had begun on the Oystermouth Railway in South Wales. Yet these were only tentative steps. The first major passenger service was the Stockton to Darlington Railway which commenced operations in 1825. Yet despite the opening day of this service seeing passengers hauled by George Stephenson's *Locomotion*, for some years afterwards horses were used to pull passenger carriages.

The opening of the Liverpool to Manchester Railway in 1830 marked the railways' coming of age in Britain, linking two major industrial areas for both freight and passengers. The country was soon gripped by railway fever, with parliamentary authorization for lines radiating out of London in all directions, as well as additional cross-country routes, such as the Manchester to Sheffield Railway. By 1837 some 2,400 km (1,500 miles) of railway had been sanctioned, and by the end of 1843, around 3,200 km (2,000 miles) were in operation.

Mine and quarry owners were now freed from their dependence on canal and seaborne transportation, while factory owners no longer had to rely on local sources of fuel, allowing Britain's industrial areas to expand still further. The South Durham and Lancashire Union Railway was built to take Furness iron ore to the Cleveland ironworks, while London, the biggest market for domestic fuel, sucked in ever greater quantities of rail-borne coal. In 1845–50 only 1.6 per cent of London's requirements had been met by coal hauled by canal or rail (the remainder being seaborne), but in 1867 rail-transported coal exceeded that carried by sea to the capital for the first time. By 1869, the North Eastern Railway was carrying 4.7 million tons of coal and 1.7 million tons of coke each year, dwarfing the 1.6 million tons of coal and 117,000 tons of coke it had transported twenty years earlier.

The railway system grew in an unplanned fashion, with private companies gaining authorization from parliament and funding the building of lines through shareholder subscriptions. The most successful companies such as the Great Western (1835) grew by extending their network, or, as with the Great Northern (1846) through amalgamation. But the competing companies had little interest in a national route map as opposed to prospectus maps of their own networks designed to pull in new shareholders. Zachary Macaulay of the Railway Clearing House – established in 1842 as a way of facilitating through-ticketing on lines owned by different companies – created one in 1851, but the most iconic maps of Britain's railway age were created by George Bradshaw, a Lancashire engraver.

In 1830, Bradshaw produced his first map of Britain's canals and railways and, once he began publishing railway timetables in 1842, he accompanied these with basic maps of the various regional networks. He created larger maps of the British railway system in 1839 and in 1852, the latter showing the density of the network, with great spines criss-crossing the country and branches radiating through the industrial heartlands. It is a fitting memorial both to the role played by railways in the Industrial Revolution and to Bradshaw himself, who died of cholera in 1853 while on a visit to Norway.

THE INDUSTRIAL REVOLUTION

George Bradshaw's 1852 highly detailed map of the railways of Great Britain and Ireland, including their stations. Mapseeker.co.uk.

Wigton
Glenluce
Gatehouse
Kirkcudbright
Sorby
Whithorn
Luce Bay
Wigton B.
Dalston
Curthwaite
Wigton
Leegate
Brisco
Kirkhaugh
Alston
Stanhope
Cold Rowley
Southwaite
Arkleby
Plumpton
Maryport
Dearham
Flimby
Workington
Harrington
Parton
Cockermouth
Whitehaven
Penrith
Clifton
Appleby
Barnard Castle
Kirby Stephen
Shap
Rossthwaite
Egremont
St. Bees
Nethertown
Braystones
Sellafield
Seascale
Drigg
Ravenglass
Eskmeals
Bootle
Silecroft
Holborn Hill
Whitehaven & Furness
Furness
Troon to Fleetwood 145 Miles
Glasgow to Liverpool 250 Miles
Belfast to Liverpool 156 Miles
75 Miles
138 Miles
K. Jurby
K. Michael
Ramsay
Peel
Isle of Man
Douglas
Castletown
Calf of Man
Tebay
Low Gill Stn.
Sedburg
Hawes
Winander Mere
Staveley
Burneside
Kendal
Kendal Junc.
Broughton
Kirkby
Milnthorpe
Ulverston
Walney
Barrow
Lancaster & Carlisle
Westmorland
Cumberland
Richmond
Ingleton
Clapham
Settle
Long Preston
Hellifield
Bell Busk
Gargrave
Bentham
Wennington
230
Lancaster
Galgate
Bay Horse
Scorton
Garstang
Brock
Broughton
Fleetwood
Kirkham
Blackpool
Poulton
Lytham
Preston
210
Clitheroe
Whalley
Colne
Blackburn
Burnley
Keighley
Marsden
Southport
Birkdale
Ainsdale
Formby
Croston
Rufford
Burscough
Ormskirk
Maghull
Wigan
Bolton
Bury
198
Rochdale
Oldham
Ashton
Waterloo
201
LIVERPOOL
Birkenhead
191
Bebington
Spital
Bromborough
Hooton
MANCHESTER
Warrington
Moore
Preston Brook 175
Stockport
Cheadle
Handforth
Wimslow
Alderley
Chelford
Holmes Chapel
Macclesfield
Acton 171¾
Hartford 168
CHESTER
178
Winsford 164
Minshull Vernon 162
Crewe
Sandbach
Congleton
Rushton
Leek
Basford 154½
Madeley 149½
Whitmore 146½
Standon Br. 142½
Norton Br. 138½
Stafford
132½
Penkridge
136
Spread Eagle
Four Ashes
126
Wolverhampton
Lichfield
Birmingham
Moseley
Kings Norton
Stoke
Trentham
Burslem
Etruria
Longton
Stone
North Stafford
Holyhead
Island
Isle of Anglesey
Beaumaris
Valley
Holyhead
Bangor
Aber
Conway
Colwyn
Abergele
Rhyl
Prestatyn
Mostyn
Holywell
Bagillt
Flint
Mold
Penmaenmawr
Llanrwst
Denbigh
Carnarvon
Straits of Menai
Bodorgan
Carnarvon
Crickieth
Tremadoc
Langwndle
Holls Mouth
Cape Curaig
Corwen
Llangollen Rd.
Chirk
Wrexham
Rhos
Ruabon
Gresford
Rossett
Saltney
Pulford
Waverton
Tattenhall
Beeston
Calveley
Nantwich
Whittington
Rednal
Oswestry
Baschurch
Shrewsbury
Leaton
Newport
Donnington
Hadley
Wellington
Oakengates
Shifnall
Albrighton
Codsall
Willenhall
Walcot
Upton Magna
Condover
Dorrington
Leebotwood
Church Stretton
Marsh Brook
Craven Arms
Onibury
Bromfield
Ludlow
Tenbury
Woofferton
Leominster
Kington
Presteign
Knighton
Newtown
Llanidloes
Bishops Castle
Wolshpool
Llanfyllin
Merioneth
Harlech
Balla
Dolgelly
Barmouth
Towyn
Machynlleth
R. Dovey or Dyfi
Montgomery
Aberystwith
Aberayron
Tregaron
Rhatadyr
Radnor
Cardigan Bay
Quay Bay
Irish Sea
Worcester
Droitwich
Bromsgrove

GLASGOW
Station
Monkland Canal
Gartsherrie 9.
Coatbridge 10½
Whifflet 11
Shettleton
Rutherglen 2.
Cambuslang 4
Clyde
Uddingston 7.
CLYDESDALE
Hamilton Branch
Castle
Bothwell
Blantyre
Palace
Hamilton 9½ Miles fr. Glasgow
Gartcosh 7½
Garnkirk 7.
Steps Road
Hartlepool
R. Tees
Whitby
Stockton
Middlesboro
Guisborough
Ruswarp
Sleights
Grosmont
Goathland
Levisham
Pickering
Northallerton
Otterington
Thirsk
Scarboro
Seamer
Malton
Flamboro Head
Driffield
Hornsea
York
Beverley
Cottingham
Hull
Leeds
Selby
Goole
Holme
New Holland
Withernsea
Spurn Hd
Gt. Grimsby 154¾
R. Humber
Doncaster
Barnsley
Sheffield
Gainsborough
Market Rasen
Louth 140
Lincoln
Horncastle
Spilsby
Newark
Boston 107
The Wash
Nottingham
Grantham
Derby
Burnham Market
Lynn
Fakenham 138¾
Dereham 126½
Norwich 126
Wymondham
Swaffham
Spalding 93
Peterboro 76½
March 87¾
Ely 72¼
Thetford
Bury St. Edmonds 94½
Diss 94½
Haughley Junc.
Stow Market 80
Huntingdon
Cambridge
Newmarket
Leicester
Rugby 82
Coventry 94
Wellingborough
Northampton
Bedford 64
Melton
Oakham
Stamford

Snow
CHOLERA MAP

The map which London doctor John Snow produced of an outbreak of cholera in London in 1854–5 played a key role in overturning long-held beliefs about how the disease spreads.

Cholera, a deadly disease which appeared to have originated in India, struck European countries with increasing severity after 1830. The prevailing medical orthodoxy was that it was an airborne disease spread by 'miasma' or the poisonous vapours to be found in abundance in the squalid and overcrowded cities of the time. Dr John Snow, a London anaesthetist, was uneasy with this simple equation, considering that the disease-producing vector must be ingested by the victims in some way.

In 1854 a new cholera epidemic struck London and Soho was hit particularly hard, with 500 deaths in just five days. The area had piped water for only two hours a day and as a result many of its residents depended on water from pumps. Snow undertook a detailed investigation and found that deaths from those who took their water from the Broad Street pump were far higher than those who relied on other water sources – of 137 people he could find who had definitely drunk water from the pump, 80 died.

Snow persuaded the members of the Parish Board of St James to remove the pump handle, and deaths from cholera soon fell. More crucially, he charted his findings on a commercial map, marking a series of black bars against addresses to mark the number of fatalities. He was not the first to map the spread of a disease – as early as 1832 A. Brigham had drawn a *Chart Showing the Progress of the Spasmodic Cholera* – nor the only one to map the 1854 Soho outbreak (Edmund Cooper working for the Metropolitan Commissioner of Sewers had also mapped it), but his plotting of the water pumps as a means of diagnosing the cause of the disease was revolutionary. A further refinement in a second version of the map, in which he added a boundary line around those areas where the Broad Street Pump was the closest water source, was an even more vivid demonstration of his point. Contaminated water was the source of cholera, and the scientific mapping of medical conditions was a phenomenon which was here to stay.

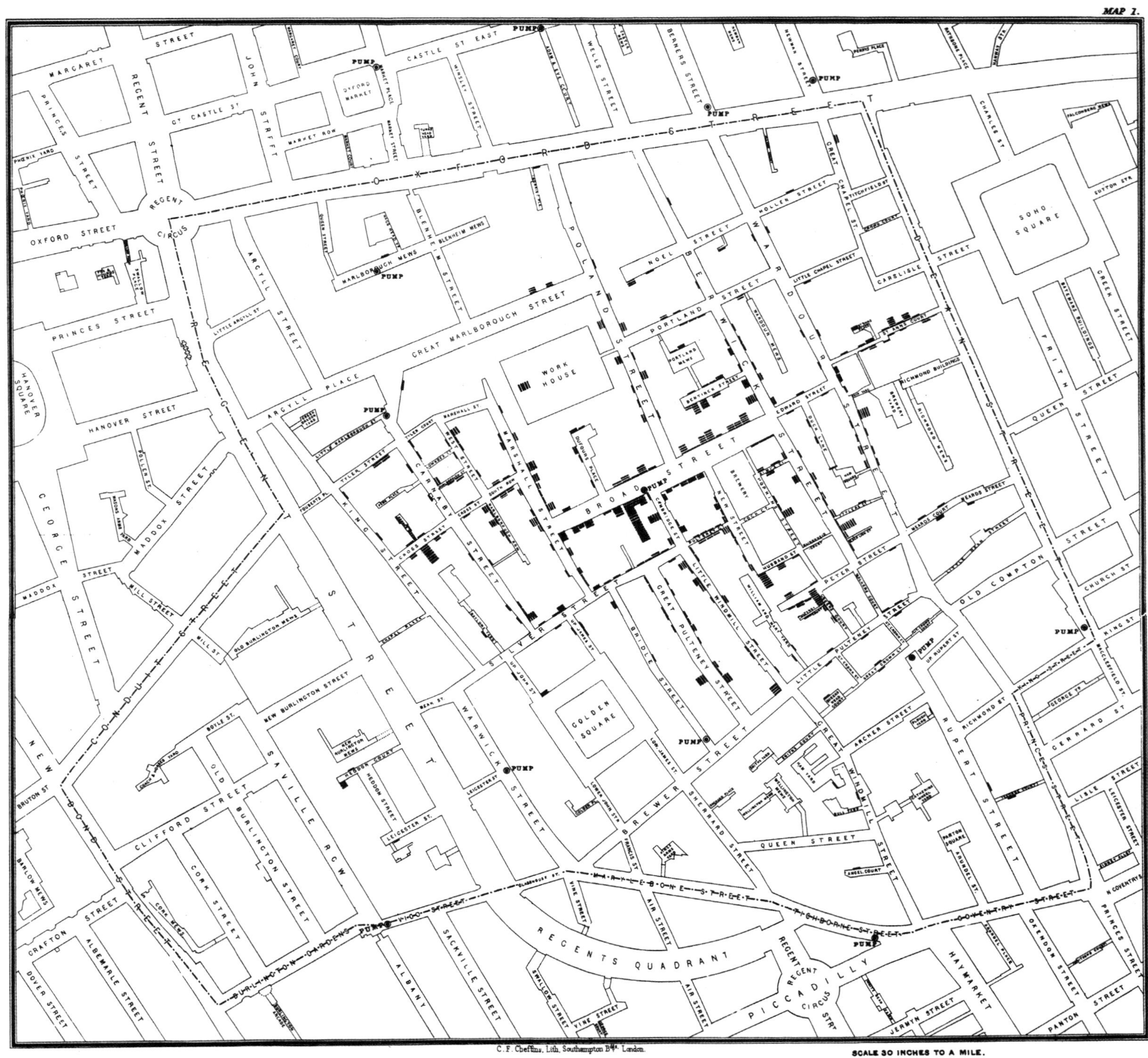

John Snow's ground-breaking thematic map of a cholera outbreak in London in 1854-5. The map led to a greater understanding of the disease, how it spread and how it could be contained. The Wellcome Trust, London, UK.

Booth POVERTY MAP

The use of mapping as a tool to visualize demographic and social trends was pioneered by Charles Booth whose maps of poverty in Victorian London crystallized unease about the plight of the urban poor.

The population of Britain's cities grew explosively in the mid-nineteenth century, with London's rising from 1.7 million in 1831 to 4.7 million by the 1881 census. These sinkholes of urban deprivation exercised legions of Victorian reformers, who pressed for health reforms, education for the poor or the wholesale renovation of the cities' blighted cores. Yet poverty wore a very personal face and scientific tools for its diagnosis and remedy were frustratingly scarce.

All this changed with the work of Charles Booth, a social scientist of herculean dedication who devoted seventeen years to a detailed survey of the plight of London's poor. Beginning in 1886, Booth and his team of enquirers, largely drawn from the ranks of the School Board visitors established by the 1868 Endowed Schools Act, visited thousands of streets and countless homes, and trawled through census returns in an effort to establish the exact level of poverty in the metropolis. The first volume (of what would become seventeen by the work's completion in 1903) of *The Life and Labour of the People* determined that thirty-five per cent of the East End – the subject of the initial enquiry – lived in poverty.

This was shocking enough, but what gave Booth's work added bite were the colour-coded maps which he included in the second (1891) and subsequent volumes. Every street in the capital was shaded in one of seven colours from yellow, denoting the prosperous quarters of the upper classes to a sombre black signifying the haunts of the 'lowest class of occasional labourers, loafers and semi-criminals'. The intermingling of rich and poor, but above all the vast swathes of London where the majority lived below Booth's chosen poverty line of eighteen shillings a week, provided a powerful visual representation of the challenges that faced reformers.

Booth's map also provided a template for future social cartography, and its scientific application has found a range of uses, from health and education planning to the setting of insurance rates by postcode.

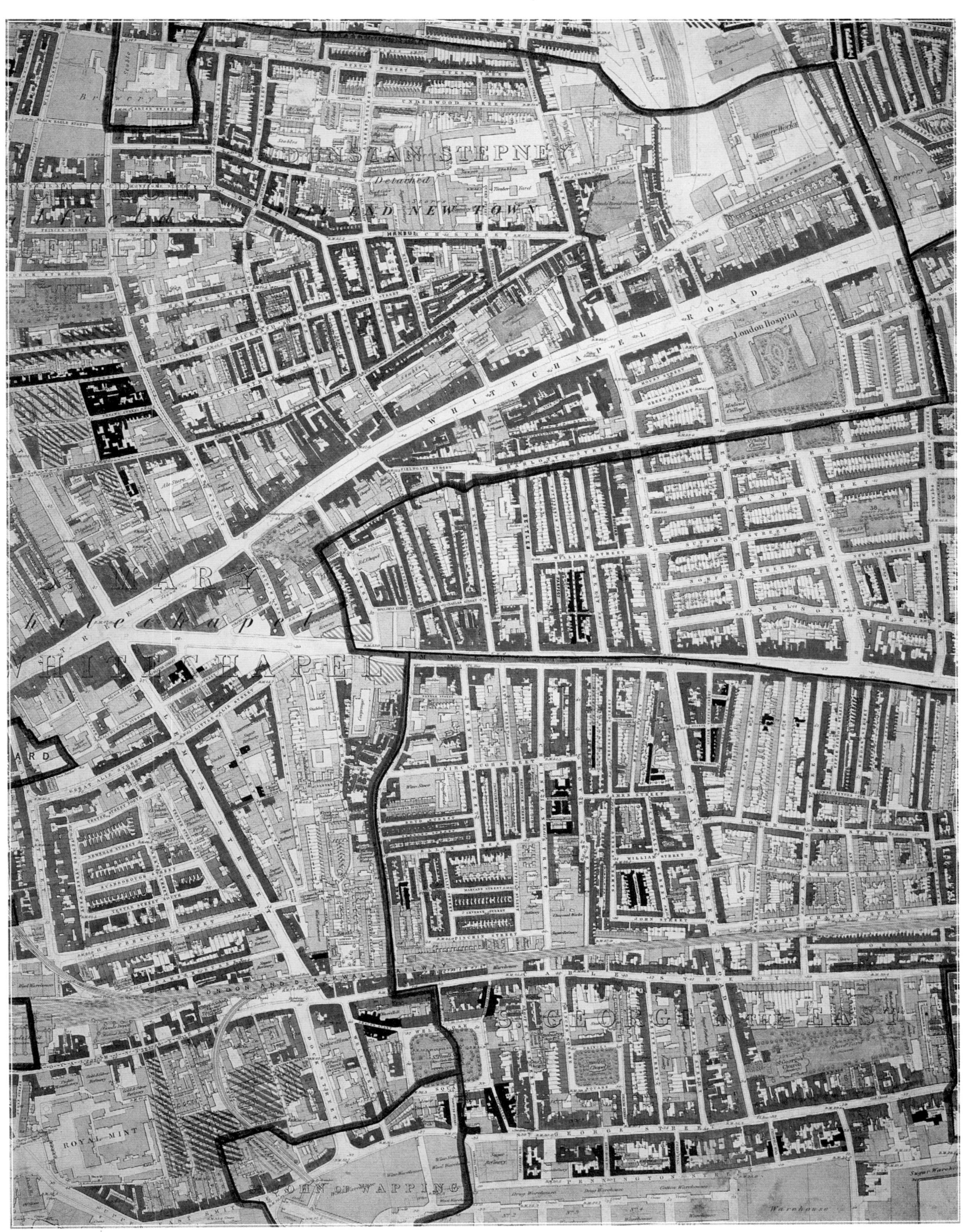

Extract from Charles Booth's map of poverty in London, 1898. Colours represent different levels of affluence, from 'Well-to-do' (red) to 'Very poor' (dark blue) and 'Semi-criminal' (black). Museum of London, London, UK.

Reynolds

Reynolds's political map of the United States, 1856, designed to show the free and slave states and the potential spread of slavery. Library of Congress, Washington D.C., USA.

SLAVERY MAP

Few maps are as emblematic of a national crisis as William C. Reynolds's 1856 *Political Map of the United States*. The map formed a part of the bitter debate between abolitionists and defenders of slavery, with states where it was permitted, and those which were free, clearly indicated, along with an ominous tide of green in those new territories which Reynolds, and many others, feared might soon be opened to slave-holding.

As the infant United States grew, particularly with the acquisition of vast territories from France through the Louisiana Purchase in 1803 and from Spain by the Treaty of Guadalupe Hidalgo in 1848, so bitter debates arose as to whether slavery, firmly entrenched in the plantations of the south, would be permitted in these newly acquired lands. The Missouri Compromise of 1820 forbade slavery in areas to the north of 36° 30'N (save in Missouri itself, which lay just to the north of that line). Partisans of each side, however, agitated for the ruling to be relaxed or strengthened, depending on their viewpoint. In 1854 Senator Stephen Douglas tried to defuse these tensions by the Kansas-Nebraska Act, which laid down that territories established from the Louisiana Purchase should be free to determine by popular vote whether slavery be allowed or outlawed.

Far from cooling tempers, the Act inflamed them and the arguments spread into the cartographic sphere, with the anti-slavery side in particular seeking to present in visual form the dangers of slavery creeping into the new territories. Although in the 1850s statistical cartography was barely known as a discipline in the United States, proponents of abolition did have access to the data gathered by the census which had taken place each decade since 1790 and which included questions on the racial composition of households. This was taken up in Europe, where the Prussian cartographer August Petermann took the 1850 census data and used it to produce a map showing the relative density of the slave population, with three levels of shading, but in the United States the 1854 final report on the census included just a simple map showing the nation's geographic regions.

Even so, in 1850 George Colton unveiled a giant map of the United States at a public anti-slavery meeting in New York, marking the areas in which slavery might be extended, and this was followed by several others, including one published in 1856 by John Jay in the *New York Tribune* which showed slavery engulfing the whole of the West, with just an island of free territory in California. William Reynolds's map, issued the same year, was a rather more sophisticated affair. The free states are drawn in red, the slave states in black, the territories potentially opened to slavery shaded in deep green, with Kansas, identified as the key to the debate, given its own, light brown hue. The map is accompanied by a wealth of statistical information comparing free to slave states, intending to show the latter in an unfavourable light in terms of their area of improved land, miles of railroad and even their number of public libraries. The purpose of the map is clear in Reynolds's warning that 'in repealing the Missouri Compromise, the institution of Slavery may be carried into ALL of the Territories – the area of which is greater than that of all the States combined.'

As war drew closer, the harnessing of mapping to make a political point became more urgent. Reconnaissance mapping carried out by the US Coast Survey from 1858 to 1861 in anticipation of a military conflict, was matched with the 1860 Census Data by Edwin Hergesheimer, in charge of the Survey's drafting division, to produce a shaded map of Virginia showing the relative density of slavery. The much lighter shading in the west of Virginia foreshadowed its breakaway as a separate state in 1863 to join the Union side of the Civil War. The map came out in June 1861, just two months after eleven states had broken away to form the Confederate States of America, committed to defending slavery and shortly after the fighting between it and the forces of the Union began. The map was so well received that in September it was extended to cover the whole of the southern states, showing, it was hoped, islands of little slavery where the Union cause might receive support.

The Reynolds map, though, had paved the way in showing how cartography might be used effectively for a particular political purpose, effortlessly showing as a single image a message which might be diluted in a thousand words of explanation. When refined by the addition of statistical information and shading, it became an immensely powerful propaganda tool.

THE ELECTION OF PRESIDENT.

States.	Pres. Electors. Reps.	Sen.	Tot'l.	Popular Vote in 1852. Scott.	Hale.	Pierce.
Maine	6	2	8	32,543	8,030	41,609
New Hampshire	3	2	5	16,147	6,695	29,997
Vermont	3	2	5	22,173	8,621	13,044
Massachusetts	11	2	13	56,062	29,993	46,880
Rhode Island	2	2	4	7,626	644	8,735
Connecticut	4	2	6	30,359	3,160	33,249
New York	33	2	35	234,882	25,329	262,083
New Jersey	5	2	7	38,556	350	44,305
Pennsylvania	25	2	27	179,122	8,524	198,568
Ohio	21	2	23	152,526	31,682	169,220
Indiana	11	2	13	80,901	6,934	95,299
Michigan	4	2	6	33,860	7,237	41,842
Illinois	9	2	11	64,934	9,966	80,597
Wisconsin	3	2	5	57,132	....	72,413
Iowa	2	2	4	15,855	1,606	17,762
California	2	2	4	34,971	100	39,665
	144	32	176	1,057,669	148,871	1,195,268
Delaware	1	2	3	6,293	62	6,318
Maryland	6	2	8	35,077	54	40,022
Virginia	13	2	15	57,132	....	72,413
North Carolina	8	2	10	39,058	59	39,744
South Carolina	6	2	8	Electors chosen by Legislature.		
Georgia	8	2	10	16,660	....	34,705
Florida	1	2	3	2,875	....	4,318
Alabama	7	2	9	15,038	....	26,881
Mississippi	5	2	7	17,548	....	26,876
Louisiana	4	2	6	17,255	....	18,647
Texas	2	2	4	4,995	....	13,552
Arkansas	2	2	4	7,404	....	12,173
Tennessee	10	2	12	58,898	....	57,018
Kentucky	10	2	12	57,068	265	53,806
Missouri	7	2	9	29,984	....	38,353
	90	30	120	365,285	440	445,094
Total	234	62	296	1,422,954	149,311	1,640,262

The Presidents and Vice-Presidents.

Terms.	Presidents.	Terms of office.	Vice-Presidents.
1.	George Washington	1789–'93	John Adams.
2.	Do. do.	1793–'97	Do. do.
3.	John Adams	1797–'01	Thomas Jefferson.
4.	Thomas Jefferson	1801–'05	Aaron Burr.
5.	Do. do.	1805–'09	George Clinton.
6.	James Madison	1809–'13	Do. do.*
7.	Do. do.	1813–'17	Elbridge Gerry.*
8.	James Monroe	1817–'21	Dan'l D. Tompkins.
9.	Do. do.	1821–'25	Do. do.
10.	John Quincy Adams	1825–'29	John C. Calhoun.
11.	Andrew Jackson	1829–'33	Do. do.
12.	Do. do.	1833–'37	Martin Van Buren.
13.	Martin Van Buren	1837–'41	Rich'd M. Johnson.
14.	Wm. H. Harrison*	1841	John Tyler.
—	John Tyler	1841–'45	———
15.	James K. Polk	1845–'49	George M. Dallas.
16.	Zachary Taylor*	1849–'50	Millard Fillmore.
—	Millard Fillmore	1850–'53	———
17.	Franklin Pierce	1853–'57	Wm. Rufus King.*

* Died before the expiration of their terms of office.

N.B.—By the DEMOCRATIC (?) legislation of 1854, in repealing the Missouri Compromise, the institution of Slavery *may be carried* into ALL the Territories—the area of which is greater than that of all the States combined.

FREEDOM VS. SLAVERY: COMPARISON OF THE CHIEF STATISTICS OF THE FREE

The Free States.	Area, in sq. miles.	Population. Whites.	Free Col'd.	Slaves.	Total.	Pop'n to sq. m.	Valuation of Property.	Cash Value of Farms.	Acres of Improved Land.	Value of Live Stock.
1. Maine	31,766	581,813	1,356	The Census erroneously reports 236 persons in New Jersey under the caption of "Slaves." In that State slavery was provisionally abolished in 1784; all children born of a slave after 1804, were free in 1820. Twenty-six (26) "slaves" were reported from Utah, "on their way to California."	583,169	18.36	$122,777,571	$54,861,748	2,039,596	$9,705,726
2. New Hampshire	9,280	317,456	520		317,976	34.26	103,652,835	55,245,997	2,251,488	8,871,901
3. Vermont	10,212	313,402	718		314,120	30.76	92,205,049	63,367,227	2,601,409	12,643,228
4. Massachusetts	7,800	985,450	9,064		994,514	127.50	573,342,286	109,076,347	2,133,436	9,647,710
5. Rhode Island	1,306	143,875	3,670		147,545	112.97	80,508,794	17,070,802	356,487	1,532,637
6. Connecticut	4,674	363,099	7,693		370,792	79.33	155,707,980	72,726,422	1,768,178	7,467,490
7. New York	47,000	3,048,325	49,069		3,097,394	65.90	1,080,309,216	554,546,642	12,408,964	73,570,499
8. New Jersey	8,320	465,509	23,810		489,555	58.84	153,151,619	120,237,511	1,767,991	10,679,291
9. Pennsylvania	46,000	2,258,160	53,626		2,311,786	50.26	729,144,998	407,876,099	8,623,619	41,500,053
10. Ohio	39,964	1,955,050	25,279		1,980,329	49.55	504,726,120	358,758,603	9,851,493	44,121,741
11. Indiana	33,809	977,154	11,262		988,416	29.24	202,650,264	136,385,173	5,046,543	22,478,555
12. Michigan	56,243	395.071	2,583		397,754	7.07	59,787,255	51,872,446	1,929,110	8,008,734
13. Illinois	55,405	846,934	5,436		851,470	15.37	156,265,006	96,133,290	5,039,545	24,209,258
14. Wisconsin	53,924	304,756	635		304,391	5.66	42,056,595	28,528,563	1,045,499	4,897,385
15. Iowa	50,914	191,851	333		192,214	3.78	23,714,638	16,657,567	824,682	3,689,275
16. California	155,980	91,635	962		92,597	.59	22,161,872	3,874,041	32,454	3,351,058
Total	612,597	13,238,670	196,016	236	13,434,922	21.91	$4,102,162,098	$2,147,218,478	57,720,494	$286,374,541
Slave States & D. C.	851,508	6,222,418	238,187	3,204,051	9,664,656	11.35	2,952,814,356	1,119,380,109	54,991,694	253,795,330
The Territories	1,472,061	91,980	292	26	92.298	.06	11,586,512	4,976,839	320,426	4,010,645
Total of U. S.	2,936,166	19,553,068	434,495	3,204,313	23,191,876	7.90	$7,066,562,966	$3.271,575,426	113,032,614	$544,180,516

The Free States.	Educational Income.	Newspapers, etc. Number.	Yearly Circ'n.	Public Libraries. Number.	Volumes.	Value of Churches.	Capital in Manufactures.	Miles of Canal.	Miles of Railr'd. In 1850.	In 1856.
1. Maine	$401,347	49	4,203,064	236	121,969	$1,794,209	$14,700,452	50	175	494
2. New Hampshire	231,529	38	3,067,552	129	85,759	1,433,266	18,242,114	11	309	66
3. Vermont	256,898	35	2,567,662	96	64,641	1,251,655	5,001,377	2	243	51
4. Massachusetts	1,486,796	209	64,820,564	1,462	684,015	10,504,888	83,357,642	100	1,095	1,40
5. Rhode Island	160,904	19	2,756,950	96	104,342	1,293,600	12,923,176	....	50	14
6. Connecticut	436,979	46	4,267,932	164	165,318	3,599,330	23,890,348	62	434	69
7. New York	2,718,939	428	115,385,473	11,013	1,760,820	21,539,561	99,904,405	989	1,070	2,79
8. New Jersey	600,282	51	4,098,678	128	80,885	3,712,863	22,184,730	147	231	50
9. Pennsylvania	2,251,520	310	84,898,672	393	363,400	11,853,291	94,473,810	936	981	1,74
10. Ohio	1,097,945	261	30,473,407	352	186,826	5,860,059	29,019,538	921	299	2,72
11. Indiana	445,664	107	4,316,828	151	68,403	1,568,906	7,941,602	367	86	1,78
12. Michigan	214,717	58	3,247,736	417	107,943	793,180	6,534,250	....	344	59
13. Illinois	419,483	107	5,102,276	152	62,486	1,532,305	6,385,387	100	22	2,21
14. Wisconsin	138,473	46	2,665,487	72	21,020	512,552	3,382,148	....	....	46
15. Iowa	65,800	29	1,512,800	32	5,790	235,412	1,292,875	....	....	6
16. California	35,092	7	761,200	....	....	288,400	1,006,197	....	....	8
Total	$10,971,768	1,800	334,146,281	14,893	3,883,617	$66,773,517	$431,290,351	3,685	5,339	16,82
Slave States & D. C.	6,819,808	722	91,166,029	722	752,794	23,038,541	95,918,842	1,113	2,016	6,41
The Territories	42,155	4	97,768	....	....	171,970	1,050,300	....	....	..
Total of U. S.	$17,824,331	2,526	426,409,978	15,615	4.636.411	$89,983,028	$527,209,193	4,798	7,355	23,24

SLAVEHOLDERS.—Of the 6,222,418 white inhabitants of the South, only 347,525 are owners of slaves. And yet this faction controls every branch of the Federal Government, and wields its influence for the increase and perpetuation of Slavery. Classification of the Slaveholders in 1850:

Holders of	1 slave	68,820
Holders of	1 and under 5	105,683
Holders of	5 and under 10	80,765
Holders of	10 and under 20	54,595
Holders of	20 and under 50	29,733
Holders of	50 and under 100	6,196
Holders of	100 and under 200	1,479
Holders of	200 and under 300	187
Holders of	300 and under 500	56
Holders of	500 and under 1,000	9
Holders of	1,000 and over	2
Total number of Slaveholders		347,525

CONGRESSIONAL REPRESENTATION.

HOUSE OF REPRESENTATIVES.

The Free States have a total of 144 members.

The Slave States have a total of 90 members.

One Free State Representative represents 91,935 white men and women.

One Slave State Representative represents 68,725 white men and women.

Slave Representation gives to Slavery an advantage over Freedom of 30 votes in the House of Representatives.

UNITED STATES SENATE.

16 Free States, with a white population of 13,238,670 have 32 Senators.

15 Slave States, with a white population of 6,186,477, have 30 Senators.

So that 413,708 Free Men of the North enjoy but the same political privileges in the U. S. Senate as is given to 206,215 Slave Propagandists.

POST-OFFICE STATISTICS FOR A SINGLE YEAR.

Chief Items of the Accounts.	Free States.	Slave States.
Amount received for Postages	$4,391,860 80	$1,486,984 06
Paid for Transportation of Mails	2,381,607 16	2,087,266 05
	$2,010,253 64	$600,281 99

Showing that there is received yearly in the Free States, Two Millions of Dollars ($2,000,000) MORE THAN IS EXPENDED, while in the Slave States the EXPENDITURES EXCEED THE RECEIPTS over Six Hundred Thousand Dollars ($600,000).

ATES AND OF THE SLAVE STATES, ACCORDING TO THE U. S. CENSUS OF 1850.

The Slave States.	Area, in sq. miles.	Population. Whites.	Free Col'd.	Slaves.	Total.	Pop'n to sq. m.	Valuation of Property.	Cash Value of Farms.	Acres of Improved Land.	Value of Live Stock.	Educational Income.	Newspapers, etc. Number.	Yearly Circ'n.	Public Libraries. Number.	Volumes.	Value of Churches.	Capital in Manufactures.	Miles of Canal.	Miles of Railr'd. In 1850.	In 1856.
Delaware	2,120	71,169	18,073	2,290	91,532	43.18	$18,855,863	$18,880,031	580,862	$1,849,281	$114,599	10	421,200	17	17,950	$340,345	$2,978,945	14	16	86
Maryland	11,124	417,943	74,723	90,368	583,034	52.41	219,217,364	87,178,545	2,797,905	7,997,634	583,303	68	19,612,724	124	125,042	3,974,116	14,753,143	184	324	466
Virginia	61,352	894,800	54,333	472,528	1,421,661	23.17	391,646,438	216,401,543	10,360,135	33,656,659	854,860	87	9,223,068	54	88,462	2,902,220	18,109,993	189	303	1,295
North Carolina	50,704	553,028	27,463	288,548	869,039	17.14	226,800,472	67,891,766	5,453,975	17,717,647	421,959	51	2,020,564	38	29,592	907,785	7,252,225	13	302	631
South Carolina	29,385	274,563	8,960	384,984	668,507	22.75	288,257,694	82,431,684	4,072,551	15,060,015	510,879	46	7,145,930	26	107,472	2,181,476	[illegible],056,865	50	241	846
Georgia	58,000	521,572	2,931	381,682	906,185	15.62	335,425,714	95,753,445	6,378,479	25,728,416	480,514	51	4,070,866	38	31,788	1,327,112	5,460,483	28	609	1,013
Florida	59,268	47,203	932	39,310	87,445	1.48	23,198,734	6,323,109	349,049	2,880,058	54,519	10	319,800	7	2,660	192,600	547,060	....	54	26
Alabama	50,722	426,514	2,265	342,844	771,623	15.21	228,204,332	64,323,224	4,435,614	21,690,112	663,798	60	2,662,741	56	20,623	1,244,741	3,450,606	51	113	467
Mississippi	47,156	295,718	930	309,878	606,526	12.86	228,951,130	54,738,634	3,444,358	19,403,662	460,205	50	1,752,504	117	21,737	832,622	1,833,420	....	60	296
Louisiana	41,255	255,491	17,462	244,809	517,762	12,55	233,998,764	75,814,398	1,590,025	11,152,275	731,165	55	12,416,224	10	26,800	1,940,495	5,318,074	101	66	[illegible]
Texas	237,504	154,034	397	58,161	212,592	0.89	55,362,340	16,550,008	643,976	10,412,927	178,411	34	1,296,924	12	4,230	408,944	539,290	....	....	36
Arkansas	52,198	162,189	608	47,100	209,897	4.02	39,841,025	15,265,245	781,530	6,647,969	105,819	9	377,000	3	420	149,686	324,065	....	....	37
Tennessee	45,600	756,836	6,422	239,459	1,002,717	21.99	207,454,704	97,851,212	5,175,173	29,978,016	443,868	50	6,940,750	34	22,896	1,246,951	6,975,279	....	....	455
Kentucky	37,680	761,413	10,011	210,981	982,405	26.07	301,628,456	155,021,262	5,968,270	29,661,436	653,036	62	6,582,838	80	79,466	2,295,353	12,350,734	486	28	284
Missouri	67,380	592,004	2,618	87,422	682,044	10.12	137,247,707	63,225,543	2,938,425	19,887,580	440,641	61	6,195,560	97	75,056	1,730,135	9,079,695	....	....	139
Total	851,508	6,184,477	228,128	3,200,364	9,612,969	11.29	$2,936,090,737	$1,117,649,649	54,975,427	$253,723,687	$6,697,536	704	80,038,793	713	654,194	$22,675,541	$95,029,877	1,113	2,016	6,414
District of Columbia	60	37,941	10,059	3,687	51,687	861.45	16,723,619	1,730,460	16,267	71,643	122,272	18	11,127,236	9	98,600	363,000	888,965	(Inclu'd with Mary'd.)		
Free States	612,597	13,238,670	196,016	(236)	13,434,922	21.91	4,102,162,098	2,147,218,478	57,720,494	286,374,541	10,971,768	1,800	334,146,281	14,893	3,883,617	66,773,517	431,290,351	3,685	5,339	16,828
Territories	1,472,061	91,980	292	(26)	92.298	.06	11,586,512	4,976,839	320,426	4,010,645	42,155	4	97,768	....	....	171,970	1,050,300	....	....	..
Total of U. S.	2,936,166	19,553,068	434,495	3,204,313	23,191,876	7.90	$7,066,562,966	$3,271,575,426	113,032,614	$544,180,516	$17,824,331	2,526	426,409,978	15,615	4,636,411	$89,983,028	$527,209,193	4,798	7,355	23,242

Gettysburg

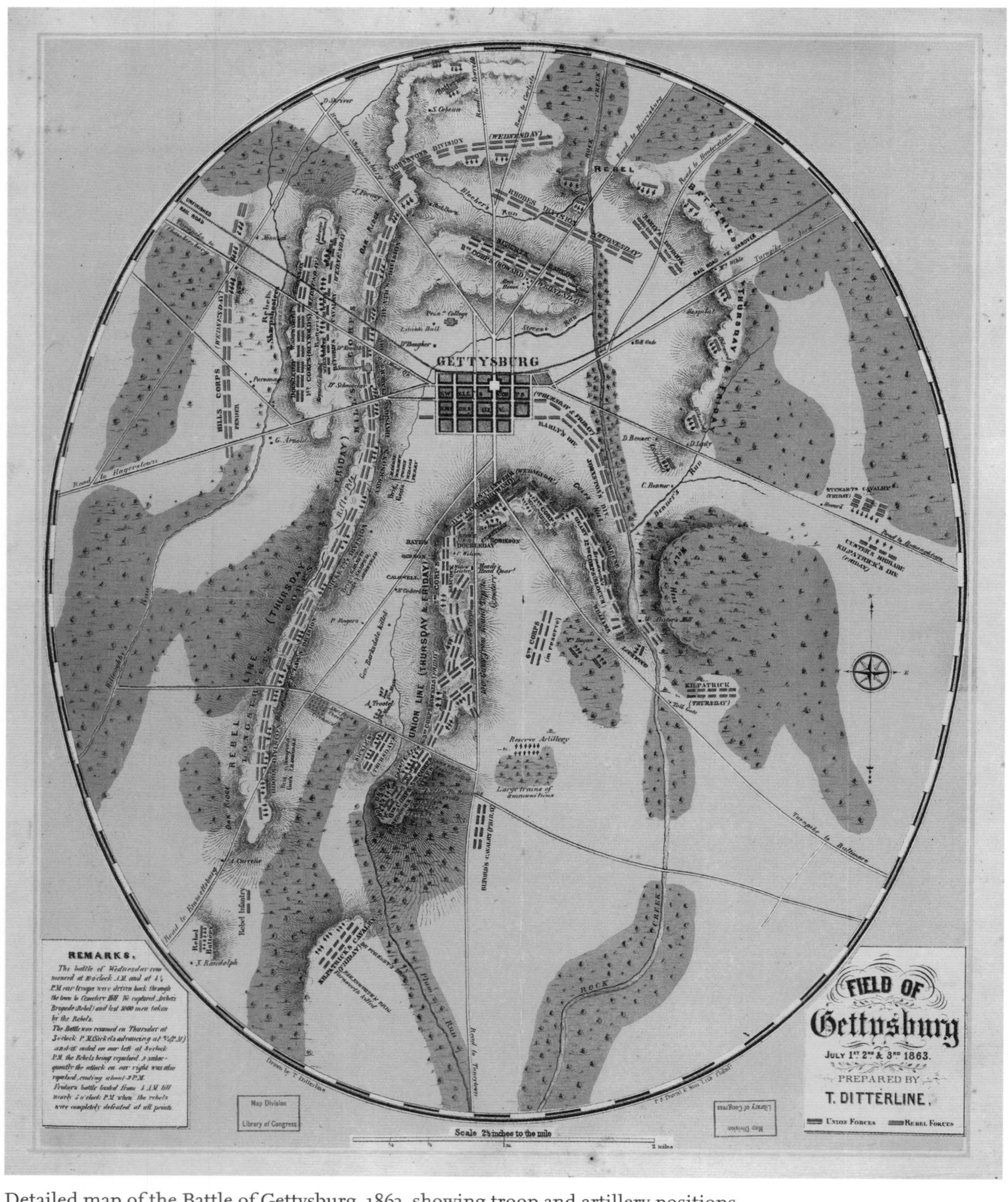

Detailed map of the Battle of Gettysburg, 1863, showing troop and artillery positions.
Library of Congress, Washington D.C., USA.

BATTLEFIELD MAP

The American Civil War raged between the northern states and southern states from 1861 to 1865. The southern states seceded from the United States in 1861, forming the Confederate States of America (CSA), also known as the Confederacy. Confederate forces won most of the early battles, but a fateful encounter near the small rural town of Gettysburg changed the course of the war. Theodore Ditterline's oval-shaped map, *Field of Gettysburg*, is considered the first map published of this pivotal battle, in which Union (US) forces stopped an attack by a Confederate army.

Gettysburg, the bloodiest battle of the Civil War, resulted in some 51,000 casualties (killed, missing, wounded). The Confederate 'Army of Northern Virginia' commanded by General Robert E. Lee, invaded the northern state of Pennsylvania with some 70,000 troops. Union troops encountered elements of the Confederate army near the town of Gettysburg, and Union forces ultimately swelled to some 90,000 soldiers as reinforcements rushed to the battlefield. The Union 'Army of the Potomac' was led by Major General George Meade, who had just been given command of that army.

The map by Ditterline shows topography, roads, railroads, troop and artillery locations, and troop movements from 1 July to 3 July 1863. Ditterline, a northern cartographer, compiled the map based on eyewitness accounts of the battle. George W. Childs, a Philadelphia newspaper publisher during the Civil War, endorsed the map and added that his opinion was shared by several officers in the battle.

On the first day of the battle (Wednesday, 1 July), the more numerous Confederate forces (in red on the map) made the Union defenders (in blue) retreat from their positions north and west of Gettysburg. The Union army fell back through Gettysburg and took a defensive position south of town on Culp's Hill and Cemetery Hill.

By Thursday, 2 July, the Union soldiers (called 'Yankees' by the Confederates) were gaining in numbers and strengthening their positions on the series of hills south of Gettysburg. Union reinforcements were streaming in along the Baltimore turnpike. Confederate troops (called 'Rebels' by Union soldiers), spearheaded by General Longstreet's corps, attacked the Union left flank (southernmost position on Ditterline's map). Lee's plan was to have the rest of the Confederate line assault the Union's centre and right flanks, but these attacks were ineffective. While Longstreet's attack almost broke the Union's left flank, thousands of Confederate soldiers died in the attempt, but the Union positions held.

The Union and Confederate lines remained about the same on Friday, 3 July. Lee decided to hit the Union centre, because he thought the flanks had been reinforced at the expense of the centre. He ordered a massive artillery bombardment to precede a Confederate infantry charge of some 12,000 soldiers, which becomes known as 'Pickett's Charge,' after Major General George Pickett. Pickett's Division is labeled on Ditterline's map. However, Meade anticipated the assault against the Union centre, and the Confederate attack failed with huge loss of life.

After the battle, Lee pulled his army out of Pennsylvania, and the rebels retreated back to Virginia. Many believe that Gettysburg marked the beginning of the end of the Confederacy. Lee's army never recovered from the loss of thousands of soldiers during the three days of battle.

Theodore Ditterline captured a critical turning point in US history, with an attractive map design that effectively tells the story of a complex battle. Collectors covet this rare map, which is 40 cm x 49 cm (15¾ in. x 19¼ in.), and a copy sold for $5,400 in 2013.

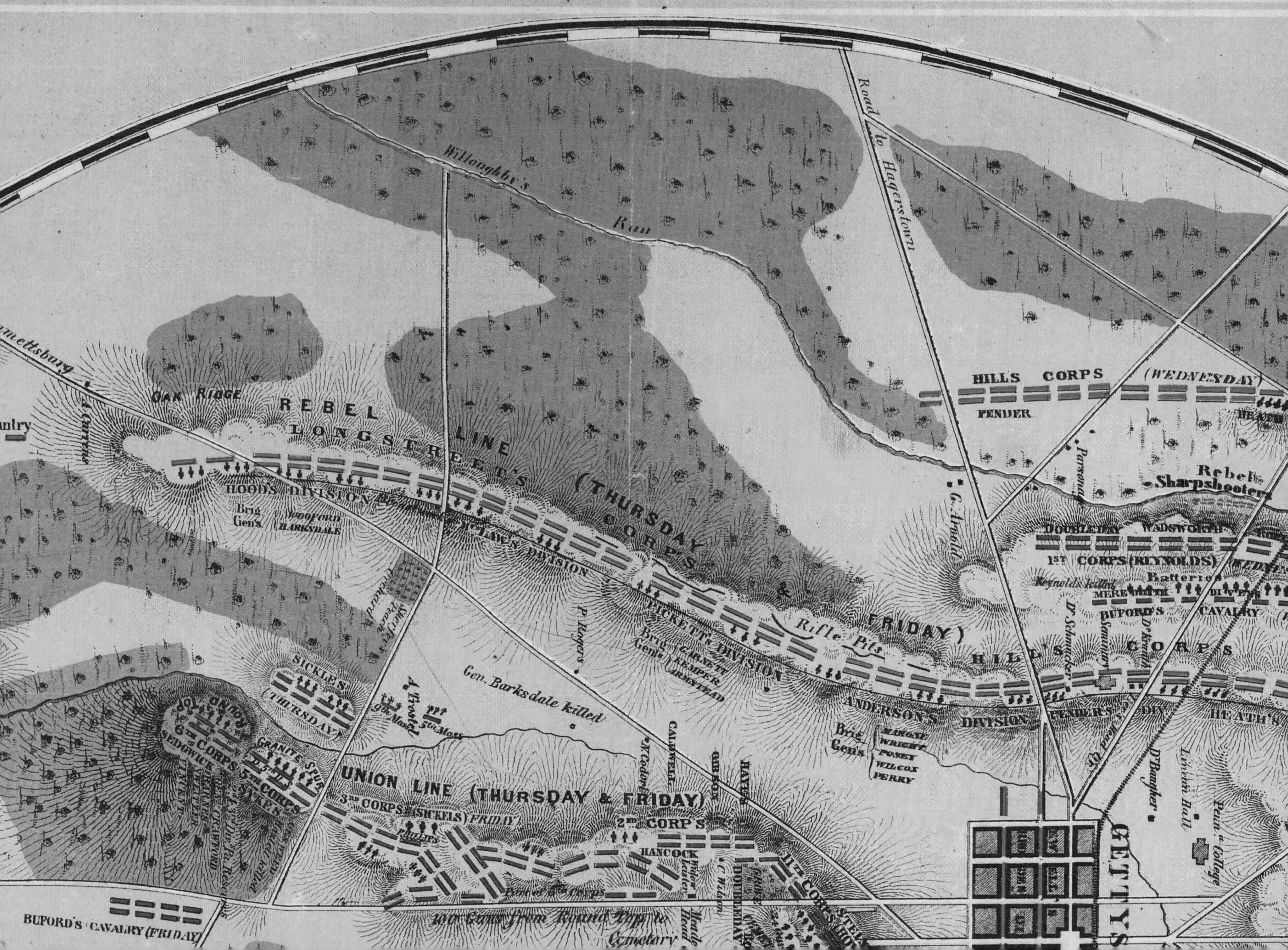

Willoughby's Run
Road to Hagerstown
HILLS CORPS (WEDNESDAY)
PENDER
Rebel Sharpshooters
Parsonage
G. Arnold
OAK RIDGE
REBEL LINE
LONGSTREET'S CORPS (THURSDAY & FRIDAY)
HOOD'S DIVISION
Brig Gen's WOOFORD BARKSDALE
McLAW'S DIVISION
PICKETT'S DIVISION
Brig Gen's GARNETT KEMPER ARMSTEAD
Rifle Pits
HILL'S CORPS
ANDERSON'S DIVISION
Brig Gen's MAHONE WRIGHT POSEY WILCOX PERRY
PENDER'S DIV
HEATH'S
DOUBLEDAY
WADSWORTH
1ST CORPS (REYNOLDS)
Reynolds killed
Batteries
MEREDITH
BUFORD'S CAVALRY
Dr Schmucker
Seminary
Dr Krauth
Dr Baugher
Lineau Hall
Penn^a College
GETTYS
Sherfey's Peach Orchard
P. Rogers
Gen. Barksdale killed
A. Trostel
SICKLES (THURSDAY)
ROUND TOP
6TH CORPS
SEDGWICK
5TH CORPS
SYKES
GRANITE SPUR
VINCENT
UNION LINE (THURSDAY & FRIDAY)
3RD CORPS (SICKELS) FRIDAY
PHILLIPS
2ND CORPS
HANCOCK
N. Codori
CALDWELL
GIBSON
HAYES
Widow Leister
C. Wilson
DOUBLEDAY
11TH CORPS
Meade Head
100 Guns from Round Top to Cemetery
BUFORD'S CAVALRY (FRIDAY)
A. Currens
mettsburg

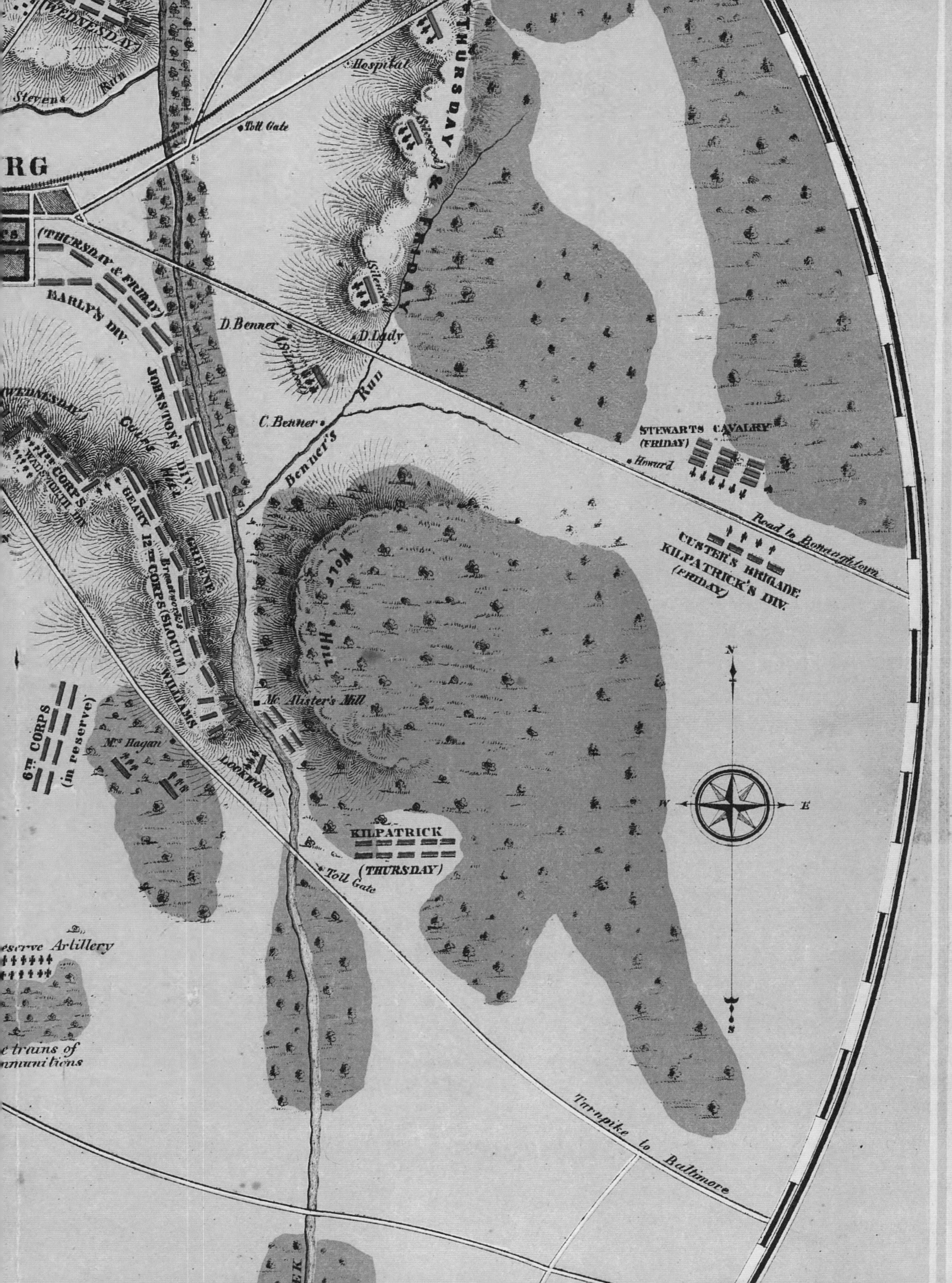

(WEDNESDAY)
THURSDAY & FRIDAY
Hospital
Stevens Run
Toll Gate
RG
(THURSDAY & FRIDAY)
EARLY'S DIV.
JOHNSTON'S DIV.
D. Benner
D. Lady
Run
C. Benner
Benner's Run
(WEDNESDAY)
Culp's Hill
GEARY
12TH CORPS (SLOCUM)
GREENE
Breastworks
WILLIAMS
Mc Alister's Mill
Wolf Hill
6TH CORPS (in reserve)
Mrs Hagan
LOOKWOOD
STEWARTS CAVALRY (FRIDAY)
Howard
Road to Bonaughtown
CUSTER'S BRIGADE
KILPATRICK'S DIV.
(FRIDAY)
N
W
E
S
KILPATRICK
(THURSDAY)
Toll Gate
Reserve Artillery
trains of munitions
Turnpike to Baltimore

SCRAMBLE FOR AFRICA

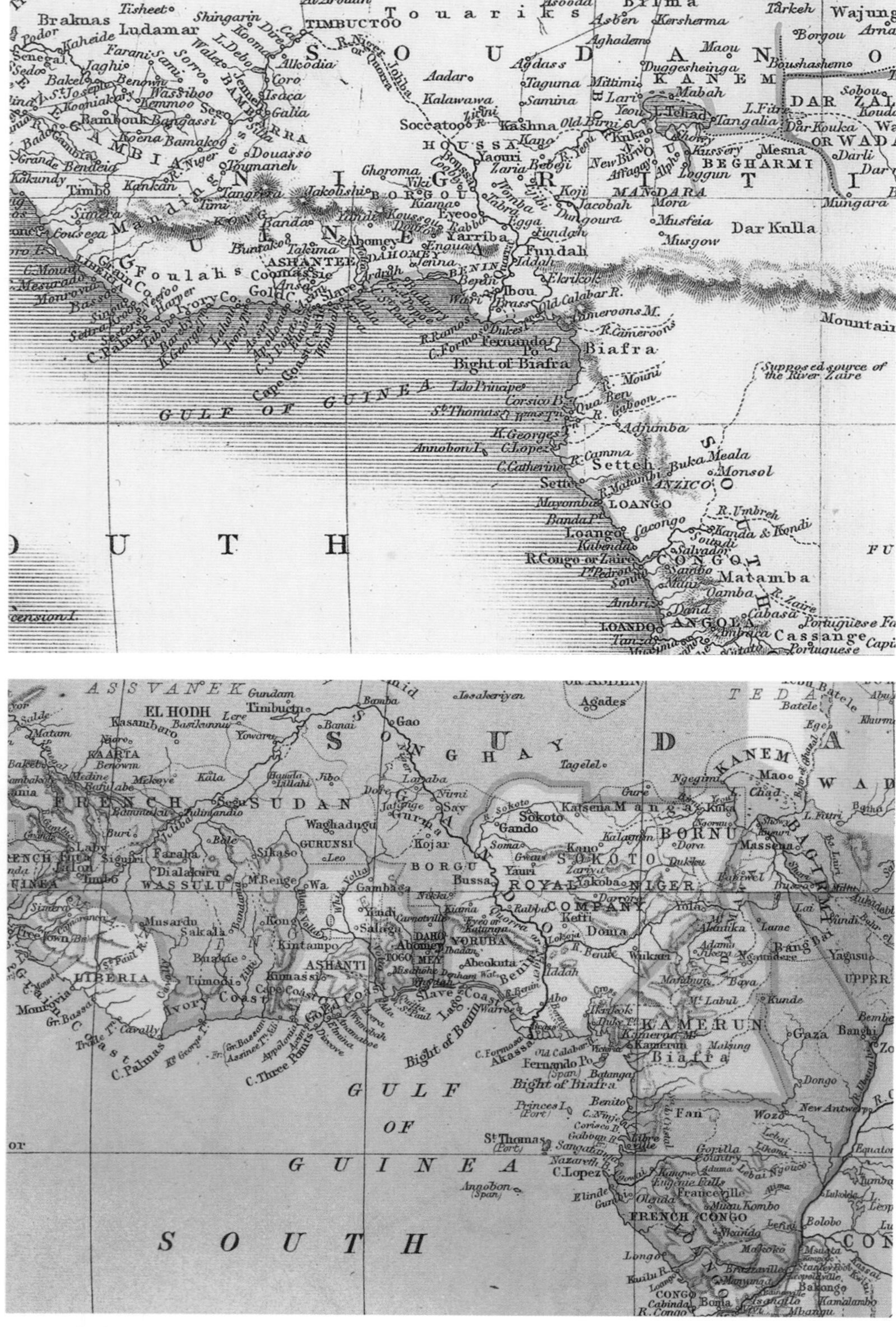

Maps of Africa showing the growing influence of colonial powers. From Collins *College Atlas for Schools and Families* (1852) and Bartholomew *Citizen's Atlas* (1898).

Two maps sharply illustrate the massive expansion of European colonial territory in Africa in the second half of the nineteenth century. In 1852 the interior was little known and European possessions clung precariously to the coast. By 1898, in contrast, the 'Scramble for Africa,' which had been set off by the Congress of Berlin fourteen years earlier, had resulted in the survival of only two independent states in Africa – Ethiopia and Liberia.

Save in the northern and southern extremities of Africa, European cartographic knowledge of the continent in 1850 was limited largely to the coastlines, which had been extensively explored since Portuguese navigators sailed along those of Senegal and Guinea in the 1440s. The great age of European exploration of the African interior was yet to come at the time of the 1852 map which still shows enormous blank areas. It also includes speculative labels such as 'supposed source of the River Zaire' and downright fallacies, such as the Mountains of the Moon, which Ptolemy (see page 16) had placed in the centre of Africa, and which for almost 2,000 years had proved surprisingly persistent among cartographers ignorant of the true topography of the continent. The greatest level of detail lay in areas controlled by Europeans, such as the British Cape Colony, the French territory of Algeria and pockets along the coast in west Africa.

Throughout the 1850s, expeditions led by Richard Burton, John Speke and James Grant discovered Lake Tanganyika, Lake Victoria and the source of the River Nile, extending British knowledge northwards from southern Africa towards Egypt and firing a long-standing ambition to link Cairo to the Cape as a means of supplying India, the jewel in Britain's colonial crown. The French, meanwhile, concentrated on north and west Africa, where René-August Caillié reached Timbuktu in 1827 and the Italian-French explorer Pierre Savorgnan de Brazza ventured up the Congo in 1879–82, setting markers for future colonial control from Paris.

These explorations vastly increased the level of knowledge of the African interior: the American explorer Henry Morton Stanley who explored the Congo from the east right to the Atlantic coast included ten maps in *Through the Dark Continent*, his account of the journey. The successive exploratory forays of the Scottish missionary Dr David Livingstone – who was the first European to see Victoria Falls in 1855 – inspired the London cartographer John Arrowsmith to produce a map of South Africa in 1857 which showed 'the routes of the Reverend Dr Livingstone between the years 1849 and 1856' and was for many years the best available for the region.

By the 1880s, European colonial expansion in Africa had become more active The French expanded from their main bases in Senegal and Algeria into central Africa and the Sahel, while the British pushed into Kenya and Uganda, extended their control in Nigeria and took political control of Egypt in 1882. A newly united Germany did not wish to be excluded from the colonial table and, under the capable direction of Chancellor Otto von Bismarck, began to acquire territory, starting with parts of Cameroon and Togo in July 1884.

The spheres of colonial ambitions of the major powers – with the Spanish, Portuguese and Italians all also making new claims or re-asserting old ones – were now in danger of colliding, and the Germans called a Conference in Berlin in November 1884 to try to head off any confrontations. The conference was ostensibly tasked with ensuring free trade in central Africa, and freedom of navigation on the Congo and Niger rivers. It also espoused the principle of 'effective occupation' – that colonial powers could not merely plant a flag to claim possession, but must have garrisons or treaties with local rulers – which had the effect of sparking a rush to claim the remaining uncolonized areas before rival European powers could do so.

One particular winner from the conference was King Leopold of Belgium, who had sponsored a series of expeditions in the Congo Basin led by the veteran explorer Henry Morton Stanley. Leopold succeeded in persuading the conference to recognize the claims of his Association internationale du Congo over a vast area of the region and he established the 'Congo Free State' there, which operated in effect as his own private colony.

The French pushed their sphere of influence east, trying to link Senegal with French Somaliland, while the British sought to cement the Cape to Cairo link by subduing the Mahdist Islamist regime in Sudan. A clash, or even war, between Britain and France seemed inevitable. On 10 July 1898 a French expeditionary force under Major Marchand reached Fashoda (in modern-day South Sudan) on the White Nile and concluded an agreement with the local ruler to acknowledge French suzerainty. The British, however, decisively defeated the Mahdists at the Battle of Omdurman in September 1898 and were able to deploy a force 100 times greater than Marchand's tiny 150-strong garrison. The French agreed to pull out and on 12 December 1898 the British flag was raised at Fashoda.

The Scramble for Africa was complete, and European administrators and cartographers set to drawing the straight lines on the map which would separate their respective colonies. The 1898 map does not yet show these pleasingly deceptive lines, but it portrays a world in which indigenous rule had been almost totally shut out. Native rulers remained in control only in Abyssinia (Ethiopia), where the defeat of the Italians at Adowa in 1896 kept the Ethiopian emperors on their throne, and in Liberia, set up in 1847 as a colony for freed African-American slaves. Compared to the blanks on the map, the tentative guesswork and the tenuous European colonies clinging to the coast in 1852, it is a stunning transformation.

Comparative lengths of Europe & Africa

Gibraltar — Europe 3,500 Miles – Area 3,754,000 — Karskain Gulf

Tunis — Africa 5,100 Miles – Area 11,750,000 — Cape of Good Hope

AFRICA.

Scale of English Miles.

200 0 200 400 600 800 1000

Longitude West 20 from Greenwich.

Longitude East 40 from Greenwich.

Drawn & Engraved by J. Archer

London Published by H. G. Collins, 22 Paternoster Row

AFRICA

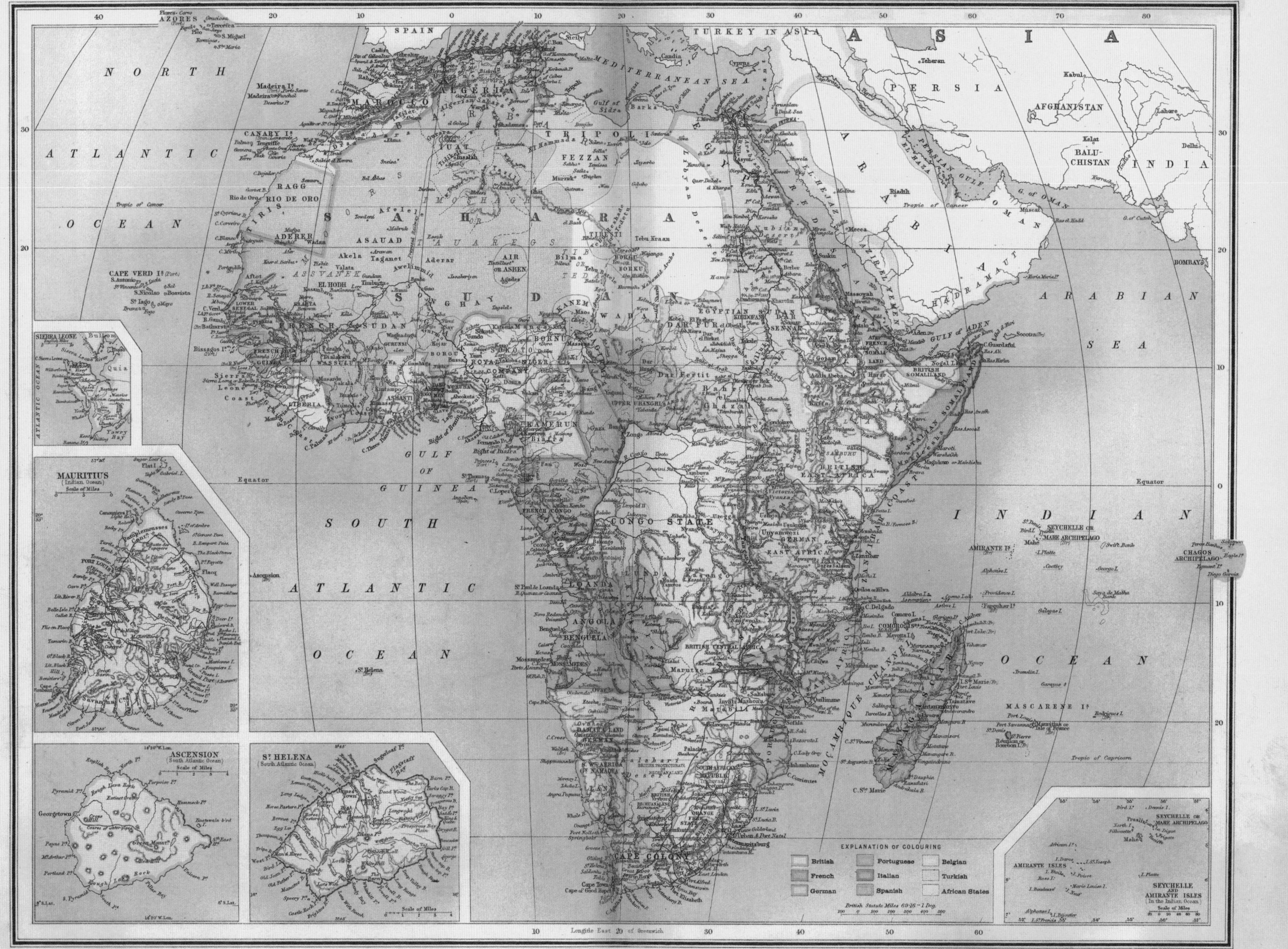

NORTH ATLANTIC OCEAN
SOUTH ATLANTIC OCEAN
INDIAN OCEAN
ARABIAN SEA
GULF OF GUINEA
MEDITERRANEAN SEA
A S I A
TURKEY IN ASIA
PERSIA
AFGHANISTAN
BALUCHISTAN
INDIA
ARABIA
SPAIN
MAROCCO
ALGERIA
TRIPOLI
FEZZAN
EGYPT
SAHARA
SUDAN
FRENCH SUDAN
LIBERIA
ASHANTI
KAMERUN
FRENCH CONGO
CONGO STATE
ANGOLA
BENGUELA
GERMAN S.W. AFRICA
DAMARA LAND
CAPE COLONY
ORANGE FREE STATE
SOUTH AFRICAN REPUBLIC
BRITISH CENTRAL AFRICA
GERMAN EAST AFRICA
BRITISH EAST AFRICA
ABYSSINIA
BRITISH SOMALILAND
ITALIAN SOMALILAND
EGYPTIAN SUDAN
MADAGASCAR
MOÇAMBIQUE CHANNEL
GULF OF ADEN
RED SEA
Equator
Tropic of Cancer
Tropic of Capricorn
MASCARENE Iˢ
SEYCHELLE OR MAHE ARCHIPELAGO
AMIRANTE Iˢ
CHAGOS ARCHIPELAGO
AZORES
Madeira Iˢ
CANARY Iˢ
CAPE VERD Iˢ (Port.)
Ascension
Sᵗ Helena
SIERRA LEONE
MAURITIUS (Indian Ocean)
ASCENSION (South Atlantic Ocean)
Sᵗ HELENA (South Atlantic Ocean)
SEYCHELLE AND AMIRANTE ISLES (In the Indian Ocean)
EXPLANATION OF COLOURING
British
French
German
Portuguese
Italian
Spanish
Belgian
Turkish
African States
British Statute Miles 69·16 = 1 Deg.
Longitude East 20 of Greenwich
John Bartholomew & Co. Edinʳ

1871 OPENING AMERICA

The completion of a transcontinental railroad across the United States in 1869 helped weld the country together as one nation rather than a disparate collection of territories which it had previously taken weeks to cross by waggon-train. Before long, maps began to appear, such as Gaylord Watson's 1871 *Rail-Road and Distance map of the United States and Canada,* which showed a dense network of tracks connecting the east and west coasts.

The first commercial railway in the United States may have been built in 1810 in Pennsylvania to transport stone from Thomas Leiper's quarry to the Delaware River, but the first substantial stretch was the 219 km (136-mile) line between Charleston and Hamburg, Georgia, which carried passengers for the first time on Christmas Day 1830. Soon railway companies were springing up all over New England, and these gradually extended their operations westward.

The idea of a railroad extending to the Pacific was suggested as early as 1844 by Asa Whitney, a Connecticut businessman, who proposed the line be funded by selling land along its route (which he would buy from the government at the bargain rate of 10 cents a mile). Progress was bedevilled by arguments over whether it should take a northern route or a more southerly track made possible by the acquisition from Mexico of huge tracts of the south-west in 1849. Four surveying parties were sent out in 1853–4 but no decision was made on a route before the outbreak of the US Civil War in 1861 halted any further work.

The Republican Party platform in the 1860 elections included a promise of a Pacific Railroad. President Lincoln made good on this pledge by authorizing a new railroad company, the Union Pacific, which would build track westward from Missouri to meet the line of the Central Pacific, which was given the responsibility of constructing a line eastward from Sacramento. All talk of a southern route had gone, with the South's defeat in the civil war leaving its interests marginalized. The 1862 Pacific Railroad Act also set out a system of land grants, five square-mile lots on each side of the track for each mile laid, which the railroad companies could sell to fund the construction. Thirty-year bonds were issued at rates of $16,000 to $48,000 per mile of track (depending on the terrain). As the actual cost turned out to be $10,600 a mile, the directors of the companies made substantial windfall profits. The Union Pacific's ground breaking ceremony took place at its Omaha railhead in December 1863, but it took three years to cover the first 64 km (40 miles) west.

Thereafter progress was far more rapid, with the tracklayers – largely civil war veterans – operating in separate blasting, grading and laying teams which enabled them to reach Cheyenne in July 1867. The Central Pacific pushed eastwards in parallel and by 1869 the two lines were close to meeting. Both companies had all the while issued regular prospectuses with maps designed to show the enormous progress they were making in the hopes of attracting fresh investment funds. The alarming prospect of their going their separate ways east and west was avoided by Congress's choice of Promontory Point, Utah as the meeting point of the two lines, and in an elaborate ceremony on 10 May 1869 a last golden spike was driven into the rails. The Union Pacific had laid 1,706 km (1,060 miles) of the track, with the Central Pacific contributing 1,110 km (690 miles).

Passenger services began from Omaha just five days later, with a fare of $111 for First Class and $40 for a distinctly less comfortable Third Class ticket. It now took just four days and four hours to get from Omaha to California, a revolutionary change in travel possibilities for the United States. New lines proliferated, with the length of lines in operation growing from 97,030 km (60,300 miles) in 1871 to 300,640 km (186,800 miles) by the end of the century. Further transcontinental routes were opened, too; in 1870 the last spike on the Kansas Pacific was hammered in, opening a line which allowed the first traverse of the continent without changing trains (the Union Pacific required a crossing of the Missouri by ferry until 1872).

It is the country at the height of this railway fever which Gaylord Watson portrays in his 1871 map, published just two years after the opening of the Union-Central Pacific line. A New York publisher, he had begun to produce maps and travellers' guides to the West in 1865, and the railway map was the culmination of this work. It shows a country bound together by the simple arithmetic of distance and travel times, and a dense network of tracks in the east which feathers out and thins to a few bold lines linking the east and west coasts – across a genuinely 'United' States.

RAILROADS ACROSS AMERICA

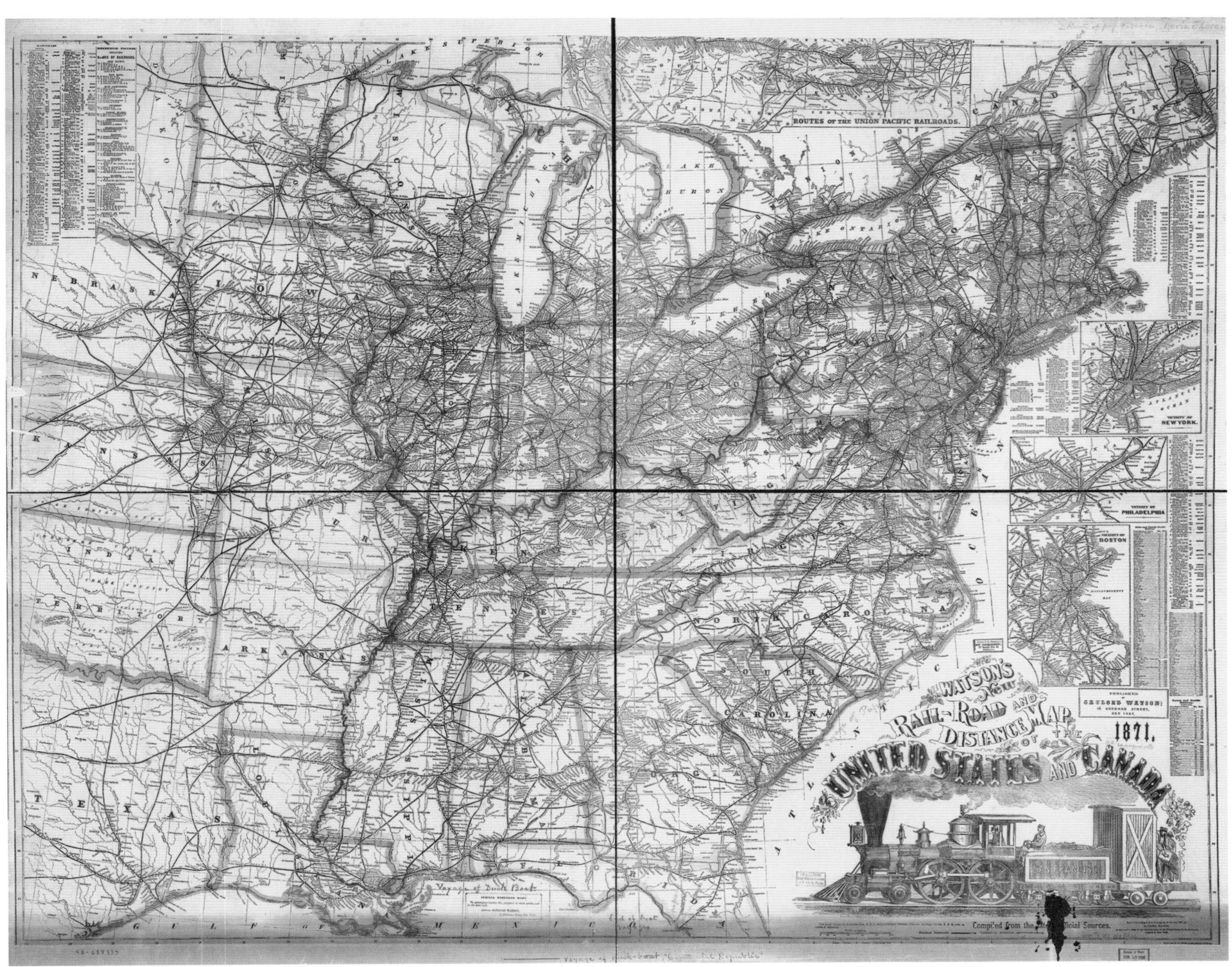

Gaylord Watson's highly detailed map from the height of railroad fever in the United States. Insets show routes of the Union Pacific Railroads in the New York Philadelphia and Boston areas. Library of Congress, Washington D.C., USA.

LAKE ERIE
NEW
PENNSYLVANIA
OHIO
MARYLAND
WEST VIRGINIA
VIRGINIA
CLEVELAND
DUNKIRK
PITTSBURGH
WHEELING
PARKERSBURG
GRAFTON
HARRISBURG
BALTIMORE
WASHINGTON
ALEXANDRIA
RICHMOND
PETERSBURG
LYNCHBURG
DANVILLE
CHARLOTTESVILLE
WINCHESTER
ELMIRA
CHATHAM
PORT STANLEY
AKRON
ALLIANCE
ROCHESTER
CONNELLSVILLE
CHAMBERSBURG
GETTYSBURG
ALTOONA
HUNTINGDON
BELLEFONTE
WILLIAMSPORT
ANNAPOLIS
COVINGTON
BRISTOL
SALTVILLE
CLARKSVILLE
Charleston
Staunton
Lexington
Fredericksburg
Hagerstown
Cumberland
Harrisonburg
Woodstock
Warrenton
Culpepper
Gordonsville
Clarion R.
James R.
Roanoke R.
Ohio River

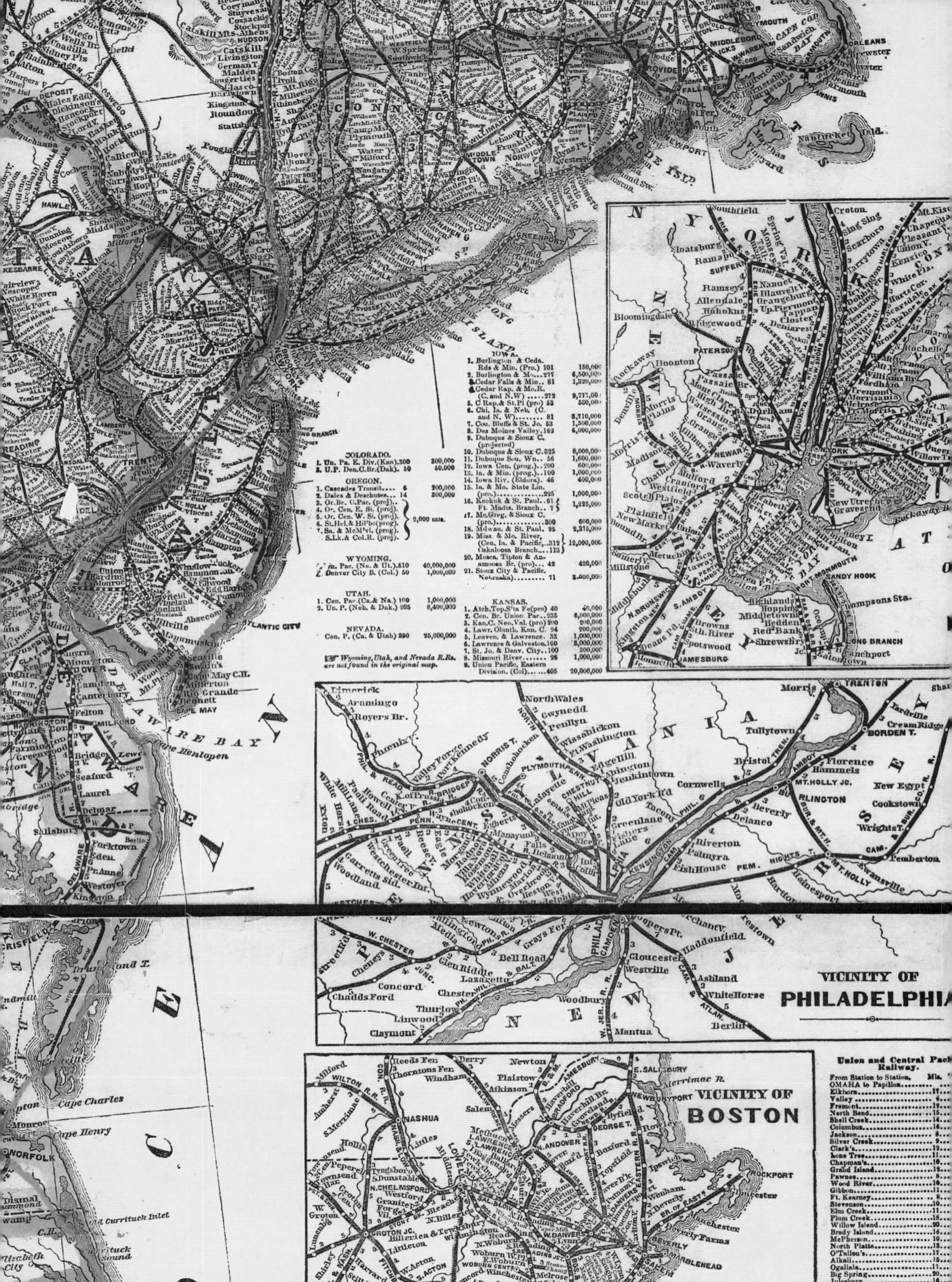

IOWA.
1. Burlington & Ceda. Rds & Min. (Pro.) 101 — 150,000
2. Burlington & Mo...277 — 6,500,000
3. Cedar Falls & Min.. 81 — 1,320,000
4. Cedar Rap. & Mo.R. (C. and N.W) .....272 — 9,777,000
5. C Rap.& St.Pl (pro) 53 — 530,000
6. Chi. Ia. & Neb. (C. and N. W)........ 81 — 3,710,000
7. Cou. Bluffs & St. Jo. 53 — 1,500,000
8. Des Moines Valley.162 — 6,000,000
9. Dubuque & Sioux C. (projected)
10. Dubuque & Sioux C.325 — 8,000,000
11. Dubuque Sou. Wn.. 56 — 1,600,000
12. Iowa Cen. (prog.)..200 — 600,000
13. Ia. & Min. (prog.)..100 — 1,000,000
14. Iowa Riv. (Eldora). 46 — 400,000
15. Ia. & Mo. State Lin. (pro.)..............225 — 1,000,000
16. Keokuk & St. Paul..61 / Ft. Madis. Branch.. 7 — 1,225,000
17. Mc.Greg. & Sioux C. (pro.)..............300 — 600,000
18. Milwau. & St. Paul. 95 — 2,375,000
19. Miss. & Mo. River, (Cen. Ia. & Pacific,..312 / Oskaloosa Branch....113 — 12,000,000
20. Musca. Tipton & Anamoosa Br. (pro)... 42 — 420,000
21. Sioux City & Pacific. Nebraska)......... 71 — 3,000,000
COLORADO.
1. Un. Pa. E. Div.(Kan).300 — 300,000
2. U.P. Den.C.Br.(Dak). 50 — 50,000
OREGON.
1. Cascades Transit.... 6 — 200,000
2. Dales & Deschutes... 14 — 300,000
3. Or.Br. C.Pac. (proj)..
4. Or. Cen. E. Si. (proj).
5. Or. Cen. W. Si. (proj).
6. St.Hel.& Hil'bo(prog).
7. Sa. & McM'vl. (prog).
S.Lk.& Col.R. (proj). — 2,000 mls.
WYOMING.
1. Un. Pac. (Ne. & Ut.).510 — 40,000,000
2. Denver City B. (Col.) 50 — 1,000,000
UTAH.
1. Cen. Pac.(Ca.& Na.) 100 — 1,000,000
2. Un. P. (Neb. & Dak.) 205 — 8,400,000
NEVADA.
Cen. P. (Ca. & Utah) 390 — 25,000,000
Wyoming, Utah, and Nevada R.Rs. are not found in the original map.
KANSAS.
1. Atch.Top.S'ta Fe(pro) 40 — 40,000
2. Cen. Br. Union Pac..235 — 6,000,000
3. Kan.C. Neo.Val. (pro) 200 — 200,000
4. Lawr. Olanth. Kan. C. 24 — 200,000
5. Leaven. & Lawrence. 33 — 1,000,000
6. Lawrence & Galveston.160 — 3,000,000
7. St. Jo. & Denv. City..100 — 300,000
8. Missouri River....... 26 — 1,000,000
9. Union Pacific, Eastern Division. (Col)......405 — 20,000,000
VICINITY OF PHILADELPHIA
VICINITY OF BOSTON
Union and Central Pac[ific] Railway.
From Station to Station. Mls.
OMAHA to Papillon.........
Elkhorn..........17
Valley..........4
Fremont..........12
North Bend..........15
Shell Creek..........14
Columbus..........16
Jackson..........8
Silver Creek..........9
Clark's..........12
Lone Tree..........11
Chapman's..........10
Grand Island..........12
Pawnee..........8
Wood River..........10
Gibbon..........11
Ft. Kearney..........8
Stevenson..........10
Elm Creek..........11
Plum Creek..........18
Willow Island..........20
Brady Island..........18
McPherson..........10
North Platte..........13
O'Tallon's..........17
Alkali..........15
Ogallala..........17
Big Spring..........20
Julesburg..........16

MAPPING THE OCEAN FLOOR

In the mid-nineteenth century, the mapping of the world seemed complete. Continents were outlined, routes across the oceans had been established, and various expeditions were delving deeper into the unknown hearts of the Americas, Africa, and Australia. However, exploration of over 70 per cent of our planet's surface – the seafloor – had not yet even begun. Findings from the Challenger Expedition (1872–6), including this map created from them, were to radically change our understanding of the oceans.

Scientific exploration of the seas began in the mid-nineteenth century, though tentatively at first, with a sounding here, a dredging there. In 1843, the first great hypothesis concerning ocean life emerged from the dredgings carried out in the Aegean Sea by the English naturalist Edward Forbes. His theory was that life in the sea was extinguished at a depth of 300 fathoms. This theory fanned the flames of scientific inquiry and scientists in the United States, Great Britain, and other nations began actively seeking to either verify or disprove this concept. Further impetus to explore and map the ocean came from the commercial development of the telegraph and the desire to 'put a girdle "round the earth"' and connect continents by laying telegraphic cables on the seabed. The technology of the times was woefully inadequate for conducting the necessary inquiries into Forbes's hypothesis and over two decades were to pass before it was disproved.

On 22 July 1869, a dredge was lowered from the deck of HMS *Porcupine* at a point southwest of Ireland to the depth of 2,435 fathoms and returned to the surface with numerous life forms. The scientist directing this operation was Wyville Thomson who carried this success into the great Challenger Expedition. HMS *Challenger* left Great Britain in December 1872, and after a three-year cruise around the world returned in 1876 having made 492 dredge hauls and 362 soundings.

Within a year of its return, Wyville Thomson had produced a two volume book for popular consumption: *The Voyage of the Challenger: The Atlantic*. What is perhaps one of the most important maps in the history of the earth sciences graced the frontispiece of Volume II of this publication. This map – *Contour Map of the Atlantic* – is reproduced here. Prominent on the map is the lighter coloured area roughly following the centre line between the Americas and Europe and Africa – this was the first map to show the continuity of the Mid-Atlantic Ridge.

There were two factors leading to Thomson's interpretation of the Mid-Atlantic Ridge. The first was related to ocean bottom temperatures. On the outward bound leg of the expedition, it was observed that the bottom temperatures on the west side of the Atlantic Ocean were lower than those on the east side. This led to the hypothesis that a mid-ocean barrier existed which prevented the mixing of water masses. The second key was the serendipitous choice of a sounding line location on the return leg in the South Atlantic Ocean which ran north from Tristan da Cunha (Tristan D'Acunha on map) Island to Ascension Island and followed the spine of the ridge. This line was the key to understanding and proving the continuity of the ridge and could be considered the single most important sounding line in the history of seafloor mapping. With the continuity of the ridge established in the South Atlantic, Thomson correctly surmised the existence of a continuous ridge in the North Atlantic. Thomson's vision of this great mountain range led to a new view of the earth, one in which the ocean basins had topography as magnificent and diverse as the continents. It was also a major step towards a view of our planet in which seafloor and continents are not static features on a fixed globe, but are ever shifting as part of great plates which make up the earth's surface – the theory of continental drift or plate tectonics.

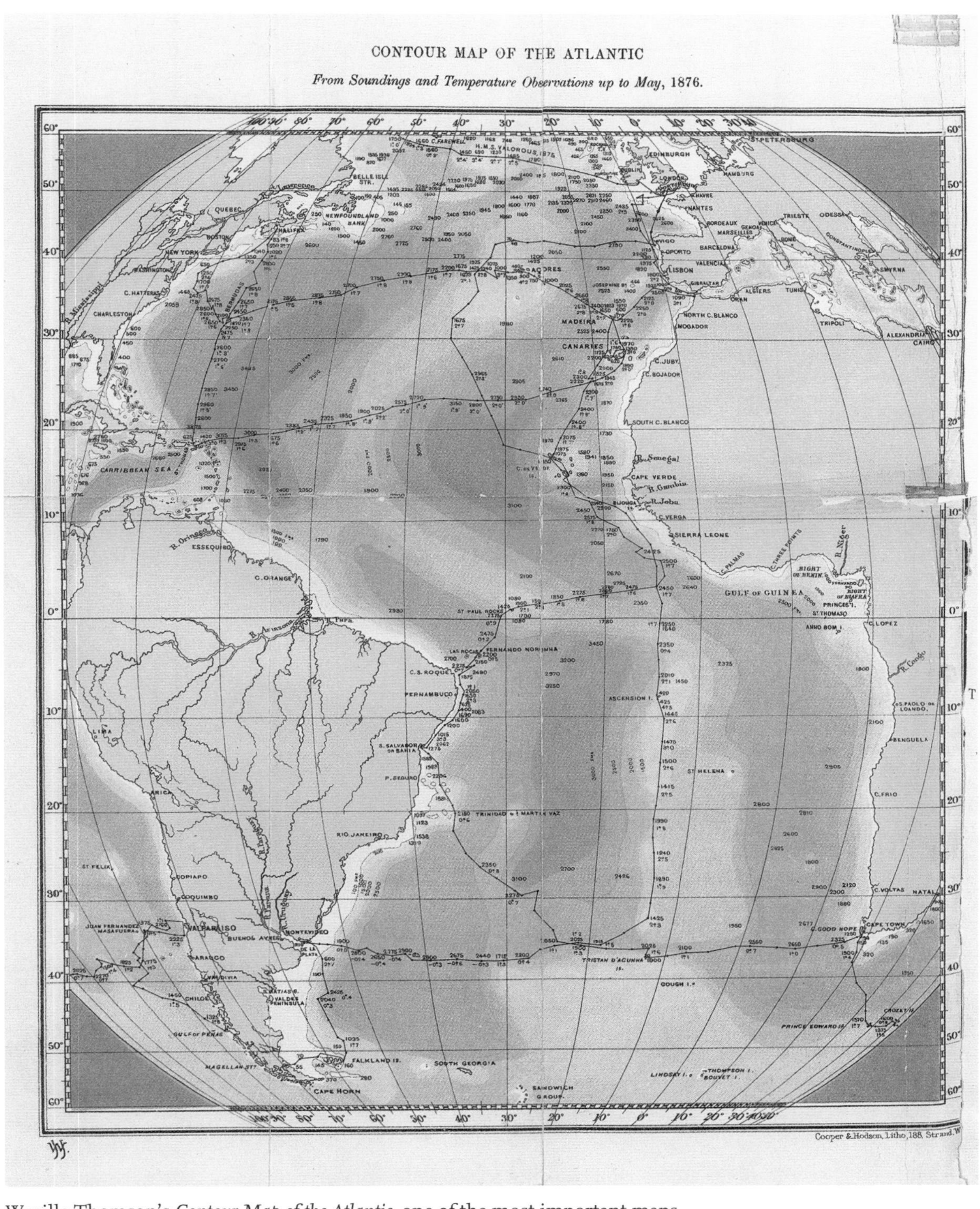

Wyville Thomson's *Contour Map of the Atlantic*, one of the most important maps of the earth sciences for its depiction of the Mid-Atlantic Ridge.
NOAA Photo Library, Silver Spring, MD, USA.

CONTOUR MAP OF THE ATLANTIC

From Soundings and Temperature Observations up to May, 1876.

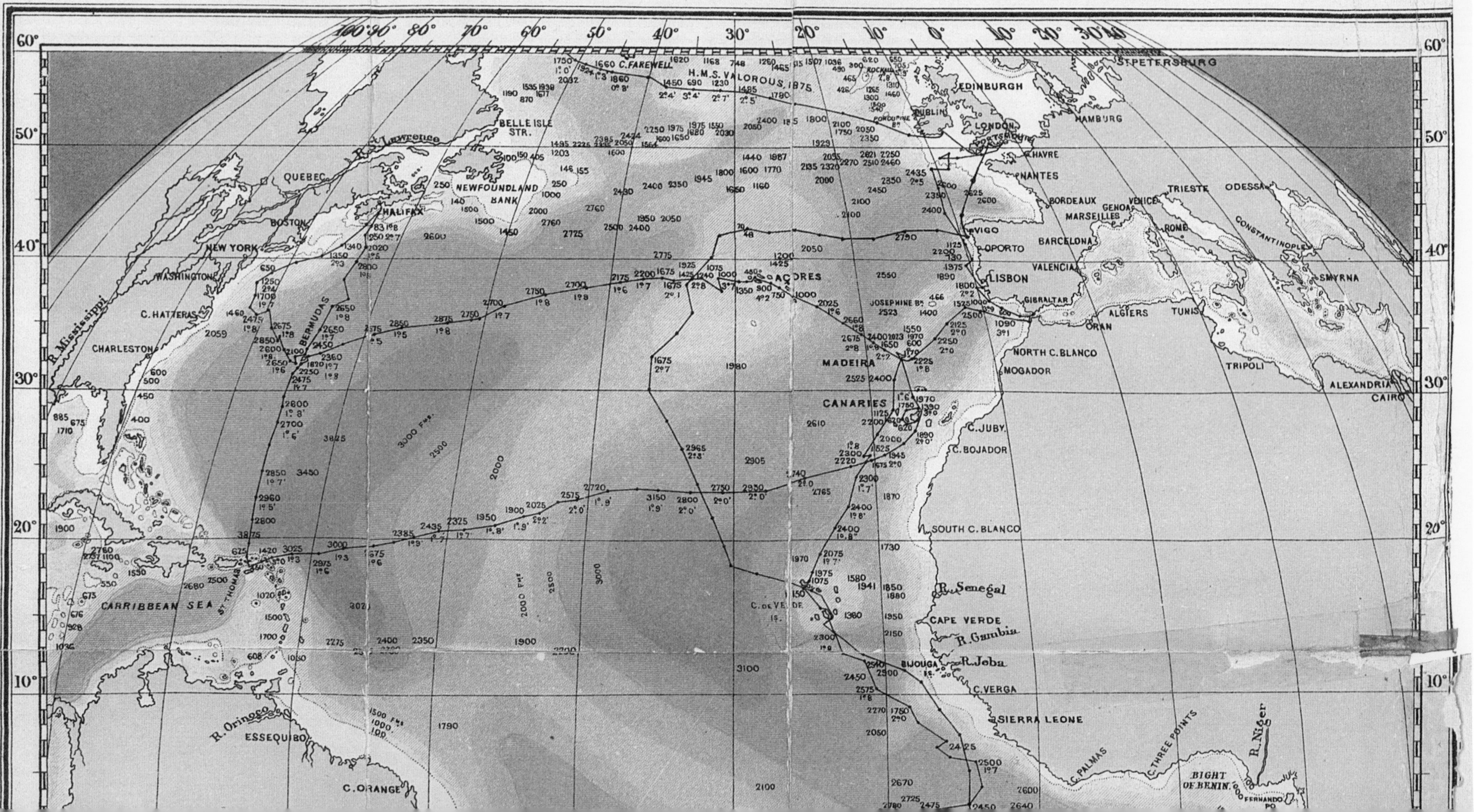

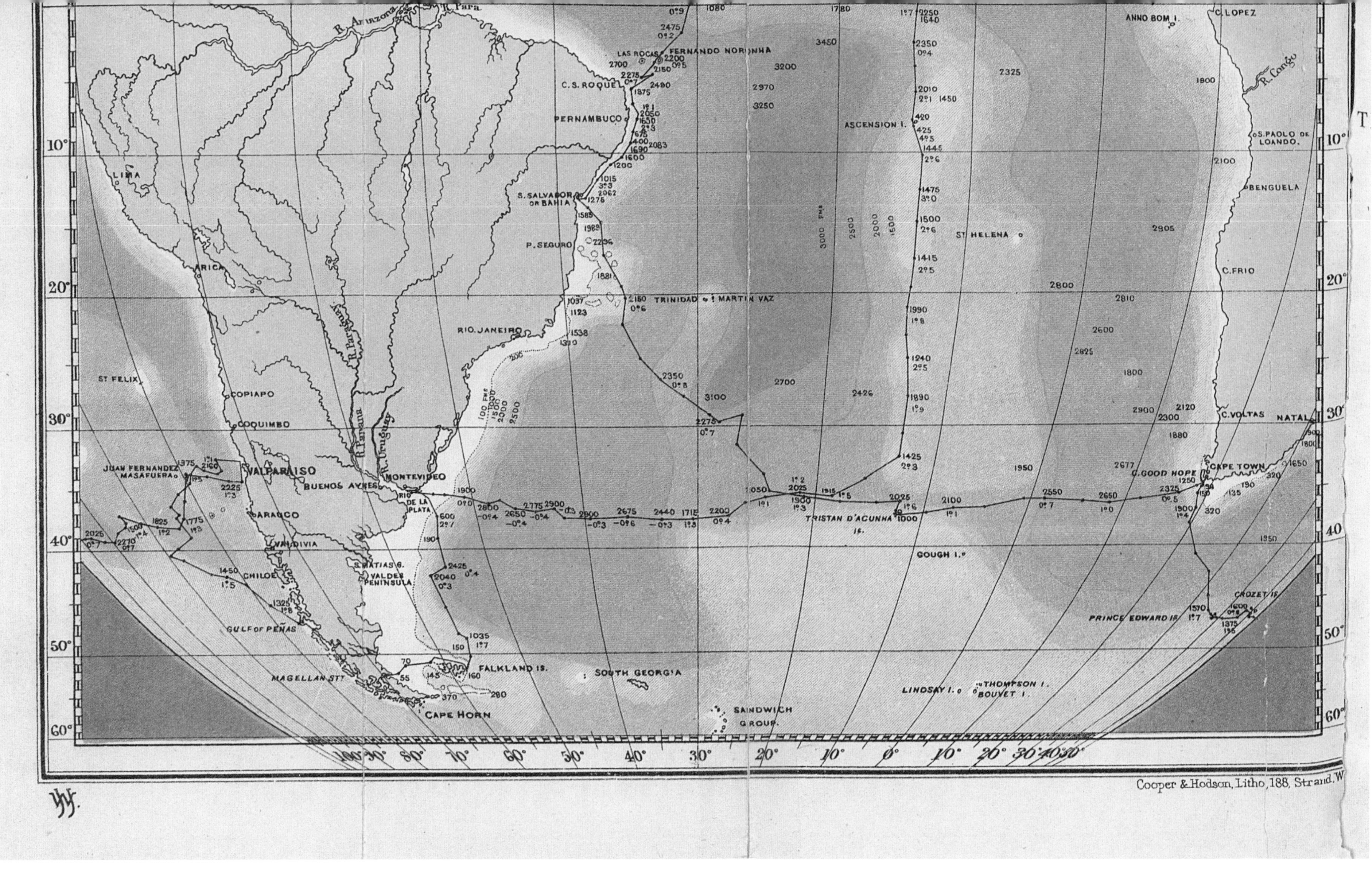

R. Amazons
R. Para
FERNANDO NORONHA
LAS ROCAS
C. S. ROQUE
PERNAMBUCO
S. SALVADOR OR BAHIA
P. SEGURO
RIO JANEIRO
TRINIDAD
MARTIN VAZ
LIMA
ARICA
ST FELIX
COPIAPO
COQUIMBO
VALPARAISO
JUAN FERNANDEZ
MASAFUERA
R. Paraguay
R. Parana
R. Uruguay
MONTEVIDEO
BUENOS AYRES
RIO DE LA PLATA
ARAUCO
VALDIVIA
CHILOE
S. MATIAS G.
VALDES PENINSULA
GULF OF PEÑAS
MAGELLAN STT
FALKLAND IS.
CAPE HORN
SOUTH GEORGIA
SANDWICH GROUP
ASCENSION I.
ST HELENA
TRISTAN D'ACUNHA IS.
GOUGH I.
LINDSAY I.
THOMPSON I.
BOUVET I.
ANNO BOM I.
C. LOPEZ
R. Congo
S. PAOLO DE LOANDO
BENGUELA
C. FRIO
C. VOLTAS
NATAL
CAPE TOWN
C. GOOD HOPE
PRINCE EDWARD IS
CROZET IS
Cooper & Hodson, Litho, 188, Strand, W.

SURVEYING INDIA

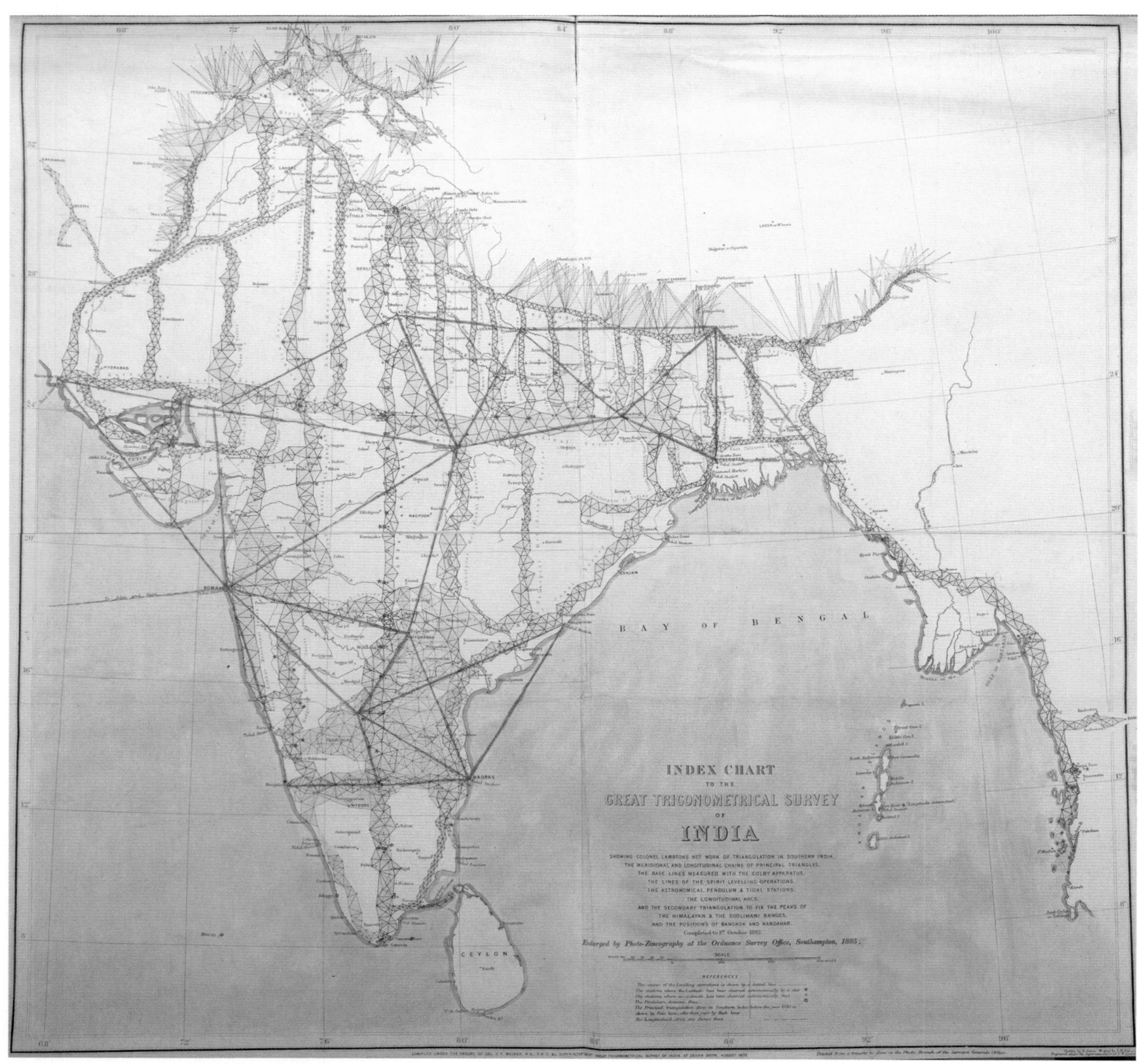

Index Chart to the Great Trigonometrical Survey (GTS) of India. The survey created an accurate framework for the subsequent detailed mapping of India. Royal Geographical Society, London, UK.

Survey Index Chart

The first maps of vast, unconnected tracts of the Indian subcontinent had been compiled by the British East India Company army by the end of the eighteenth century. As they expanded their territories, the need for standard comparable maps for strategic military needs and for the administration of occupied territories, became compelling. Thus the Great Trigonometrical Survey of India (GTS) was conceived.

The GTS began under the guidance of the Superintendent, GTS, William Lambton in 1802. Using terrestrial survey methods, this monumental exercise was completed in 1870, well after Lambton's death in 1823. The project continued under the stewardship of George Everest, the next Superintendent, and others later. Over seven decades, it created a common framework for mapping, showing accurate positions of places and heights above sea level, throughout the Indian subcontinent.

The accuracy of the surveys depended greatly on the precise measurement of a base line. The first base line was of 12.1 km (7½ miles), measured near St Thomas's Mount in Madras (present-day Chennai) in 1802. Using this line as one side, a triangle was marked on the ground. The three sides of this triangle then formed bases for further triangles to build upon. In this manner, the peninsula was covered.

Systems changed over time. For instance, while Lambton followed the method of covering the entire ground with triangles, Everest chose to make a series of triangles along selected lines of longitude (the meridional series) and lines of latitudes which joined them (the longitudinal series). Survey methods improved as instruments such as theodolites became smaller in size, more accurate, and easier to carry around. This was reflected in a high degree of accuracy in the final results of the GTS.

This 1882 Index map shown here is a simplified representation of the GTS. It is not a map in the conventional sense but an index showing the superstructure of the surveys conducted. It highlights the principal GTS triangles, and marks the places where observations were made. What is particularly interesting, but not evident from the map, is that each vertex is physically identifiable by a feature on the ground, be it a hill top, an enormous boulder, a temple turret or even a specially constructed survey tower.

As the GTS extended, a parallel project was undertaken. It involved accurate measurement of the arc of a meridian, named the Great Meridional Arc, passing through the length of India. The line of 78°E longitude, joining Cape Comorin on the southern tip of India to Mussoorie in the north, was selected for this survey. From the size of the Great Meridional Arc, it was possible to calculate the length of a meridian. The purpose of this survey was to study the shape and size of the earth, which had an important bearing on the GTS. The project measuring the Great Meridional Arc took four decades to complete.

There were many significant achievements of the GTS project. India now had a mesh of triangles which formed the framework for compiling maps of the whole country. The measurement of the geodetic anomaly, a mathematically complex process, was also undertaken successfully. This enabled the correct demarcation of boundaries of territories and the calculation of their area. When the GTS reached the Himalayas, the heights of high peaks were measured and the elevations of Mount Everest, K2, Kangchenjunga and other peaks calculated. With these height measurements in hand, it was possible to announce that Mount Everest was the highest peak in the world.

The GTS was limited to areas under British political influence. At that time, Tibet strictly forbade Europeans from entering the country. But traditionally, Indian pilgrims and traders walked across through the Himalayan passes. The British took advantage of this and identified educated Indians, '*pundits*' or learned men, trained them in survey methods and then sent them on secret missions to map Tibet. Some of these spy missions took years to complete and added greatly to the geographical knowledge of Tibet.

Today, we can only marvel at the magnitude of the GTS project. Hundreds of surveyors and native staff persevered for decades, passing through forests and deserts, and encountering floods, hostile local people, wild animals, and more. Its success can be judged by the fact that the GTS was the foundation of Indian mapping for the next century and a half.

DEHLI
Moradabad
Pillibheet
Shikerpoor
Bekaner
Boolundshuhur
Bareilly
Badaoon
Shahjuhanpoor
Sukkur
Allygurh
Ulwar
Muthra
AGRA
Mynpooree
Furukhabad
Jessulmer
Bhurtpoor
Jeypoor
LUCKNOW
Fyzabad
Oudh
Sehwen
Etawah
Cawnpoor
Ajmer
GWALIOR
Jaloun
Azimgurh
Jodpoor
Jounpoor
Humeerpoor
HYDRABAD
Bandah
Benares
Mirzapoor
Reewah
Tidal Station
Oodeypoor
Neemuch
Deesa
Pahlanpoor
Seronj Base
Seronj
Saugor
Augur
Bhopal
Ougein
Sehor
Jubbulpoor
Sohagpoor
Ahmedabad
Bhooj
Hanstal Tidal St.
GULF OF CUTCH
Okha Tidal St.
Morvi
Manoodabad
Indore
Dhar
Mhow
Rajkot
Amjerrah
Nurbudda River
Dwarka
Cambay
Baroda
Gondal
Ruttunpoor
Belaspoor
Oomrait
Seonee
Porbandar
Kattiawar Triangulation
Kompta
Taptee R.
Surat
Dhoolia
Ellichpoor
NAGPOOR
Bundara
Ryepoor
Diu
GULF OF CAMBAY
Damaun
Maligaon
Karinjah
Hingunghat
Kareealh
Dowlutabad
Aurungabad
Jalna
Chanda
Godavery R.
Ahmednuggur
Amba
Nirmul
BOMBAY
Tidal Station
To Aden and Suez
Ryepoor
Poona
Ellichpoor
Bider Base
Bider
Warungal
Satara
Nul Dg
Bolaram
HYDRABAD
Koolburga
Kummummet
Rajamundry
Coconada
Ratnagiri
Mulkhair
Nelgoondah
Beejapoor
Suggur Dg
Davurcondah
Kistna R.
Gunjoor
Masulipatam
Moodgul
Goa
Darwar
Bellary
Karwar
Tidal Station
Nellore
Chittle Droog
Cuddapah
Nuggur or Biddenoor
MADRAS
Tidal Station
Mangalore
Bangalore Base
Bangalore
Arcot
Seringapatam
MYSORE
Cananore
Salem
Pondicherry
Ootacamund
Calicut
Beypore
Tidal Station
Coimbatoor
Trichinopoly
Tanjore
Tranquebar
Ponany
Negapatam
Tidal Station
Pulney
Dindigul
Cochin
Madura
Alleppy
Ramnad
Pamban
Tidal Station
Quilon
Tuticorin
Tidal Station
Trincomalee
Tinnevelly
Minicoy
Trivandram
Cape Comorin Base
Cape Comorin
CEYLON

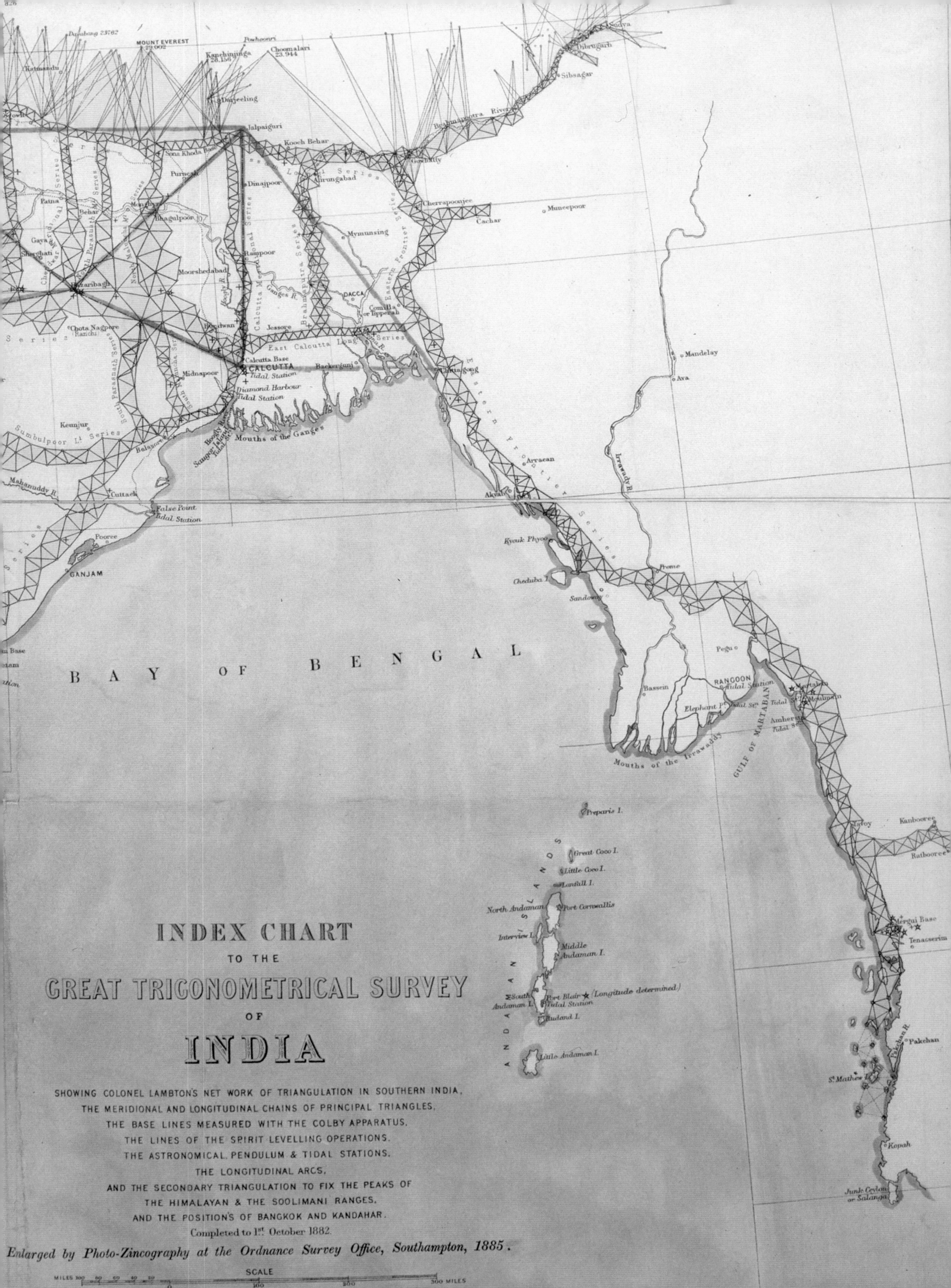

INDEX CHART
TO THE
GREAT TRIGONOMETRICAL SURVEY
OF
INDIA
SHOWING COLONEL LAMBTON'S NET WORK OF TRIANGULATION IN SOUTHERN INDIA,
THE MERIDIONAL AND LONGITUDINAL CHAINS OF PRINCIPAL TRIANGLES,
THE BASE LINES MEASURED WITH THE COLBY APPARATUS,
THE LINES OF THE SPIRIT LEVELLING OPERATIONS,
THE ASTRONOMICAL, PENDULUM & TIDAL STATIONS,
THE LONGITUDINAL ARCS,
AND THE SECONDARY TRIANGULATION TO FIX THE PEAKS OF
THE HIMALAYAN & THE SOOLIMANI RANGES,
AND THE POSITIONS OF BANGKOK AND KANDAHAR.
Completed to 1st. October 1882.
Enlarged by Photo-Zincography at the Ordnance Survey Office, Southampton, 1885.
SCALE
MILES 100 80 60 40 20 0 100 200 300 MILES
MOUNT EVEREST
Kanchinjinga
Choomalari 23.944
Darjeeling
Jalpaiguri
Kooch Behar
Sona Khoda Base
Dibrugarh
Sibsagar
Brahmapootra River
Gowhatty
Dinajpoor
Purneah
Patna
Behar
Monghyr
Bhagulpoor
Gaya
Sherghati
Hazaribagh
Rampoor
Moorshedabad
Cherrapoonjee
Muneepoor
Cachar
Mymunsing
Ganges R.
DACCA
Comilla or Tipperah
Jessore
Burdwan
Chota Nagpore (Ranchi)
East Calcutta Long! Series
Eastern Frontier Series
Calcutta Base
CALCUTTA
Tidal Station
Backergunj
Chittagong
Midnapoor
Diamond Harbour Tidal Station
Keunjur
Sumbulpoor Lt. Series
Mouths of the Ganges
Balasore
Mahanuddy R.
Cuttack
False Point Tidal Station
Pooree
GANJAM
Mandelay
Ava
Irrawady R.
Arracan
Akyab
Kyouk Phyoo
Cheduba I.
Sandoway
Prome
Pegu
RANGOON Tidal Station
Bassein
Elephant Pt. Tidal Stn.
Mouths of the Irrawaddy
GULF OF MARTABAN
Martaban
Moulmein
Amherst Tidal Stn.
BAY OF BENGAL
Preparis I.
Great Coco I.
Little Coco I.
Landfall I.
North Andaman
Port Cornwallis
Interview I.
Middle Andaman I.
South Andaman I.
Port Blair ★ (Longitude determined) Tidal Station
Rutland I.
Little Andaman I.
ANDAMAN ISLANDS
Tavoy
Kanbooree
Ratbooree
Mergui Base
Tenasserim
Pakchan R.
Pakchan
St. Mathew
Kopah
Junk Ceylon or Salanga

John Colomb

The rapid expansion of the British Empire in the mid-nineteenth century brought new problems of imperial administration and increased rivalries with other European imperial powers. Yet it also gave birth to a tradition of imperial cartography, in which Britain's extensive colonies were proudly coloured in red. Sir John Colomb's 1886 map is one of the most famous examples, yet it was also a vehicle for the promotion of a very particular theory of empire – Imperial Federalism.

The British Empire grew rapidly in the second half of the nineteenth century. The tempo of acquisitions increased in the 1880s, with the imposition of British rule in Egypt in 1882, the conquest of the remainder of Burma in 1885 and the acceleration of colonial rule following the Berlin Conference in 1884–85, called to arbitrate on differences between the European powers over their respective spheres of influence in Africa (see page 144).

Yet this growth brought new challenges, with Britain for the first time facing real competition from France in Indo-China and West Africa, from Germany as it established colonies in East Africa in the 1870s, and from Russian expansion in Central Asia. The cost of imperial defence was keenly felt, with the economy in Britain having suffered severe agricultural and industrial depressions in the 1870s. This, together with concerns over Prime Minister Gladstone's expansionist policies which were seen as ill-considered, and calls from the older established colonies in Canada and Australia for more self-rule, led to the birth of a new movement.

The Imperial Federation League was formally launched in July 1884 at a meeting in London which included leading politicians such as Lord Rosebery and interested diplomats, such as the Canadian High Commissioner Charles Tupper. Among its founders was Sir John Colomb, whose advocacy of a proper imperial defence policy based on a strong navy made him a champion of closer ties between London and the colonies. The League argued for some form of federation, whether by an Imperial Parliament or a Colonial Council, which would ensure that the empire would stick together through common interest rather than seeing the more developed colonies drift away.

Colomb published this map in 1886 to illustrate the idea of federation, with the contrast between the inset map and the main drawing showing how explosively the empire had grown in a century. The drawings of the inhabitants of the empire arrayed around the frame amply demonstrate that the proposed federation was, as another of its proponents, the historian Sir John Seeley, wrote, a matter of 'Anglo-Saxon unity' between settlers of British descent. The white settlers, uniformed, armed, or taking a rest from tilling the soil, are a stark contrast with the indigenous peoples portrayed in distinctly inferior poses – for them there was to be no 'federation'.

The medium which Colomb chose to put across his message is revealing. Although the use of globes to study geography had been advocated by the poet John Milton as early as 1644, the general state of geographical education in Britain was poor (there was no full-time university post in the subject before Oxford University established a Readership in Geography in 1887). The growth of the empire made this a pressing concern and in 1852 Joseph Guy in his *Illustrated London Geography* urged that schoolchildren be shown maps so that 'they may be shown the extent of the British Empire and the countries and islands under the sceptre of our Queen'. Maps of the empire began to appear, with the 1841 edition of Henry Teesdale's *New British Atlas* being the first to show Britain's imperial possessions in red, a practice which became generally accepted by the 1850s. Another practice commonly adopted by cartographers and map publishers (including Colomb) at the time was to construct such maps on Mercator's projection (see page 76) – a map projection which through its distortions of the relative sizes of land areas gave particular prominence to British colonial areas. Wall maps appeared in increasing numbers in classrooms and played a key role in implanting the concept of empire as the natural orders of things in Victorian minds.

Colomb's map, with the figure of Britannia, trident in hand, presiding over her empire, was firmly based on this tradition. With its wealth of statistical information about the population of the key colonies, and the trade routes which bound them to the metropolis of London, it is both a celebration of empire and a plea for political measures which would ensure its survival.

THE BRITISH EMPIRE

Imperial Federation – map of the world showing the extent of the British Empire in 1886 by Colomb. British territories are shown prominently in red. Private Collection.

BAFFIN BAY
DAVIS STRAIT
HUDSON BAY
DOMINION OF CANADA
British North America. | 1851. | Present Time. Latest Returns
Area .. Square Miles | — | 3,510,592
Population .. .. .. | 2,471,237 | 4,504,819
Total Trade {Imports, £ | 7,586,791 | 25,599,657
Total Trade {Exports, £ | 5,182,441 | 20,150,809
Shipping, Entd. & Clrd. tons | 3,234,823 | 9,211,025
Imports from the United Kingdom, 1851 £3,359,897
" " " present time 9,564,088
Exports to " " 1851 2,583,897
" " " present time 9,309,424
BRITISH COLUMBIA
VANCOUVER
Port Moody
VICTORIA
QUEBEC TO PORT MOODY 2930 MILES
Manitoba
NORTH AMERICA
NORTH PACIFIC OCEAN
QUEBEC TO LIVERPOOL 2653 MILES
BRITISH ISLANDS
NEWFOUNDLAND
St Johns
Quebec
Montreal
Boston
New York
Philadelphia
San Francisco
New Orleans
NORTH ATLANTIC OCEAN
BERMUDA I.
GIBRALTAR
MADEIRA
CANARY Is
GULF OF MEXICO
BAHAMA I.
ABACO
CAT I.
CAICOS
WEST INDIA ISLANDS
CUBA
HAYTI
ST THOMAS
ANTIGUA
DOMINICA
ST LUCIE
BARBADOES
ST VINCENT
GRENADA
TRINIDAD
ST THOMAS TO SOUTHAMPTON 3626 MILES
Tropic of Cancer
BRITH HONDURAS
Belize
JAMAICA
CARIBBEAN SEA
Panama
CAPE VERDE Is
BATHURST
SIERRA LEONE
GOLD COAST
British West Indies. | 1851. | Present Time. Latest Returns
Area .. Square Miles | — | 128,874
Population .. .. .. | 886,717 | 1,506,730
Total Trade {Imports, £ | 4,759,685 | 9,468,992
Total Trade {Exports, £ | 4,809,961 | 9,489,352
Shipping, Entd. & Clrd. tons | 1,075,190 | 6,558,828
Imports from the United Kingdom, 1851 £2,760,262
" " " present time 4,015,648
Exports to " " 1851 3,592,779
" " " present time 4,486,094
Georgetown
BRITISH GUIANA
SYDNEY TO LONDON BY PANAMA 12631 MILES
SOUTH PACIFIC OCEAN
SOUTH AMERICA
Tropic of Capricorn
HOMEWARD ROUTE AUSTRALIA TO ENGLAND BY CAPE HORN 13600 MILES
ASCENSION
SOUTH ATLANTIC
TRISTAN D
FALKLAND Is
S. GEORGIA

India, Ceylon and Straits.	1851.	Present Time. Latest Returns
Area .. Square Miles	——	895,151
Population	178,687,148	201,978,221
Total Trade {Imports, £	17,369,048	91,644,871
Total Trade {Exports, £	20,511,316	109,528,827
Shipping, Entd. & Clrd. tons	2,065,688	17,104,897
Imports from the United Kingdom, 1851		£9,065,035
" " " present time		55,809,827
Exports to " 1851		8,892,023
" " present time		42,682,224

British South Africa.	1851.	Present Time. Latest Returns
Area .. Square Miles	——	298,450
Population	405,906	1,674,819
Total Trade {Imports, £	1,799,243	6,906,547
Total Trade {Exports, £	673,157	5,162,878
Shipping, Entd. & Clrd. tons	428,068	1,998,484
Imports from the United Kingdom, 1851		£1,554,114
" " " present time		5,884,271
Exports to " 1851		589,864
" " present time		4,516,094

Australasia.	1851.	Present Time. Latest Returns
Area .. Square Miles	——	3,161,842
Population	584,798	8,281,768
Total Trade {Imports, £	4,358,899	64,001,120
Total Trade {Exports, £	4,898,718	54,572,766
Shipping, Entd. & Clrd. tons	1,066,898	15,881,858
Imports from the United Kingdom, 1851		£3,071,984
" " " present time		31,875,817
Exports to " 1851		8,341,147
" " present time		27,025,996

North Atlantic Ocean

The oceans, so long the domain of traders, explorers, raiders and small groups of migrants became integrated into a global transportation system in the mid-nineteenth century, as steamships began to ply major maritime routes, cutting the time taken to cross the Atlantic from two months to two weeks. As established passenger routes grew in number, so maps became possible which showed the web of connections being made between the continents.

This map of the North Atlantic Ocean from 1907 offers a remarkable picture not just of the economic and communications links but also of the social and cultural connections which tied Britain and Europe with the USA and Canada in the early twentieth century. This was a time when shipping dominated the world of international travel.

The map shows primarily the transatlantic shipping connections between the northeastern USA and the Canadian Maritimes, and the big five British ports: Glasgow, Liverpool, Southampton, London and Cork (Cork was part of the United Kingdom until 1922). These lines appear in red, in the map's upper half.

Also plotted less prominently are lines from other countries – such as Bremerhaven to Baltimore and Naples to New York – and connections from Britain to cities beyond the map's boundaries, such as Kingston (Jamaica) and New Orleans. The map plots a contemporary picture of a period of mass migration from Europe which in 1907 alone saw over 1.25 million people enter the USA.

When viewed together with the other shipping lines linking countries and ports beyond the map's main perspective, the final impression is of an intricate and complex web of surface connections which is striking over a century later. It is a reminder that although the means of travel for passengers has changed dramatically – particularly since the growth of affordable air travel since the 1950s (see page 220) – the same network of sea and shipping lanes continues in existence today, albeit less prominently in the cultural mind.

Also plotted are the undersea cables which connected the world, trailing from the extremities of the European nations – Valentia Island, Land's End, Brest and Lisbon – to their counterparts on the Eastern Seaboard of the USA and beyond. In some cases, the cables spanned half a hemisphere.

The numbers appearing in mid-ocean, and the contours marked around the continental landmasses, show the extent and accuracy of contemporary plotting of the ocean depths. Parts of the route taken across the North Atlantic by the most famous survey ship, HMS *Challenger* on the Challenger Expedition of 1872–6 (see page 152), can be seen between Lisbon, Gibraltar, Madeira, the Canary Islands, Bermuda and the Azores. This expedition was instrumental in establishing the science and practice of oceanography.

TRANSATLANTIC SHIPPING

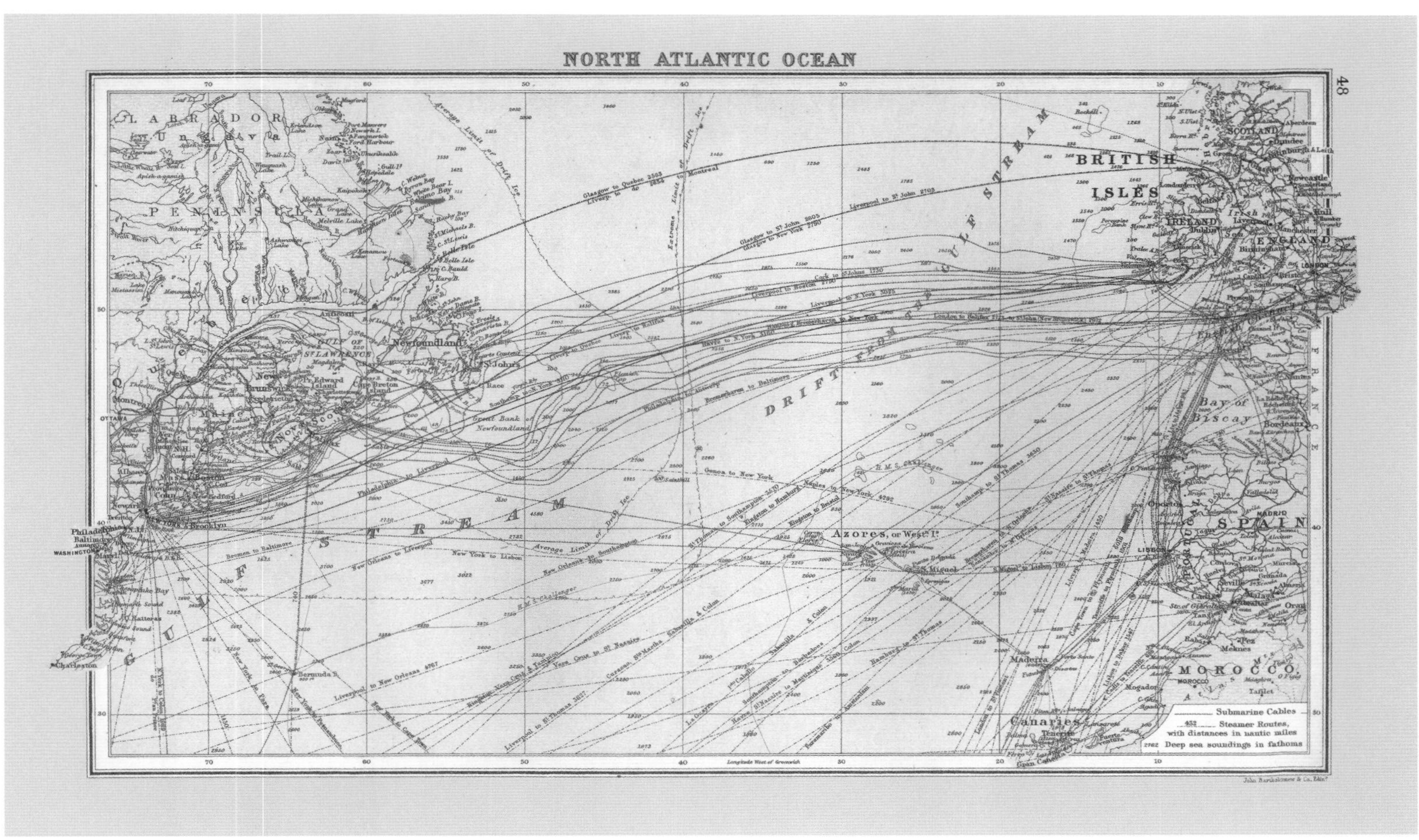

Map of the North Atlantic showing shipping routes and submarine cables. From the *Atlas of the World's Commerce*, Bartholomew/George Newnes, 1907. Collins Bartholomew.

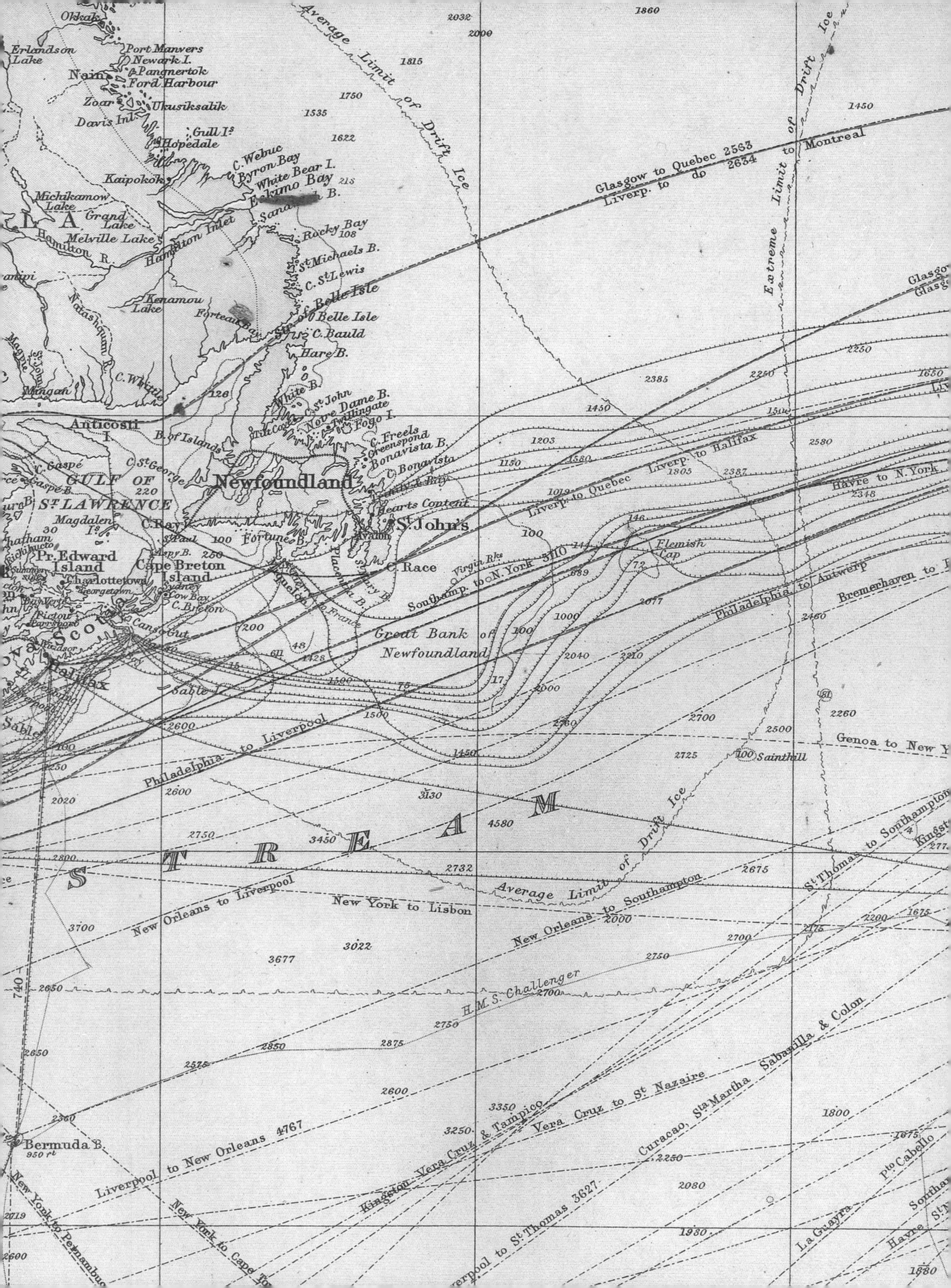

Okkak
Erlandson Lake
Port Manvers
Newark I.
Nain
Pangnertok
Ford Harbour
Zoar
Ukusiksalik
Davis Inl.
Gull Is
Hopedale
C. Webuc
Byron Bay
White Bear I.
Kaipokok
Eskimo Bay
Michikamow Lake
Grand Lake
Melville Lake
Hamilton R.
Hamilton Inlet
Sandwich B.
Rocky Bay
St Michaels B.
C. St Lewis
Kenamou Lake
Natashquan R.
Forteau Bay
Str. of Belle Isle
Belle Isle
C. Bauld
Hare B.
C. Whittle
Mingan
White B.
St John
Notre Dame B.
Twillingate
Fogo I.
Anticosti I.
B. of Islands
C. Freels
Greenspond
Bonavista B.
C. Bonavista
C. Gaspé
Gaspé B.
C. St George
GULF OF ST LAWRENCE
Newfoundland
Trinity B.
Hearts Content
Magdalen Is
C. Ray
St Paul
Fortune B.
St John's
Avalon
C. Race
Placentia B.
St Mary B.
Pr. Edward Island
Charlottetown
Georgetown
Cape Breton Island
Sydney
Cow Bay
C. Breton
Pictou
Canso Gut
Miquelon
St Pierre
Nova Scotia
Halifax
Windsor
Sable I.
Sable
Great Bank of Newfoundland
Virgin Rks
Flemish Cap
Sainthill
Average Limit of Drift Ice
Extreme Limit of Drift Ice
Glasgow to Quebec 2563
Liverp. to do 2634 to Montreal
Liverp. to Halifax
Liverp. to Quebec
Havre to N. York
Southamp. to N. York 3110
Philadelphia to Antwerp
Bremerhaven to
Philadelphia to Liverpool
Genoa to New York
GULF STREAM
New Orleans to Liverpool
New York to Lisbon
New Orleans to Southampton
St Thomas to Southampton
Kingston
H.M.S. Challenger
Bermuda B.
950 ft
Liverpool to New Orleans 4767
Vera Cruz to St Nazaire
Curacao, Sta Martha Sabanilla & Colon
Kingston Vera Cruz & Tampico
Liverpool to St Thomas 3627
New York to Pernambuco
New York to Cape Town
La Guayra
Pto Cabello
Havre
Southampton
2032
2000
1860
1815
1750
1535
1622
1450
215
108
126
220
30
100
250
200
48
1428
75
1500
1200
17
2000
2760
2600
1450
2725
2700
2500
2260
2385
2250
1650
1500
1450
1203
1150
1580
1805
2387
2580
1019
2348
146
114
72
889
1000
2077
2460
100
2040
2210
2600
2020
3130
4580
2750
3450
2800
2732
2675
3700
3022
3677
2000
2200
1675
2175
2700
2750
740
2650
2700
2750
2850
2875
2575
2650
2600
2360
3350
3250
1800
1675
2250
2080
1930
2719
2600
1880

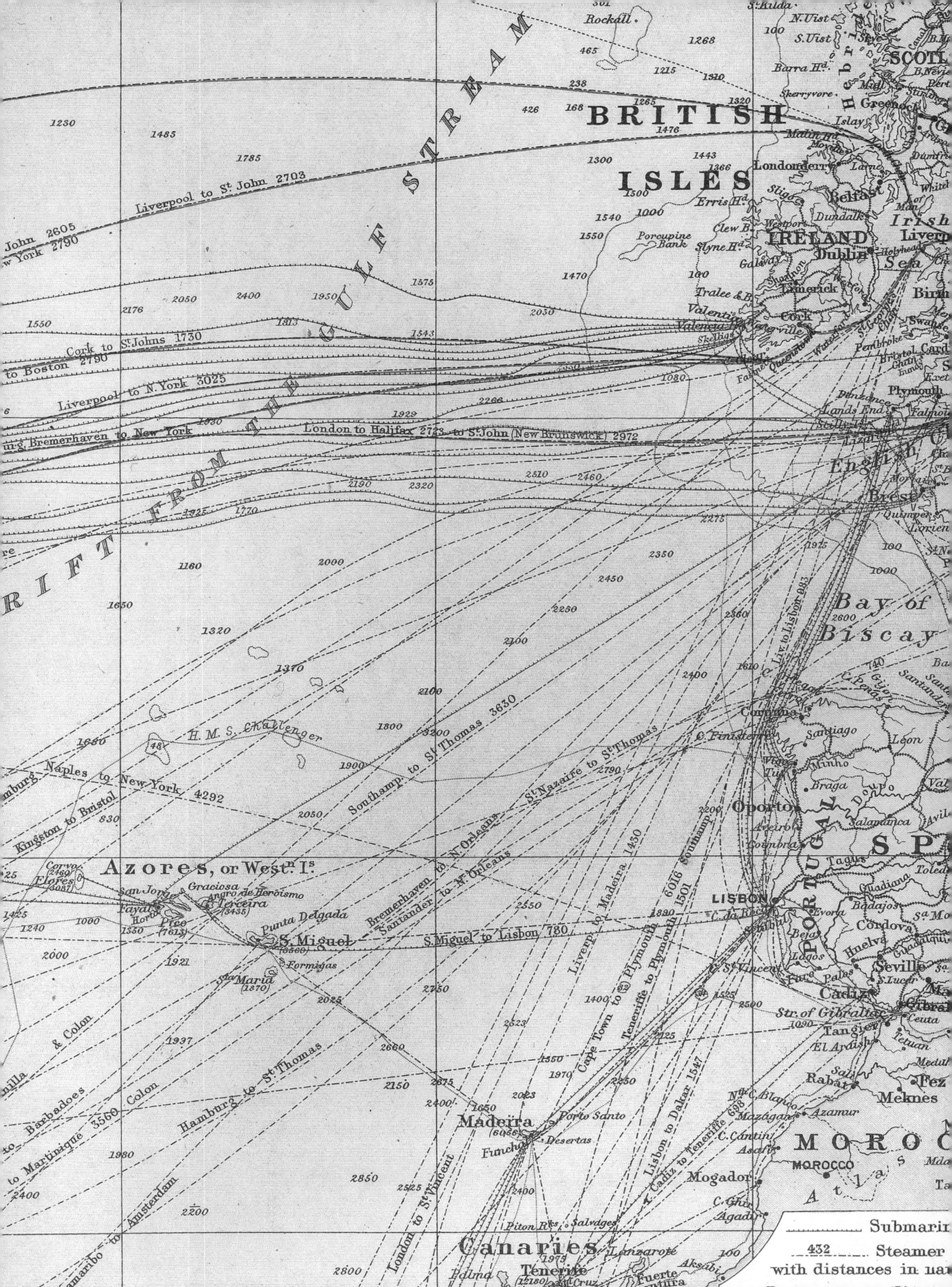
Rockall
St. Kilda
N. Uist
S. Uist
Barra Hd.
Skerryvore
Islay
Greenock
BRITISH
ISLES
Liverpool to St. John 2703
John 2605
York 2790
STREAM
GULF
THE
FROM
DRIFT
Londonderry
Belfast
Sligo
Erris Hd.
Clew B.
Westport
IRELAND
Dublin
Dundalk
Irish
Sea
Porcupine Bank
Slyne Hd.
Galway
Limerick
Tralee B.
Valentia
Cork
Queenstown
Waterford
Skelligs
Pembroke
Bristol Chan.
Cork to St. Johns 1730
to Boston 2790
Liverpool to N. York 3025
Plymouth
Penzance
Lands End
Scilly Is.
Lizard
Bremerhaven to New York
London to Halifax 2723 to St. John (New Brunswick) 2972
English
Brest
Quimper
Lorient
Bay of
Biscay
C. Liv. to Lisbon 983
Gijon
Santander
Corunna
Santiago
Leon
C. Finisterre
Vigo
Minho
Braga
Douro
Oporto
Aveiro
Coimbra
Salamanca
PORTUGAL
SP
Tagus
LISBON
C. da Roca
Evora
Badajoz
Cordova
Guadiana
Huelva
Seville
Lagos
C. St. Vincent
Cadiz
Str. of Gibraltar
Tangier
Ceuta
Tetuan
El Araish
Rabat
Fez
Mekines
Azamur
Mazagan
C. Cantin
Mogador
MOROCCO
Atlas
C. Ghir
Agadir
H. M. S. Challenger
Naples to New York 4292
Kingston to Bristol
Southamp. to St. Thomas 3630
St. Nazaire to St. Thomas
Bremerhaven to N. Orleans
Santander to N. Orleans
Azores, or Westn. Is.
Corvo
Flores
San Jorge
Graciosa
Angra de Heroismo
Terceira
Fayal
Horta
Pico
Punta Delgada
S. Miguel
Formigas
Sta. Maria
S. Miguel to Lisbon 780
Liverp. to Madeira 1430
Southamp. 6016
Cape Town to Plymouth
Teneriffe to Plymouth 1501
Lisbon to Dakar 1547
Cadiz to Teneriffe 698
& Colon
Barbadoes
to Martinique 3560 Colon
Hamburg to St. Thomas
Paramaribo to Amsterdam
London to St. Vincent
Madeira
Porto Santo
Funchal
Desertas
Piton Rks.
Salvages
Canaries
Lanzarote
Tenerife
Palma
Fuerte ventura
Submarine
432 Steamer
with distances in

India

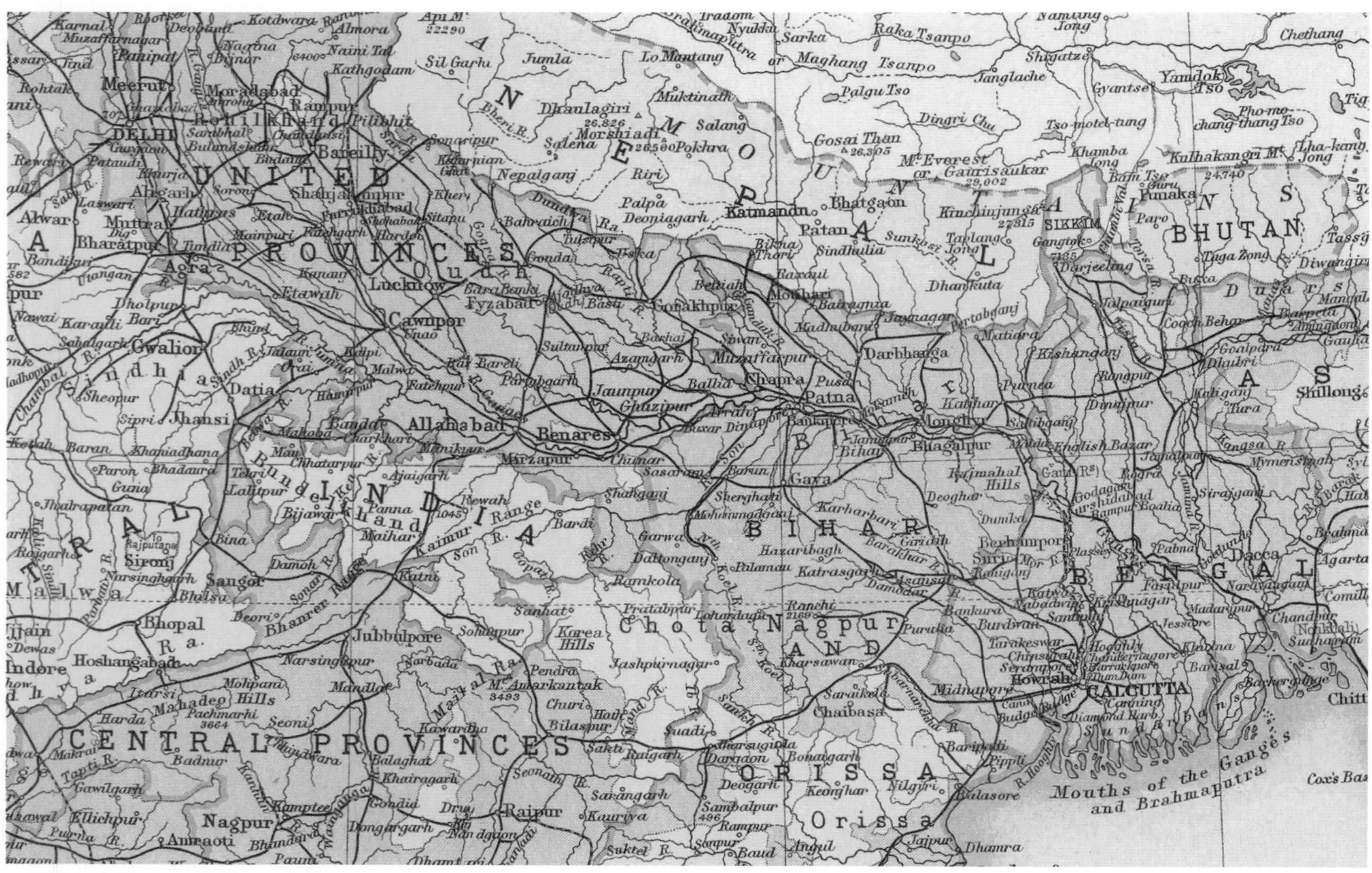

Map of the Indian Empire by J.G. Bartholomew, from the *International Reference Atlas of the World*, Bartholomew/George Newnes, 1914. Collins Bartholomew.

BRITISH IMPERIAL POWER

Dating from immediately before the outbreak of the First World War, this map graphically illustrates the reach of British imperial power on the Indian subcontinent at that time. The map demonstrates the complex interplay of direct rule and local autonomy which was essential to the smooth administration of such a vast and diverse territory.

The map projects the writ of the King Emperor George V across a territory that encompasses the present-day states of India, Pakistan, Bangladesh and Myanmar (Burma). Although not part of British India, the strategically vital Crown colony of Ceylon (now Sri Lanka) and the northern kingdoms of Nepal and Bhutan also fell under its influence.

In the second half of the eighteenth century, the British East India Company occupied extensive territories in India. Having monopolized trade within India, the Company subsequently acquired military and administrative responsibilities in the region, until it became the de facto ruler.

Areas coloured in pink on the map were referred to as British India. These provinces were administered directly by London through its principal representative, the Governor-General, the head of the British Administration in India.

By contrast, the areas in yellow mark out the clusters of Princely States – those nominally sovereign territories outside the direct control of the British Government whose rulers had entered into personal treaties with the Crown. These treaties allowed each state a degree of local autonomy and freedom to issue its own laws and set its own government. However, each was obligated to provide military assistance to the British Raj (the British government in India) in return for protection. The monarch's personal representative to these Princely States was the Viceroy.

As a consequence of the monarch's influence over this large population, the British Indian Army supported Britain's First World War effort with 1.4 million people being sent to fight alongside British soldiers in the Middle East and Iraq. Having served honourably, once the war ended, Indians expected in return a greater role in the governance of India. A groundswell of opinion in favour of self-rule and Dominion status finally led to Independence at midnight on 14 August 1947.

Also evident from the map is the extent of India's rail network. This had been started in 1853 under private enterprise, and by 1907, when all rail companies were brought under government control, the network extended for over 14,500 km (9,000 miles), mostly spreading inland to the British administered provinces from the major ports and termini of Calcutta (Kolkata), Bombay (Mumbai) and Madras (Chennai).

Burma (Myanmar), which was annexed on 1 January 1886 following the three Anglo-Burmese wars of the nineteenth century, is shown as part of the Indian Empire. In the northeast, the McMahon line was delineated, having been agreed upon by Tibet and Britain in accordance with the Simla Accord of 1914. A century later, this boundary line remains contentious for the countries on either side of it, China and India.

Earlier, in 1893, the northwestern boundary of the Empire was defined by the Durand Line which cut through Pashtun tribal lands in the traditionally undefined frontier between Afghanistan and Hindoostan. The Durand Line now separates Afghanistan and Pakistan.

Indian Empire highlights the extent of British political control in 1914 – from Baluchistan in the west to Burma in the east, Kashmir in the north to the Indian Ocean in the south. In less than thirty-five years this influence faded away and the age of empire came to an end.

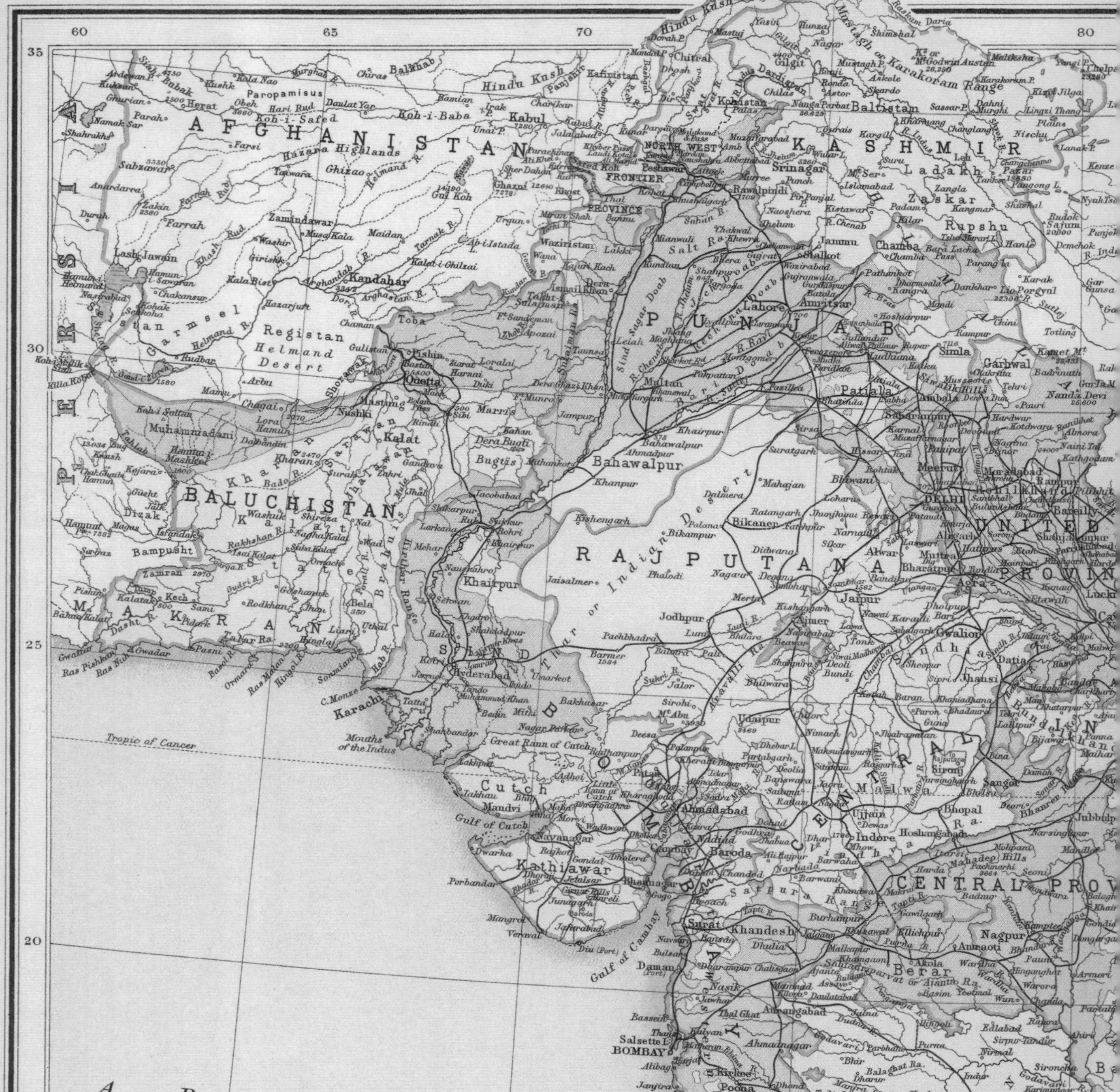

INDIAN EMPIRE

British Statute Miles 69·16 = 1 Degree

100 50 0 100 200

NOTE TO COLOURING

British Possessions

French Possessions

Subordinate Native States

Portuguese Possessions

Railways open —— constructing —— Canals ------

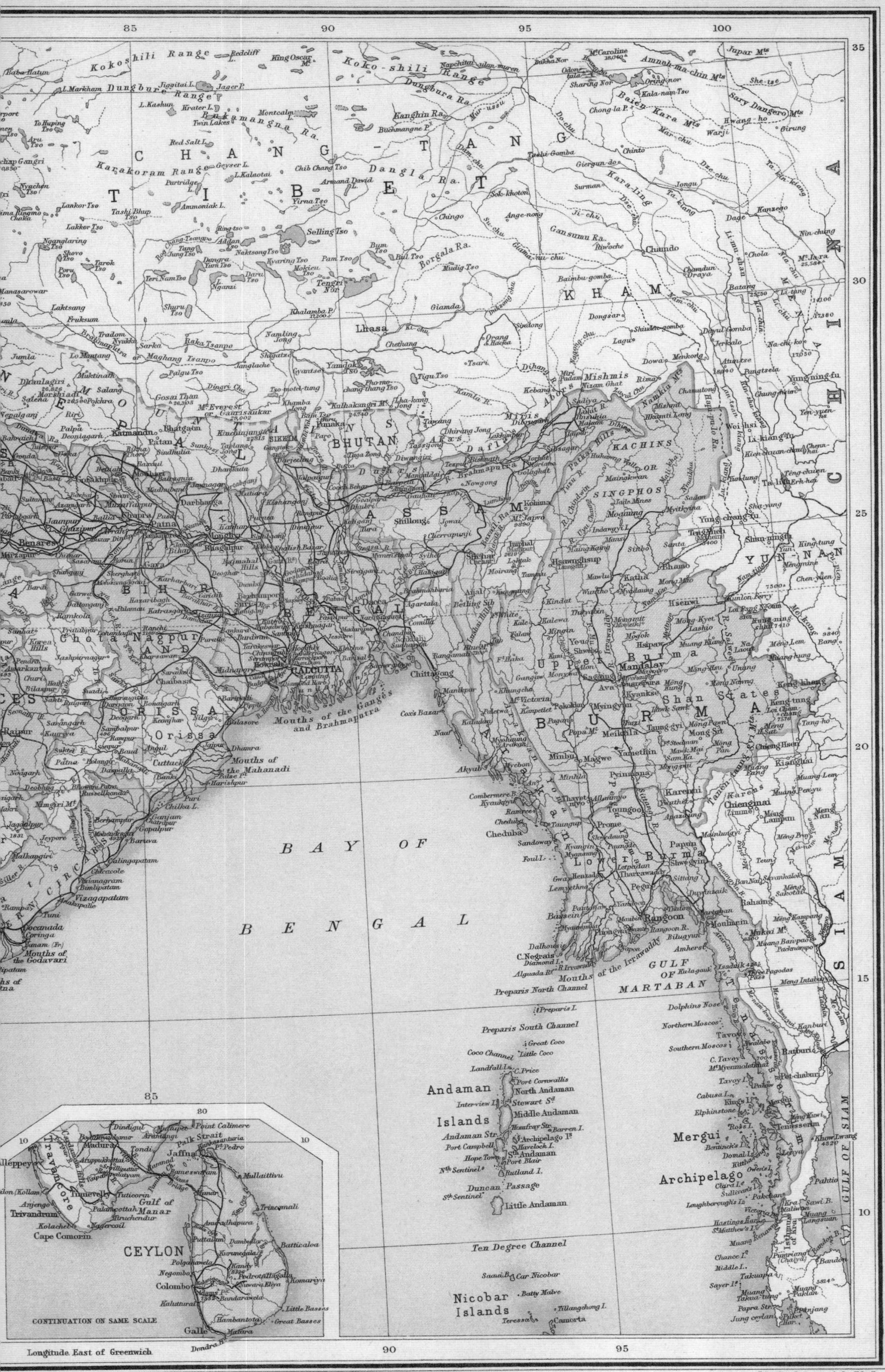

John Bartholomew & Co. Edinr

DIVIDING THE MIDDLE EAST

Map of the Sykes-Picot Agreement of 1916. Hand colouring and annotations refer to the proposed administration of the region, including British and French spheres of influence, independent Arab States, and the 'Sykes-Picot Line'. The National Archives, London, UK (MPK 1/426).

As the First World war progressed, the Allies embarked on a strategy of defeating the Ottoman Empire (which had allied itself to Germany) by fomenting a revolt among its Arab subjects. Dismemberment of the Ottoman empire became a war aim, and in 1916 the French and British agreed a partition of the Arab lands into areas directly occupied by them, and larger spheres of influence. The map drawn by the negotiators, Sykes for the British and Picot for the French, cast a shadow over the Middle East for a century.

On 28 October 1914, the Ottoman Empire entered the First World War as one of the Central Powers (with Germany, Austro-Hungary and Bulgaria). As part of their strategy for defeating the Turkish armies, the Allies sought to attract local Arab rulers to their side with promises of independence after the war was over. Arab forces led by Hussein and Faisal, the sons of Sherif Hussein of Mecca, proved particular thorns in the Ottoman side. Their British liaison officer, T.E. Lawrence earned undying fame as 'Lawrence of Arabia'.

Vague, and often contradictory promises of Arab autonomy had to be balanced with other needs, and this proved next to impossible. The primary need was to ensure that Turkey did not collapse completely, thereby removing a buffer against Russian expansionism and against the increasingly vocal demands of Zionists for a Jewish homeland in Palestine. A long-running correspondence which began in July 1915 between Hussein and Sir Henry McMahon, the British High Commissioner in Egypt, promised an independent Arab state south of the thirty-seventh parallel. However, it excluded the Syrian coastal strip and the holy places in Palestine– the former of which was to remain French, while the status of the latter was left somewhat unclear.

Yet fear of French ambitions in the region kept Hussein from joining the Allies, while in turn Britain needed to ensure France felt rewarded for the huge sacrifices it had made in the war on the Western Front. As a result Britain appointed the politician and diplomat Sir Mark Sykes to negotiate with his French counterpart François Georges Picot over what was, in effect, a carve-up of the Middle East. While the French wanted to keep the Ottoman Empire intact, Picot soon recognized this was impossible, but wished to secure direct French rule over Lebanon and Syria. Sykes sought to cement British influence in what is now Iraq and Jordan, as well as over Palestine. The compromise which finally emerged in May 1916 allocated northern Syria and Lebanon to direct French rule, with a similar area of British control in southern Iraq. The remainder of Iraq and Jordan were designated as 'Area A' – in which French influence would dominate – and 'Area B' (mainly modern Jordan) – which would be under British influence. In Palestine, Britain was to get Acre and Haifa and a strip of land connecting it to Area B, while the rest was to fall under international administration.

The agreement in the end satisfied nobody. It was not long before the Balfour Declaration (in 1917) seemed to supersede the Sykes-Picot provisions in Palestine by promising a Jewish national homeland there. And the Arabs felt – rightly as it turned out – that 'Area A' and 'Area B' were merely proxies for Franco-British colonial control. After the Bolsheviks were brought to power by the Russian Revolution in 1917, they leaked the terms of the agreement (which had hitherto been secret), further undermining it. The 'Fourteen Points' issued by US President Wilson in January 1918 recognizing the right to self-determination of nations, also seemed to contradict the fundamental basis of Sykes-Picot, which was that this independence would be denied.

The Paris Peace Conference of 1919, which met to decide the terms for peace, had therefore to deal with several conflicting currents when it came to deciding the post-war borders of what had been the Ottoman Empire. In the north, the 1920 Treaty of Sèvres delineated an independent Armenia and an area of French influence in Kurdish majority areas of southeastern Anatolia. However, its harsh terms were rejected by Turkish revolutionaries under Kemal Ataturk who ultimately obtained the far more favourable Treaty of Lausanne in 1922. The settlement for the Levant was agreed at the San Remo Conference in 1920, which determined that Palestine, Mesopotamia (Iraq) and Transjordan (modern Jordan) were to be ruled as League of Nations Mandates by Britain, and Lebanon and Syria were similarly mandated to France. As ever in the Middle East, seemingly simple solutions engendered complex problems and a war broke out in Syria, where the British had tacitly supported the declaration of an independent kingdom by Sharif Hussein's son Faisal. The French won out and the British similarly crushed a rebellion in Mesopotamia, where they then installed Faisal as king as a consolation for his expulsion from Syria.

Mesopotamia received limited autonomy, but the rest of the Arab lands which had formed part of the Sykes-Picot agreement remained under French or British rule. The reward for the Arabs' support in defeating the Ottoman empire turned out to be the almost complete betrayal of their hopes for independence.

MAP OF EASTERN TURKEY IN

BLACK SEA
BAHR-ES-SIAH OR KARA DENIZ

MEDITERRANEAN SEA
(BAHR-ES-SUFEID OR AK DENIZ)

CYPRUS

SYRIAN DESERT

A

IA, SYRIA AND WESTERN PERSIA
398
CASPIAN
SEA
LAKE VAN
BAGHDAD
BUSRA
PERSIAN
GULF
Reference
Scale 1: 2,000,000 or 1 Inch 31·56 Stat. Miles
Heights in Feet.
rected to November 1915.
F. Georges Picot

Ypres

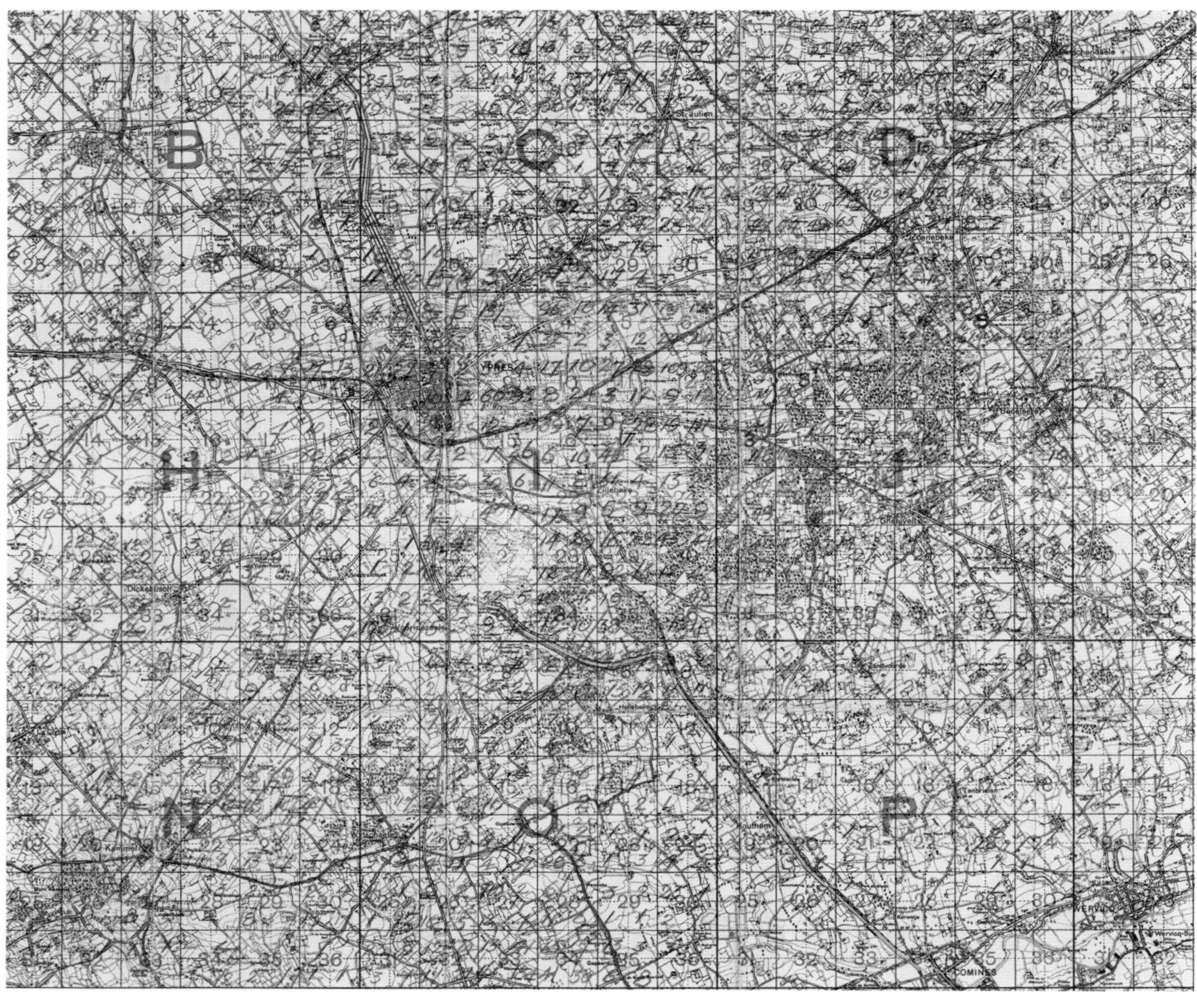

Extract from a British 1:40,000 military map of Ypres, annotated in each 500-yard grid square with the number of bodies collected after the war. Imperial War Museum/Western Front Association.

FIRST WORLD WAR AFTERMATH

The static warfare of the First World War trenches produced casualties on an unimaginable scale. Between eight and ten million soldiers lost their lives between 1914 and 1918, with almost 20,000 British troops dying on 1 July, the first day of the Battle of the Somme alone. The Red Cross brought some sense of respect for the multitude of fallen, from 1914 cataloguing deaths and looking after war graves. Its map of the body count at Ypres is cartographic testament to its calm and methodical work in the face of appalling horrors.

By 1915 the work of the Red Cross was regularized under the Graves Registration Commission and in May 1917 this evolved into the Imperial War Graves Commission. Grave registration units combed the battlefields searching for burials and by 1918 some 587,000 graves had been identified and a further 559,000 casualties were registered as having no known grave. Many bodies had been blown apart by shellfire or had sunk into the mud. They are still being found. Those located were exhumed and moved to the established military cemeteries, some of which dated from the first months of the war. The names of those still missing were inscribed on memorial panels, for example at Ypres' Menin Gate, Tyne Cott near Passchendaele, Dud Corner Cemetery near Loos and the Thiepval memorial on the Somme.

Millions of casualties had been suffered during the first three years of the First World War, in East Prussia and Russian Poland, in the territories of the Austro-Hungarian Empire, the Balkans and Romania, in France and Belgium, in northern Italy and in the Ottoman Empire, largely proving the point that a well-dug-in defence, equipped with modern weapons, could bring to a halt the heaviest assault. The resultant stalemate of trench warfare, which set in on all fronts where there was a great density of men relative to space, could only be broken, the generals argued, by even greater concentrations of artillery and ammunition.

By the time the Third Battle of Ypres (Passchendaele) opened, at the end of July 1917, the Western Front had seen monstrous and inconclusive battles fought in Artois and Champagne (1915), at Verdun and on the Somme (1916), and at Vimy Ridge, Arras, the Chemin-des-Dames and Messines (1917). For the Allies (France, Russia, Britain and her Empire, Belgium, Italy, Romania) the doctrine of 'attrition', or wearing down the enemy through remorseless, grinding attacks, was based on calculations that the Central Powers (Germany, Austria-Hungary, Bulgaria, Turkey), lacking the reserves of manpower and industry available to the Allies, would eventually and inevitably succumb to this strategy.

The entry of the United States into the war on the side of the Allies in April 1917 reinforced this belief, but this was soon to be countered by the growing weakness of a feudal and inefficient Russia riven by revolution. By the autumn and winter of 1917, Russia was effectively out of the war. The need to keep pressure on the Germans, both to divert their attention from the French army and to assist the Russians, led to Sir Douglas Haig being authorized by Lloyd George's War Cabinet to launch his Flanders offensive.

The three-month 1917 battle known as Third Ypres, or Passchendaele, originated in a British pincer-movement involving tanks and infantry, cavalry movements towards Zeebrugge and Ostend, and an amphibious assault on the Flanders coast. Unfortunately for the British and their French and Belgian allies, the Germans, on the high ground, could see the obvious preparations and reinforced their defences. Hard-hit by counter-attacks, the Allied attack, launched on 31 July, ground to a halt.

The British Commander-in-Chief, Haig, initiated a change in tactics. This coincided with a period of dry weather which dried out the sodden ground, and succeeded in gaining much high ground. But in October the rains returned, and the final attacks to take the Passchendaele Ridge were made in appalling conditions. The ground became a bottomless bog, trenches were non-existent, guns and ammunition could only be moved on plank roads and tramways, and duckboards for men and mules ended far short of the front line. Canadian troops finally captured Passchendaele in early November.

The grandiose ambitions of the sweep to the coast and amphibious landings had long been forgotten. Attrition was the ghoulish key-word; the German army had to be constantly gripped in a death-struggle to keep its attention from the weakening Russian front. At Ypres, the casualty figures were, as usual, difficult to establish. British casualties had been perhaps 280,000, and German 240,000; about one-third of these were killed, the remainder wounded, missing or sick.

Those who fell, though, were not forgotten. As well as having memorials in almost every town and village in the combatant nations, they were buried close to where they fell – sites now tended to by organizations such as the War Graves Commission which looks after 1.7 million graves from the First World War and later conflicts.

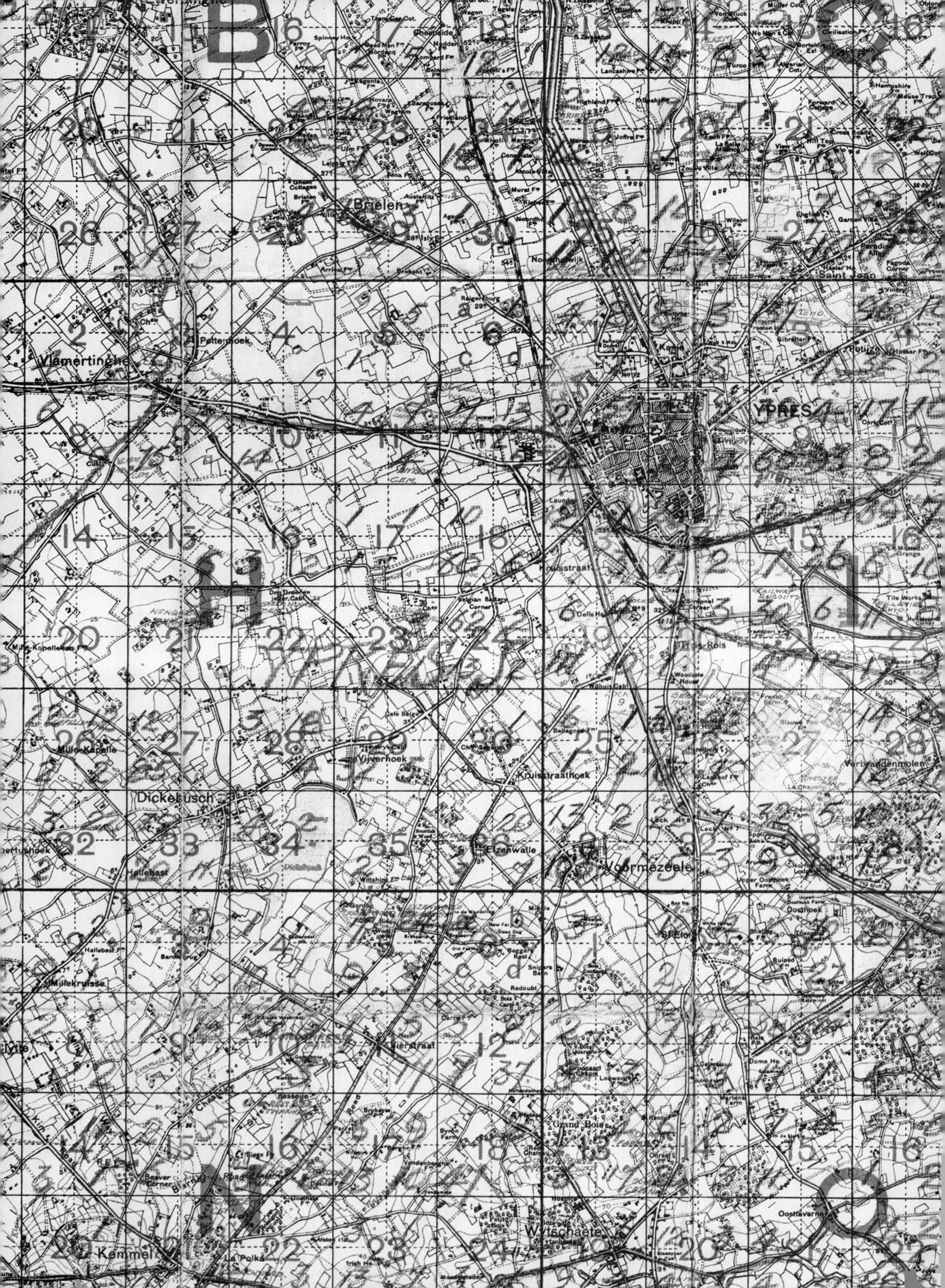

B
H
N
O
Brielen
Vlamertinghe
YPRES
Saint Jean
Potijze
Kaaie
Reigersburg
Dickebusch
Mille-Kapelle
Hallebast
Millekruisse
Vijverhoek
Cafe Belge
Belgian Battery Corner
Kruisstraat
Kruisstraathoek
Elzenwalle
Voormezeele
Dolls Ho
Shrapnel Corner
Trois Rois
Woodcote House
Verbrandenmolen
St Eloi
Vierstraat
Grand Bois
Kemmel
La Polka
Wytschaete
Irish Ho
Spinney Ho
Lancashire Fm
Highland Fm
Wilson Fm
Garden Villa
Cross Roads
Hill Top
Lancer Fm
Gibraltar Fm
Cork Cot
Tile Works

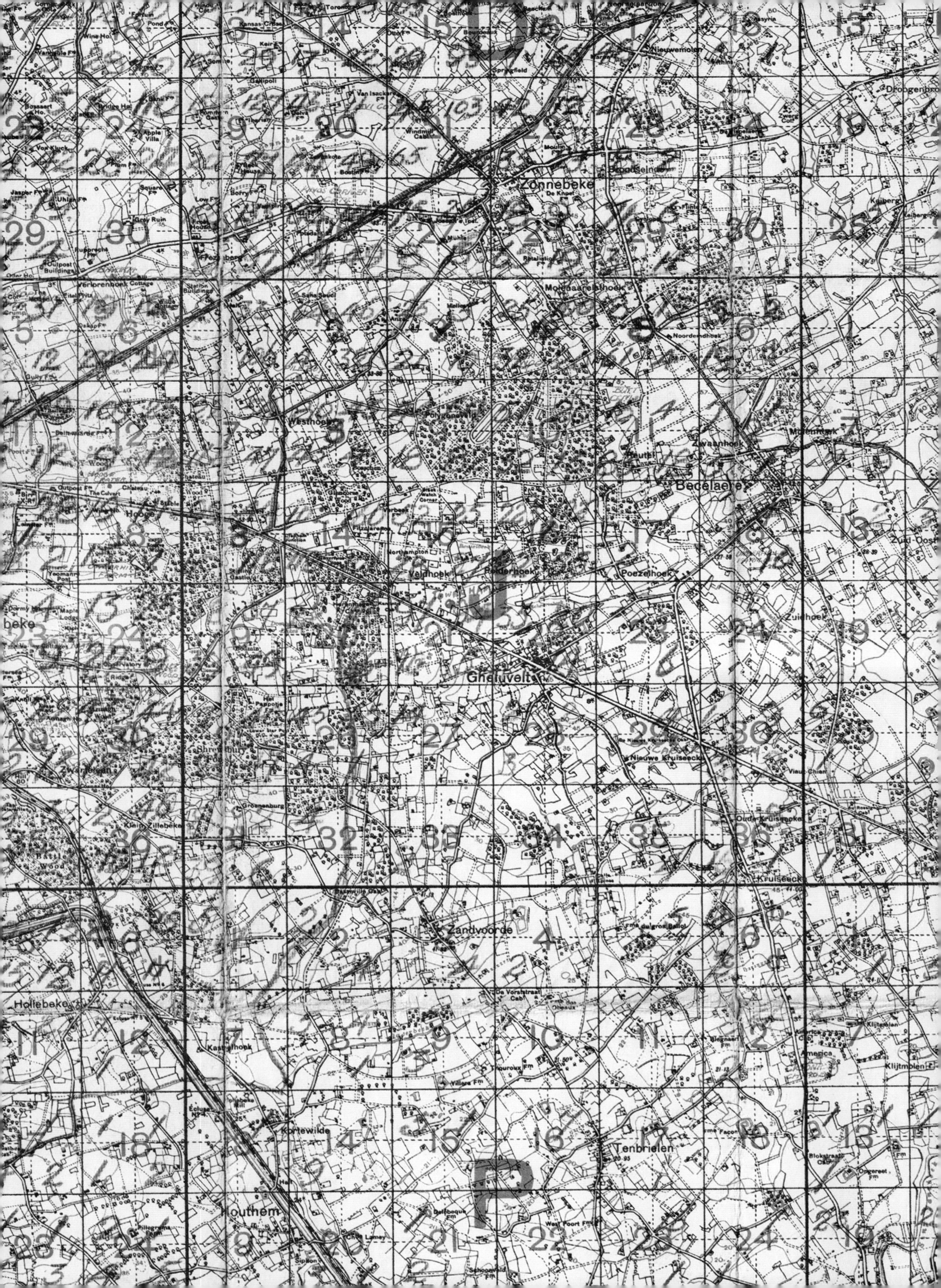

Zonnebeke
Broodseinde
Nieuwemolen
Verlorenhoek
Gray Ruin
Westhoek
Reutel
Zwaanhoek
Becelaere
Poezelhoek
Veldhoek
Zuidhoek
Gheluvelt
Nieuwe Kruiseecke
Kruiseecke
Zwarteleen
Groenenburg
Zandvoorde
Hollebeke
Klijtmolen
Kortewilde
Tenbrielen
Houthem

Germany

The positioning of boundaries on a map, and name forms used, have long been tools of state propaganda. The art of mapping for political purposes was refined during the period of Nazi rule in Germany after 1933, when maps were produced incorporating each successive German takeover of neighbouring territory, within the boundaries of a newly enlarged Gross Deutsches Reich (Greater German State).

German nationalists chafed at the provisions of the 1919 Treaty of Versailles which ended the First World War, particularly those which stripped Germany of control of the Saarland, demilitarized the Rhineland and placed stringent controls on the strength of its armed forces. The feeling that Germany had been betrayed and the discontent caused by soaring unemployment levels (which reached six million in 1933) contributed greatly to the rise of Adolf Hitler's National Socialist German Workers' Party (the Nationalsozialistische Deutsche Arbeiterpartei, or NSDAP) – the Nazi Party. In 1932 the Nazis became the largest party in the German Reichstag and in January 1933 Hitler was appointed Chancellor. He soon embarked on a programme of anti-Jewish legislation and military rearmament at home and an aggressively nationalist foreign policy which sought to absorb into Germany those neighbouring territories which had substantial German-speaking populations.

There were around thirty million Germans outside the borders of the Reich and Hitler set about increasing Germany's influence among them, in particular in Austria where he vigorously undermined the government of Chancellor Dolfuss (who was murdered in July 1934 and replaced with the substantially more pliable Schuschnigg). The first major annexation, though, was in the Saarland. This area had been awarded as a League of Nations Mandate to France under the terms of Versailles, with the provision that a referendum on its future status would be held after fifteen years. Intimidation by the Nazis and heavy-handed French attempts to expunge German influence, however, resulted in a plebiscite being held in January 1935 and a 91 per cent vote to rejoin Germany.

By 1936 Hitler had repudiated the Locarno Treaties (which guaranteed borders in western Europe) and left the League of Nations, so freeing himself from controls on further foreign adventures. In March 1936 German forces marched into the Rhineland, which had been demilitarized under the Treaty of Versailles. The reaction of other European powers was muted and so Hitler was emboldened to turn his attention to Austria, a German-speaking land which some nationalists had always considered should have become part of Germany at the time of unification in 1871. When Schuschnigg rejected Hitler's blandishments, the German Chancellor retorted, 'The whole history of Austria is just one uninterrupted act of high treason' and on 12 March 1938 he sent the German army across the border. The *Anschluss* – the annexation of Austria – was confirmed by another plebiscite, and it soon became 'Ostmark', another province of Gross Deutschland.

Czechoslovakia was the next to be engulfed by Hitler's ambitions. Its large ethnic minority of three million Germans was largely concentrated in Sudetenland in the west of the country. After fomenting political agitation among then, Hitler succeeded in getting France and Britain to pressure the Czechoslovak government into ceding Sudetenland to Germany in October 1938. The Munich Agreement which sanctioned this betrayal of the Czechs, though, was not enough to save the country and in March 1939, German troops marched in to occupy the rest.

This map shows the situation in summer 1939 after Germany had also seized a strip of territory which Versailles had awarded to Lithuania. The new German territories are shown in deep orange, with a lighter shade indicating the rump of Czechoslovakia which had been organized as the 'Protectorate of Bohemia and Moravia' and not yet fully integrated into the German Reich. A legend to the right details the stages of Germany's enlargement, carefully listing the area and population the country gained by each successive act of annexation. Cartography was carefully controlled by the Nazi government and it encouraged the publication of celebratory maps after the Austrian *Anschluss*. Between July 1934 and June 1944 it issued about sixty regulations related to maps, including one in 1937 which decreed that place names in territories lost by the Treaty of Versailles were to be printed first in German (in bold), with local names only in smaller type. On world maps, the German (rather than the British) Empire was to be indicated in red. To ensure compliance by map-makers, the Ministry of Popular Enlightenment and Propaganda, headed by Dr Joseph Goebbels, was given the task of supervising all national, linguistic and political maps. Hitler viewed the gathering-in of German-speaking populations into the Reich as part of Germany's destiny.

Mapping, and the image of the Greater Germany which its effective manipulation could present, was a highly effective tool for enabling and consolidating this goal.

THE THIRD REICH

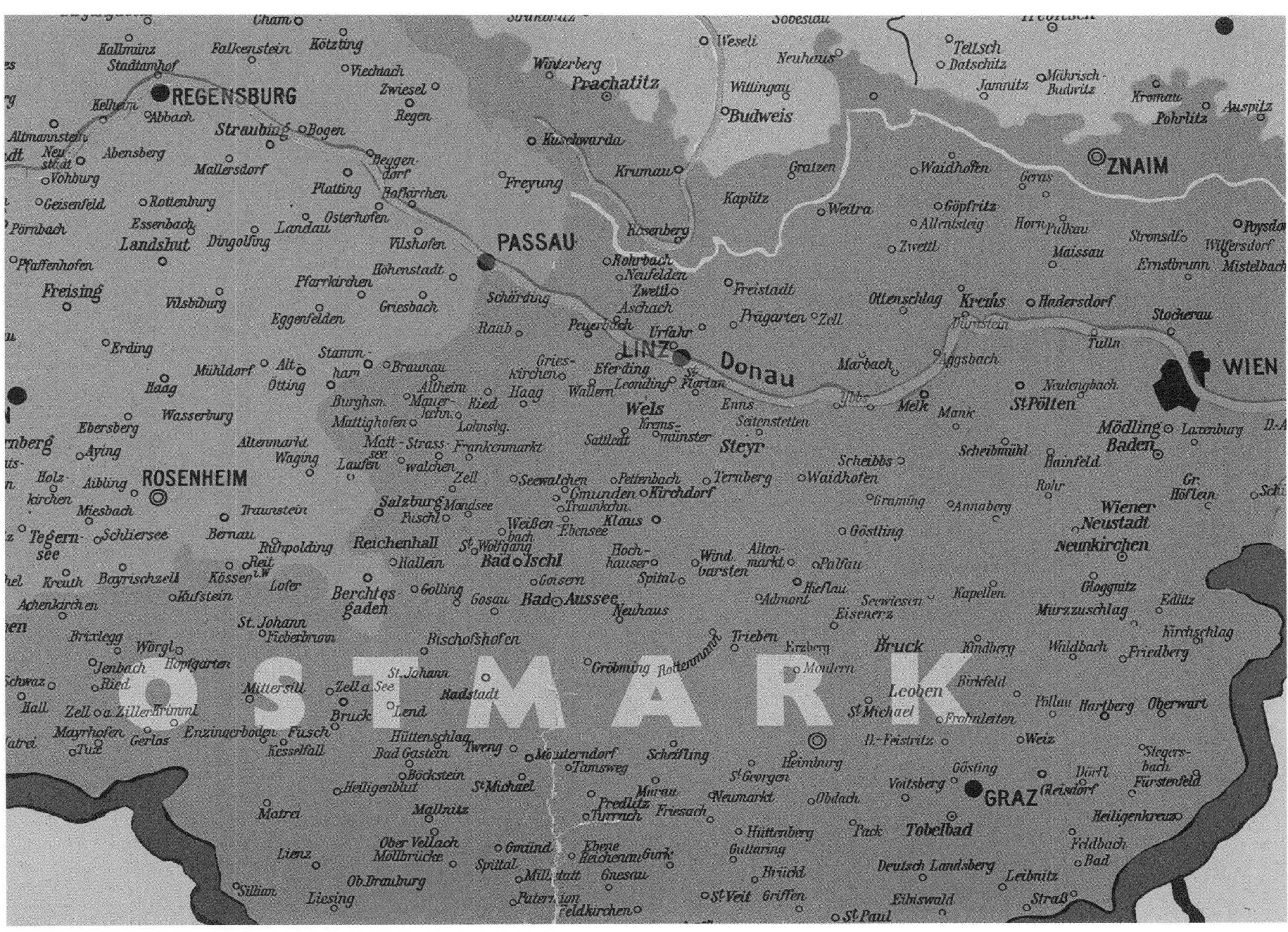

Map of the German Reich published in Berlin after the annexations of Austria and Sudetenland. National Library of Scotland, Edinburgh, UK.

Nord-
See
DÄNE-
MARK
NIEDERLANDE
BELGIEN
LUXEMBURG
FRANKREICH
SCHWEIZ
ITALIEN
GROSS
DEUTSCHES
Ostfriesische Ins.
Nord friesische Ins.
KOPENHAGEN
AMSTERDAM
GRONINGEN
LÜTTICH
ZÜRICH
BERN
HAMBURG
BREMEN
HANNOVER
BERLIN
KASSEL
KÖLN
MÜNSTER
OSNABRÜCK
DÜSSELDORF
FRANKFURT
MAINZ
WIESBADEN
KOBLENZ
TRIER
SAARBRÜCKEN
MANNHEIM
LUDWIGSHAFEN
KARLSRUHE
STUTTGART
FREIBURG
KONSTANZ
AUGSBURG
MÜNCHEN
NÜRNBERG
WÜRZBURG
REGENSBURG
INNSBRUCK
LEIPZIG
HALLE
ERFURT
MAGDEBURG
CHEMNITZ
PLAUEN
COBURG
ROSTOCK
LÜBECK
KIEL
HUSUM
EMDEN
WILHELMSHAVEN
Rhein
Main
Mosel
Donau
Elbe
Weser

50 PFENNIG

Gedruckt im Deutschen Verlag Berlin

East Asia

As Japan's empire in East Asia expanded in the 1930s, Japanese leaders sought to develop an ideology which would affirm the country's pre-eminent place in East Asia and its moral superiority to the European colonial powers which it had displaced. This dogma found political expression in the 'Greater East Asia Co-Prosperity Sphere' within which Japan's new subordinate territories found themselves grouped, while Japanese propaganda-inspired maps and atlases portrayed the new order as a profoundly positive development for Asia.

Japan had found itself a colonial power late, beginning with the acquisition of Taiwan in 1895 and of Korea in 1905 after its stunning victory against Tsarist Russia in the Russo-Japanese War. A Japanese League of Nations Mandate over the former German territories in the South Pacific (the Marianas, Palau and the Marshall Islands) followed in 1919, but the pace of expansion accelerated in the 1930s with the seizure of Manchuria (northeast China) in 1932 and large parts of China by 1940. At first, Japanese ideology had mirrored that of the European colonial powers, but a growing feeling that Japan should distance itself from western notions of empire and instead search for a specifically Asian concept of empire, led to the growth of movements such as *Kominka*, which argued for the complete Japanization of the country's colonial possessions amid mass campaigns to inspire pro-Japanese feelings among their populations.

After the Japanese joined the Second World War to fight the Allies as an Axis power in 1941 and swept through the British and Dutch colonial possessions in the Pacific, the need to promote a political agenda which would, at least superficially, appeal to Asian nationalists, grew more urgent. In a radio broadcast in June 1940, Japan's Foreign Minister Hachiro Arita had floated the idea of a 'Co-Prosperity Sphere' uniting friendly nations under an economic and political umbrella. This was given theoretical rigour in a series of lectures in summer 1942 by economist Nagao Sakuro, who promoted the idea of a racially-based 'co-prosperous' economic bloc as the new basis of international political power.

By 1943 the Japanese had achieved most of their objectives in the Pacific, with the British hanging on grimly to a strip of Burma and the Allies fighting desperately to keep hold of New Guinea. In November of that year, just after, as it turned out, the tide had irrevocably turned against them, the Japanese held a Greater East Asia Conference in Tokyo. This affirmed that independent (but pro-Japanese) governments would be installed in Burma, the Philippines, Indonesia, India and China. These would commit to mutual political co-operation, respect of each other's sovereignty and culture, and a programme of joint economic development. Premier Tojo contrasted what he characterized as Japan's tolerant agenda with that of the British Empire which, he proclaimed had 'through fraud and aggression, acquired vast territories throughout the world'.

As part of the propaganda effort, the Japanese government issued the *Declaration for Greater East Asia Co-operation* in 1942, a textbook intended to be used by Japan's Asian colonies and allies. It was filled with attractive images of children and other locals across east Asia appearing happy to participate in Japan's great new venture. In the map which accompanied it, Roosevelt and Churchill are shown glowering disconsolately from the corners as their own forces are depicted being chased by the 'liberating' Japanese into the sea. The text reads, for those not taking in the visual message of the Burmese soldier joyfully waving a flag as a British trooper retreats with a Japanese bayonet at his back, 'Now with ringing footsteps, let us all advance together'. Subsequent pages bore images of ecstatic Javanese, Malay and Burmese people, rejoicing at the driving out of their colonial masters and the motivational message, 'From now, each country of Greater East Asia will grow great and live in friendship with one another'.

But the truth was very different. Japan needed the economic resources of the Co-Prosperity Sphere – iron and coal from Manchuria, oil and rubber from Indonesia – and any notion of political liberation for the peoples of the conquered territories was strictly subordinate to the needs of Japan's military machine. Indeed as *An Investigation of Global Policy with the Yamato Race as Nucleus*, a secret Japanese military document of 1943 made clear, Japan was to take the leading role with other nations being subordinate. It spoke of *shujin minzoku* ('master peoples'), in other words the Japanese, whose race made them superior to the *yukin minzoku* (or 'friendly peoples') who made up the bulk of the Japanese Empire.

The sincerity of Japan's commitment to the Greater East Asia Co-Prosperity Sphere was never put to the long-term test. Wartime conditions for the inhabitants of its new possessions were if anything harsher than those they had experienced before, and the final defeat of Japan in summer 1945 meant that its vision of pan-Asian nationalism vanished. Instead it was to be replaced in the region by a new ferment of nationalist uprising against the Dutch colonial powers in Indonesia and by communist revolution in China.

JAPAN'S CO-PROSPERITY SPHERE

Propaganda map from the 1942 publication Declaration for Greater East Asia Co-operation illustrating the Japanese vision for the region. CPA Media, Chiang Mai, Thailand.

チュウクヮミンコク
ヂュウケイ
インド
ビルマ
ラングーン
タイ
バンコク
インドシナ
ハノイ
サイゴン
セウナン
マライ
セウナン
スマトラ
パレンバン
ボルネオ
バンジェルマシン
ジャカルタ
ジャワ
インドヤウ

私タチ ノ ダイトウア ヲ、私
タチ ノ 手ニ カヘサウ ト、
日本 ハ 立チ アガリマシタ。
ソシテ、ツヨイ 日本グン ハ、
テキ ヲ ダイトウア カラ
オヒハラヒマシタ
マンシウコク
ニッポン
シャンハイ
フィリピン
ダバオ
セレベス
マカッサル
ニューギネヤ
モレスビー
オーストラリヤ

Omaha Beach D-DAY

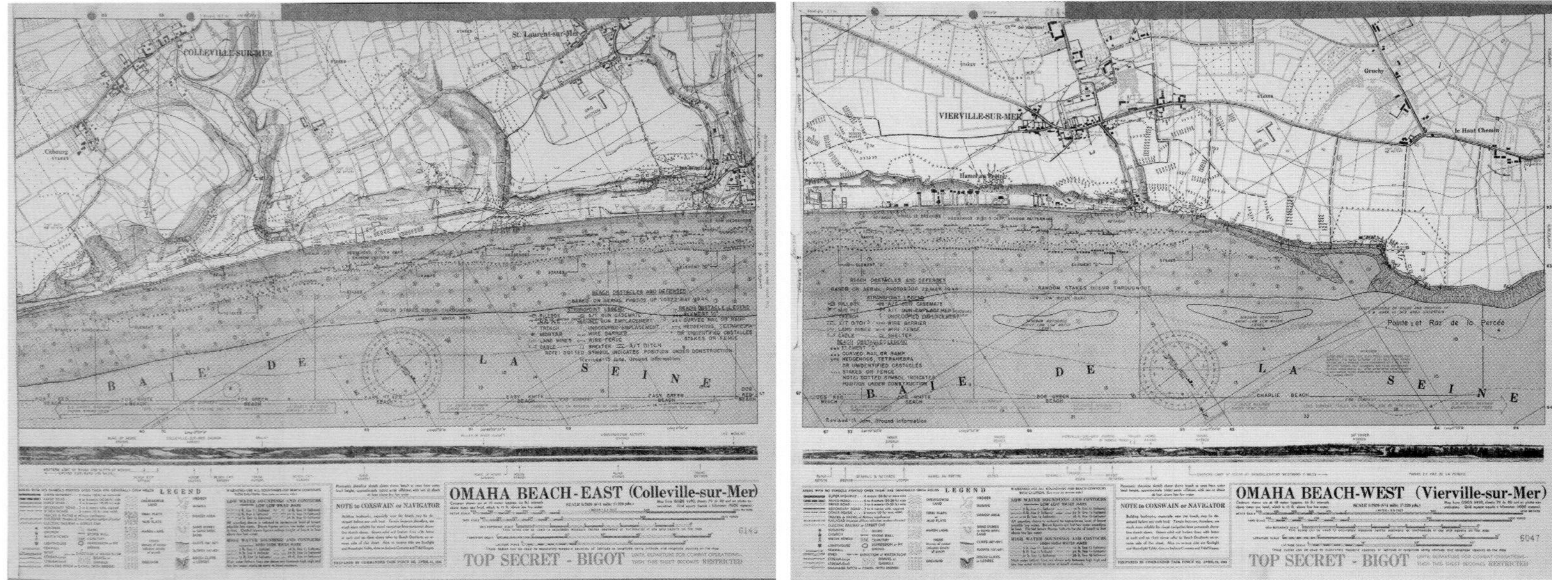

US military maps of Omaha Beach prepared for the Normandy landings on D-Day, June 1944. The maps were prepared under the highest security classification, code-named BIGOT. The National Archives, London, UK.

Operation Overlord was the largest amphibious assault in history. On D-Day – 6 June 1944 – after years of planning and benefiting from experience of similar landings in North Africa, Sicily and Italy, over 150,000 men landed from more than 4,000 ships on 80 km (50 miles) of Normandy coastline. A crucial part of the intelligence push to prepare for the invasion was the production of detailed maps, an effort which had begun the previous year.

The Second World War (1939–1945) was the result of the ruthless expansionist policies of Hitler's Germany (see page 180), Mussolini's Italy and Hirohito's Japan. The turning point of the War, prior to D-Day, came in 1942 when German forces were defeated by the Soviet Red Army at Stalingrad, and by the British and Commonwealth forces at Alamein, and the United States inflicted a crushing defeat on the Japanese at Midway Island in the Pacific. In 1943, the Germans were further defeated by the Russians in a tank battle at Kursk. The U-boat menace was largely overcome, and Allied forces landed in North Africa, Sicily and on the Italian mainland. Germany, now developing secret weapons and still a major force, had to be tackled directly by a major cross-Channel attack.

Careful examination of existing maps revealed that, because of its relatively flat terrain and lack of obvious physical obstacles, Normandy was the most suitable area for an Allied invasion. Vast quantities of new maps – many drawn from existing maps, postcards and photographs, updated using aerial photos and intelligence from various sources – had to be prepared. Despite the fact that they could not all be printed in time for the invasion, General Montgomery, commanding the ground forces in Normandy, said that 'at no time did map supply fail or prejudice the conduct of operations.'

The 'Baby Benson' programme of special 1:12,500 assault mapping of the coastal invasion area was inaugurated by COSSAC (Chief of Staff, Supreme Allied Commander) in 1943 and completed under SHAEF (Supreme Headquarters Allied Expeditionary Force). Forty maps of specially selected coastal areas were prepared from enlargements of the 'Benson' 1:25,000 sheets, with additional detail plotted from aerial photos. Before the assault, SHAEF Survey Directorate controlled the production of operational and intelligence overprints and other special maps, showing all known enemy defences and dispositions, beach obstacles and other information for assault planning. For the assault itself, the Directorate and the Theatre Intelligence Section (TIS) produced overprints for over twenty sheets showing German defences and engineer intelligence.

For use by naval craft (including landing craft) in the close bombardment, run-in and landing stages of the operation, 1:12,500 Assault Chart-Maps (Beach Landing Maps) were prepared in different styles by the British and the Americans. The British Admiralty's Hydrographic Department prepared five such sheets, one for each of the British and American landing beaches, entitled Utah, Omaha, Gold, Juno and Sword. These incorporated topographical detail from the 1:12,500 series maps, with defence intelligence and beach obstacle overprints provided by the TIS. The sheets carried the warning note: 'N.B. Underwater Obstacles of various types are being laid with great rapidity and are likely to extend along further stretches of the coast.' The hydrographic information was minimal, the sea below low water mark being coloured blue, and no soundings or compass rose were included. For detailed hydrographic information, the ships and landing craft all carried admiralty charts, while bombarding ships also carried special 1:50,000 gridded chart-maps.

The US Navy prepared its own set of four chart-maps, covering only the American beaches, two sheets for each beach: Utah Beach North and South, and Omaha Beach West and East (shown here). The Omaha Beach sheets are particularly significant because it was here that the Allied assault forces suffered their heaviest casualties in trying to fight ashore against strongpoints and to capture beach exits before exploiting inland. The maps all carried the note: 'Prepared by Commander Task Force 122, April 21, 1944,' – Admiral Kirk, US Navy, commanding the Western Task Force. The topographical information was taken from the GSGS 4490 series and 'air photo examination'. Hydrographic information included soundings, low water mark, rocks, etc., together with a detailed explanatory legend.

The result of all this preparation was that the US and other Allied forces had vital information about the topography of the landing beaches and their hinterland, a testament to the level of preparation which made the highly risky venture of D-Day ultimately a success.

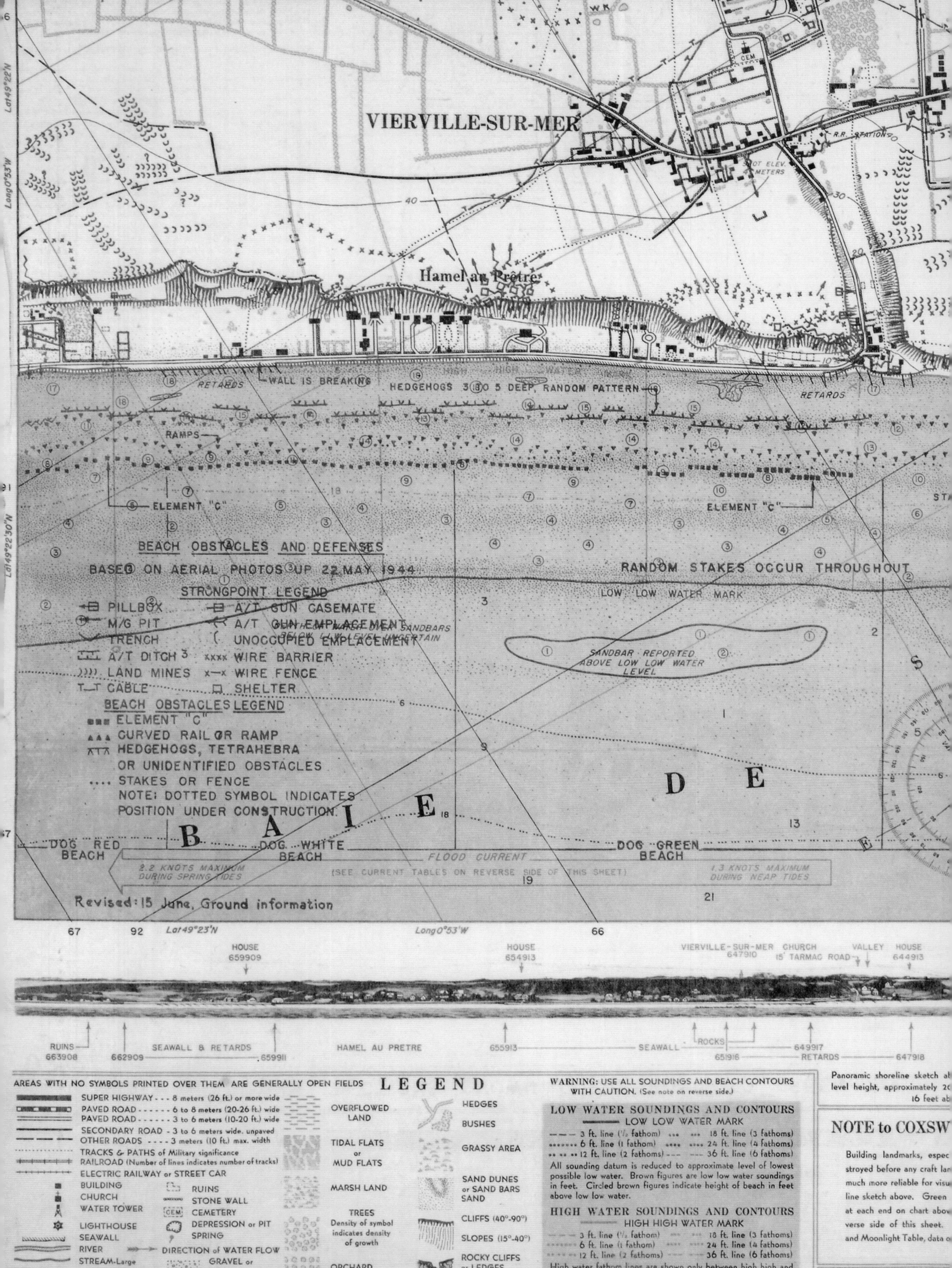

VIERVILLE-SUR-MER
CEM
R.R. STATION
SPOT ELEV. METERS
Hamel au Prêtre
WALL IS BREAKING
RETARDS
HEDGEHOGS 3 TO 5 DEEP, RANDOM PATTERN
RAMPS
ELEMENT "C"
BEACH OBSTACLES AND DEFENSES
BASED ON AERIAL PHOTOS UP 22 MAY 1944
STRONGPOINT LEGEND
PILLBOX
M/G PIT
TRENCH
A/T DITCH
LAND MINES
CABLE
A/T GUN CASEMATE
A/T GUN EMPLACEMENT
UNOCCUPIED EMPLACEMENT
WIRE BARRIER
WIRE FENCE
SHELTER
BEACH OBSTACLES LEGEND
ELEMENT "C"
CURVED RAIL OR RAMP
HEDGEHOGS, TETRAHEBRA OR UNIDENTIFIED OBSTACLES
STAKES OR FENCE
NOTE: DOTTED SYMBOL INDICATES POSITION UNDER CONSTRUCTION.
RANDOM STAKES OCCUR THROUGHOUT
LOW LOW WATER MARK
SANDBAR REPORTED ABOVE LOW LOW WATER LEVEL
B A I E D E
DOG RED BEACH
DOG WHITE BEACH
DOG GREEN BEACH
FLOOD CURRENT
2.2 KNOTS MAXIMUM DURING SPRING TIDES
(SEE CURRENT TABLES ON REVERSE SIDE OF THIS SHEET)
1.3 KNOTS MAXIMUM DURING NEAP TIDES
Revised: 15 June, Ground information
Lat 49°22'N
Long 0°53'W
Lat 49°22'30"N
Lat 49°23'N
Long 0°53'W
HOUSE 659909
HOUSE 654913
VIERVILLE-SUR-MER CHURCH 647910
15' TARMAC ROAD
VALLEY
HOUSE 644913
RUINS 663908
662909
SEAWALL & RETARDS
659911
HAMEL AU PRETRE
655913
SEAWALL
ROCKS
651916
649917
RETARDS
647918
AREAS WITH NO SYMBOLS PRINTED OVER THEM ARE GENERALLY OPEN FIELDS
LEGEND
SUPER HIGHWAY - 8 meters (26 ft.) or more wide
PAVED ROAD - 6 to 8 meters (20-26 ft.) wide
PAVED ROAD - 3 to 6 meters (10-20 ft.) wide
SECONDARY ROAD - 3 to 6 meters wide, unpaved
OTHER ROADS - 3 meters (10 ft.) max. width
TRACKS & PATHS of Military significance
RAILROAD (Number of lines indicates number of tracks)
ELECTRIC RAILWAY or STREET CAR
BUILDING
CHURCH
WATER TOWER
LIGHTHOUSE
SEAWALL
RIVER
STREAM-Large
STREAM-Small
RUINS
STONE WALL
CEMETERY
DEPRESSION or PIT
SPRING
DIRECTION of WATER FLOW
GRAVEL or SHINGLE
OVERFLOWED LAND
TIDAL FLATS or MUD FLATS
MARSH LAND
TREES Density of symbol indicates density of growth
ORCHARD
HEDGES
BUSHES
GRASSY AREA
SAND DUNES or SAND BARS SAND
CLIFFS (40°-90°)
SLOPES (15°-40°)
ROCKY CLIFFS or LEDGES
WARNING: USE ALL SOUNDINGS AND BEACH CONTOURS WITH CAUTION. (See note on reverse side.)
LOW WATER SOUNDINGS AND CONTOURS
LOW LOW WATER MARK
3 ft. line (½ fathom)
6 ft. line (1 fathom)
12 ft. line (2 fathoms)
18 ft. line (3 fathoms)
24 ft. line (4 fathoms)
36 ft. line (6 fathoms)
All sounding datum is reduced to approximate level of lowest possible low water. Brown figures are low low water soundings in feet. Circled brown figures indicate height of beach in feet above low low water.
HIGH WATER SOUNDINGS AND CONTOURS
HIGH HIGH WATER MARK
3 ft. line (½ fathom)
6 ft. line (1 fathom)
12 ft. line (2 fathoms)
18 ft. line (3 fathoms)
24 ft. line (4 fathoms)
36 ft. line (6 fathoms)
High water fathom lines are shown only between high high and low low water marks to serve as beach contours.
Panoramic shoreline sketch a
level height, approximately 2
16 feet a
NOTE to COXSW
Building landmarks, espec
stroyed before any craft la
much more reliable for visu
line sketch above. Green
at each end on chart abov
verse side of this sheet.
and Moonlight Table, data o
PREPARED BY COMMANDE

HOUSE 645920

30' TOWER 638926

EASTERN LIMIT OF ROCKS AT 646920, EXTEND WESTWARD II MILES

POINTE ET RAZ DE LA PERCEE

...s beach as seen from water ... offshore, with sea at about ...w water

OMAHA BEACH-WEST (Vierville-sur-Mer)

Contours shown are at 10 meter (approx. 33 ft.) intervals above mean sea level, which is 13 ft. above low low water.

SCALE 1:7920 (8"=1 mile; 1"=220 yds.)

Map from GSGS 4490, sheets 79 & 80 and air photo examination. Grid square equals 1 kilometer (1000 meters).

METER SCALE 500 400 300 200 100 0 500 1000 METERS (1 METER = 3.3 FEET)

YARD SCALE 500 400 300 200 100 0 500 1000 YARDS

GRID REFERENCE SCALE 0 1 2 3 4 5 6 7 8 9 10

This scale can be used to accurately measure tenths eastward or northwards in any grid square on the map.

LONGITUDE SCALE 0" 10" 20" 30" 40" 50" 60"

LATITUDE SCALE 0 10" 20" 30"

These scales can be used to accurately measure seconds of latitude or longitude using latitude and longitude squares on the map.

6047

...or NAVIGATOR

... the beach, may be de... ...in features, therefore, are ...ion from panoramic shore... ...broken lines with letter ... Beach Gradients on re... ...reverse side are Sunlight ... Currents and Tidal Stages.

...FORCE 122, APRIL 21, 1944

TOP SECRET - BIGOT

UNTIL DEPARTURE FOR COMBAT OPERATIONS-- THEN THIS SHEET BECOMES RESTRICTED

INDEPENDENCE OF ISRAEL

Maps played a crucial role in the struggle to find an equitable way to divide the British Mandate of Palestine between its Arab and Jewish communities. The 1947 United Nations Partition proposal envisaged Arab and Jewish states in an economic union, alongside an internationalized Jerusalem. Yet for all the efforts of the committee which created it, the map representing this proposal was soon superseded by a civil war which established much larger borders for the new State of Israel.

Britain acquired a League of Nations Mandate over Palestine in 1923 as part of the post-First World War settlement which divested the Ottoman Empire of most of its Arab territories (see page 172). At the time there were around 700,000 Arabs and 56,000 Jews within the mandated borders, but the growth of Zionism and the British Government's intention, as stated in the 1917 Balfour Declaration, that it would support a 'national home for the Jewish people in Palestine', served to encourage Jewish immigration.

Tensions in the Arab community were inflamed by the growth in land purchases by Jewish settlers and in 1936 this erupted into outright revolt. The rebellion was put down with the loss of 300 lives. The Peel Commission, established to determine the causes of the unrest, unexpectedly recommended the partition of Palestine into Jewish and Arab states in its final report in July 1937. Under the plan, Britain was to retain control of Jerusalem and a corridor to the sea around Jaffa as a permanent mandate, but the British government stepped back from the proposal after another report in 1938 (by the Woodhead Commission) cast doubts on the practicality of partition.

Both Arabs and Jews rejected the May 1939 White Paper on Palestine, which proposed a federative solution (and a restriction on further Jewish immigration to 75,000 over the succeeding five years), and the next eight years saw a rising level of violence, the most damaging directed by Jewish guerrilla groups such as Irgun against the British. Events such as the bombing of the King David Hotel, Jerusalem (home to the British headquarters for Palestine) on 22 July 1946, which left ninety-one dead, sapped the will of the British to keep control of their troublesome Mandate. In April 1947 they in effect washed their hands of the situation by turning the fate of Palestine over to the United Nations.

The UN Special Committee on Palestine (UNSCOP) was made up of representatives of eleven nations and deliberated for six months. It was subjected to vigorous lobbying by the two sides, and in September 1947 it produced a report containing a majority proposal that Palestine be divided into two states (Arab and Jewish), with an economic and currency union between them, and an internationalized Jerusalem. A minority report endorsed by India, Iran and Yugoslavia proposed a federal union in which greater Arab numbers would dominate. The majority plan was modified by an Ad Hoc Committee, which awarded a part of the Negev Desert and an enclave around Jaffa to the Arabs, but the proposed Jewish state was still to receive eastern Galilee, the coastal plain and much of the Negev desert, amounting to 14,200 sq km (5,500 sq miles) or around 53 per cent of the land area.

The Arabs, who got western Galilee, the hill country of Samaria and Judea and the Gaza Strip, were outraged at the 43 per cent this amounted to (despite the Arab population outnumbering that of the Jews by more than two to one) and rejected the plan outright. Despite this, the UN General Assembly voted to adopt the plan on 29 November 1947 by thirty-three votes to thirteen. On 14 May 1948, the last day of the British mandate, Jewish political leaders pre-empted the UN plan by declaring an independent State of Israel. This move provoked an invasion by Arab armies from Lebanon, Egypt, Syria and Jordan, which threatened for a short while to overwhelm the infant state But the Israelis fought back hard, advanced into Jerusalem and by the time of a final armistice in April 1949 they occupied some 20,700 sq km (8,000 sq miles) of Palestine, almost half as much again as they had been allocated under the UNSCOP plan. The Arabs were left with two large and separated tracts of territory, on the West Bank and the Gaza Strip, not enough to form a viable state. The former area ultimately came under the jurisdiction of Jordan and the latter was administered by Egypt.

These would remain the borders of Israel until 1967 when victory in the Six Day War against another Arab coalition of Syria, Egypt and Jordan led to the Israeli acquisition of the West Bank, Gaza Strip and part of the Golan Heights bordering Syria. Those new boundaries – although not internationally recognized – marked yet another stage in the troubled history of Palestine's shifting frontiers.

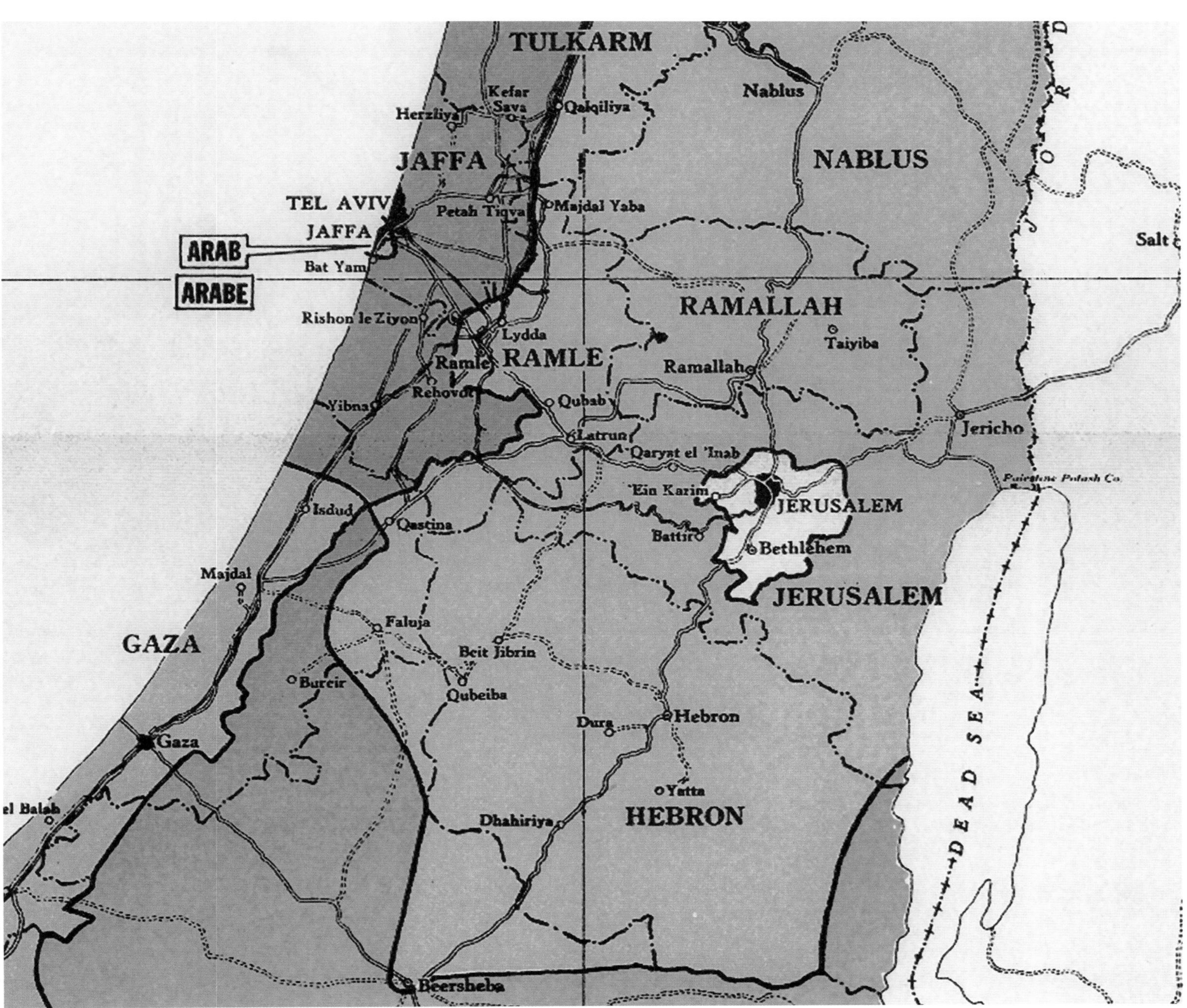

United Nations map of the UN Partition Plan for Palestine, adopted 29 Nov 1947.

Annex A
Annexe A

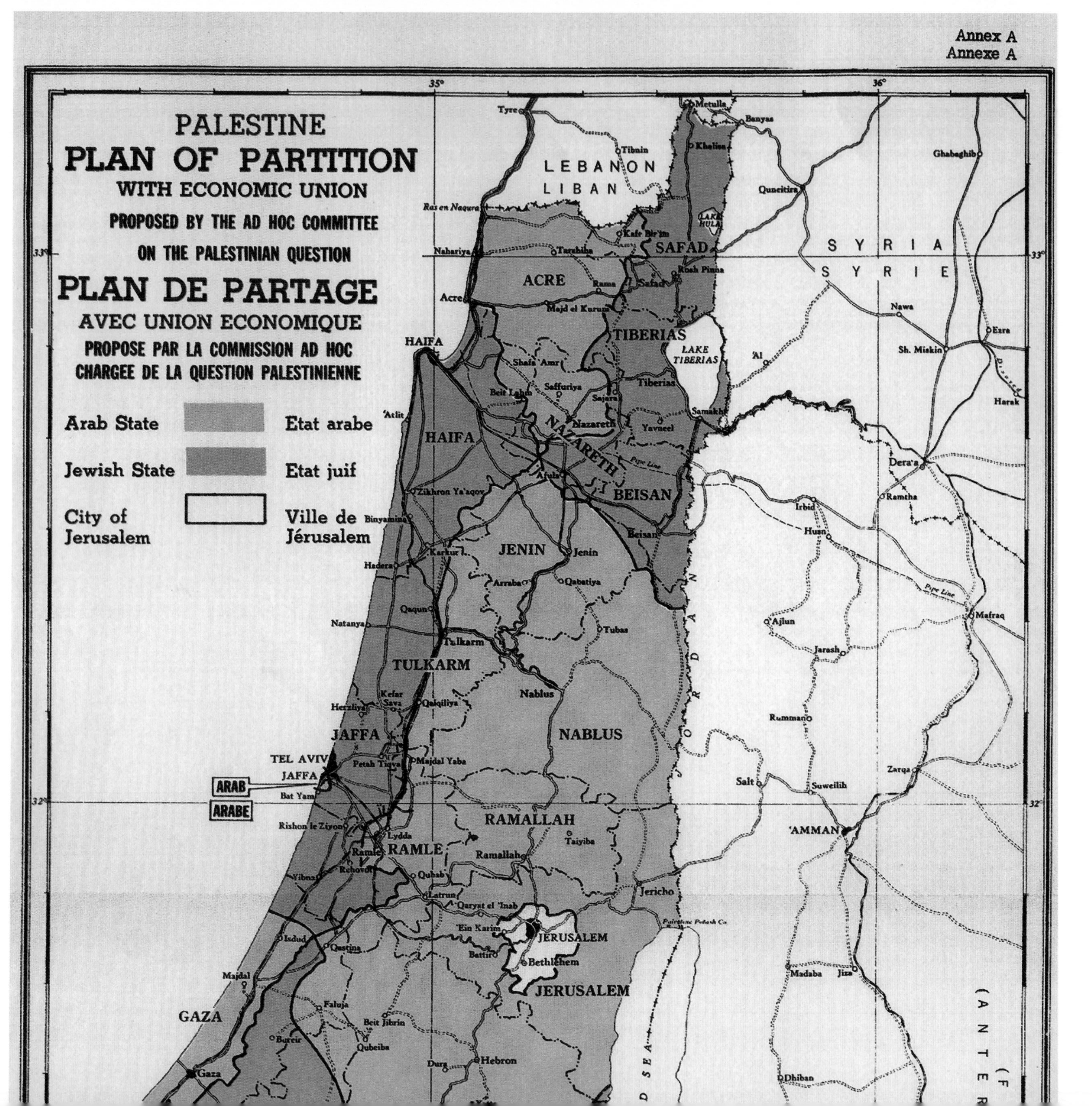

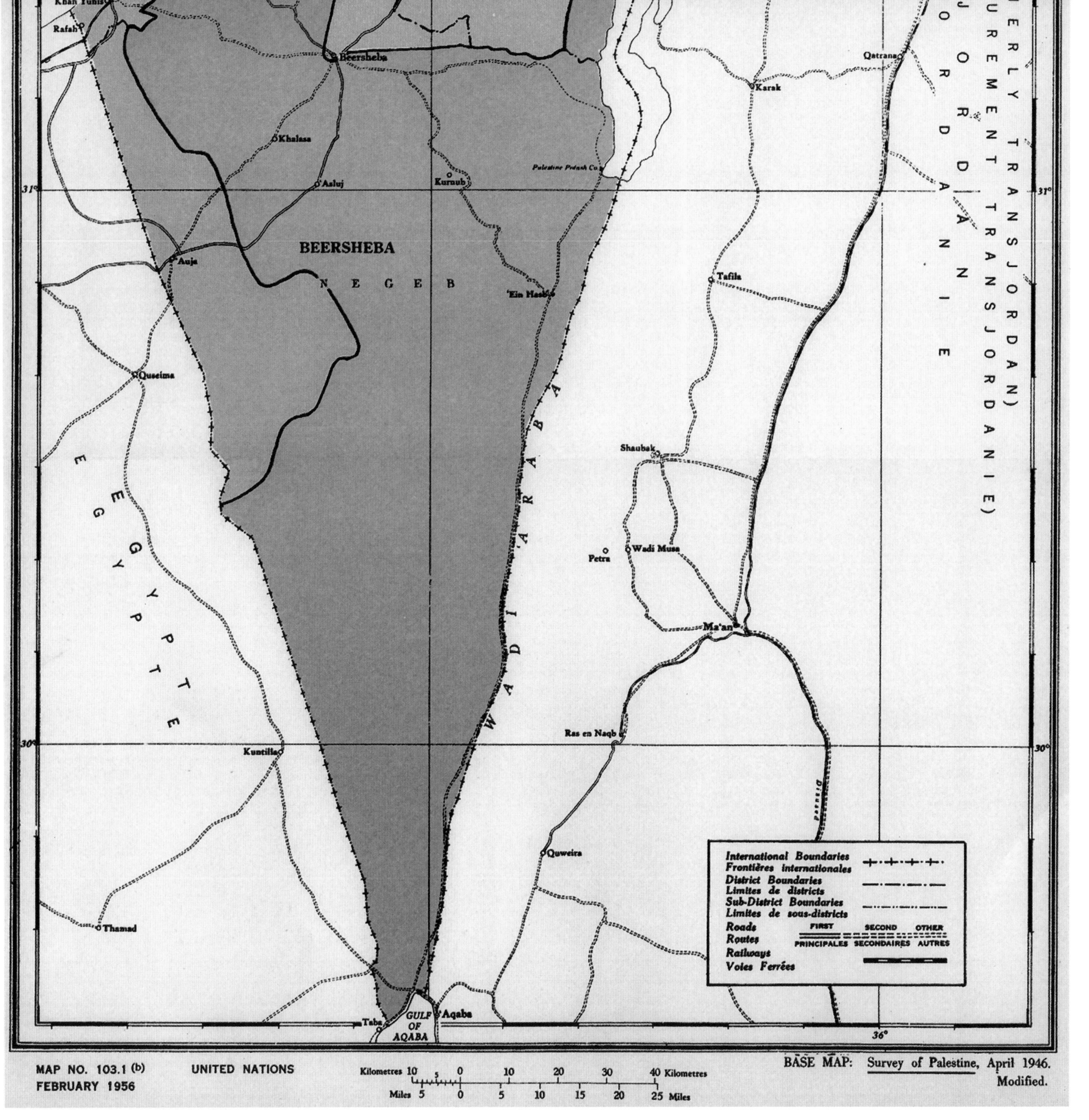

MAP NO. 103.1 (b) UNITED NATIONS
FEBRUARY 1956

BASE MAP: Survey of Palestine, April 1946. Modified.

Punjab

The line drawn on a map of British India, leading to the Partition of its territory, thereby creating the new nations of India and Pakistan, came into effect at midnight on 14 August 1947. For millions in the Indian subcontinent this momentous event resulted in 'freedom at midnight' from almost two centuries of British colonization. The impact of Partition on the lives of people was immediate and tumultuous. It is believed that the Partition of the Punjab in the west alone resulted in the migration of ten million people. It is estimated that at least a million people were killed in subsequent unrest.

Britain had ruled over progressively larger territories for almost two centuries. Demand for freedom from the Raj had become insistent after the First World War and the government in Britain announced the decision to give India independence in June 1947. Indian politicians welcomed the idea of an independent, sovereign and secular India. But in 1946, Muslim leaders had laid claim to a separate land, to be named Pakistan, for Muslim majority areas. There were two clear concentrations of Muslim majority areas in the subcontinent, one in the east (now Bangladesh, earlier East Pakistan) and another in the west (Pakistan). In Muslim-dominated areas of the Punjab, concentrations of Sikhs were apprehensive about joining Pakistan because of earlier communal disturbances.

When it became clear that at independence India was to be divided on communal lines, two Boundary Commissions were established and instructed to demarcate boundaries 'on the basis of ascertaining the contiguous majority areas of Muslims and non-Muslims'. In doing so, they were also 'to take into account other factors'. The Boundary Commissions were responsible for partitioning British India while the Princely States (see page 169) would choose whether to join Pakistan or India.

Sir Cyril Radcliffe, Chairman of both Commissions, in his report on the Boundary Commissions, admitted to the difficulty of agreeing an actual boundary line to demarcate British India. The people of the two new nations only came to know the boundaries of their countries through *The Reports of the Bengal Boundary Commission and of the Punjab Boundary Commission* two days after Independence. This gave rise to the bizarre situation in which some villages which had hoisted one country's flag on independence day then found themselves to be in another country – as defined by the map – two days later.

The creation of the maps of the Partition of India and Pakistan remains one of the greatest enigmas of modern cartography. The procedure followed to delineate this line is not documented. The maps were made in haste and without following the principles of drawing international boundaries or by marking pillars on the ground. Neither Radcliffe nor the members of the committees had any experience in demarcating boundaries. Rather it had been an exercise in negotiating for territory and for inclusion by each incipient nation.

Was the 1941 Census used to identify places of Muslim majority? Perhaps. But the Census did not publish figures of populations by religion for villages, where large numbers of agricultural people lived. Nor were all these villages marked on the topographic maps which were used for the purpose. The figures used were for *tahsil* or *thana* – aggregation of villages.

In Bengal, largely a flat riverine delta crisscrossed by numerous water channels, the international boundary line was the cause of many disputes. The Punjab was divided on the basis of majority areas of Muslims and non-Muslims as well as other factors such as administrative viability, natural boundaries, and communication, water and irrigation systems. Many 'other factors', some reasonable and others unexplained, were decisive in partitioning British India.

As well as the waves of refugees which Radcliffe's dividing line caused when it became known, the decision process by which the princely states chose which state to join proved problematic. In Jammu and Kashmir, which had a Muslim-majority population, the Maharaja, who was a Hindu, looked likely to opt to join India. To pre-empt this, pro-Pakistani fighters surged into the territory and were close to seizing the capital, Srinagar, before Indian intervention pushed them back. A ceasefire in 1948 left Kashmir partitioned between India and Pakistan, and two further wars between the two countries erupted in 1965 and 1971. Once again, neat lines on the map had fallen foul of political reality.

PARTITION OF INDIA

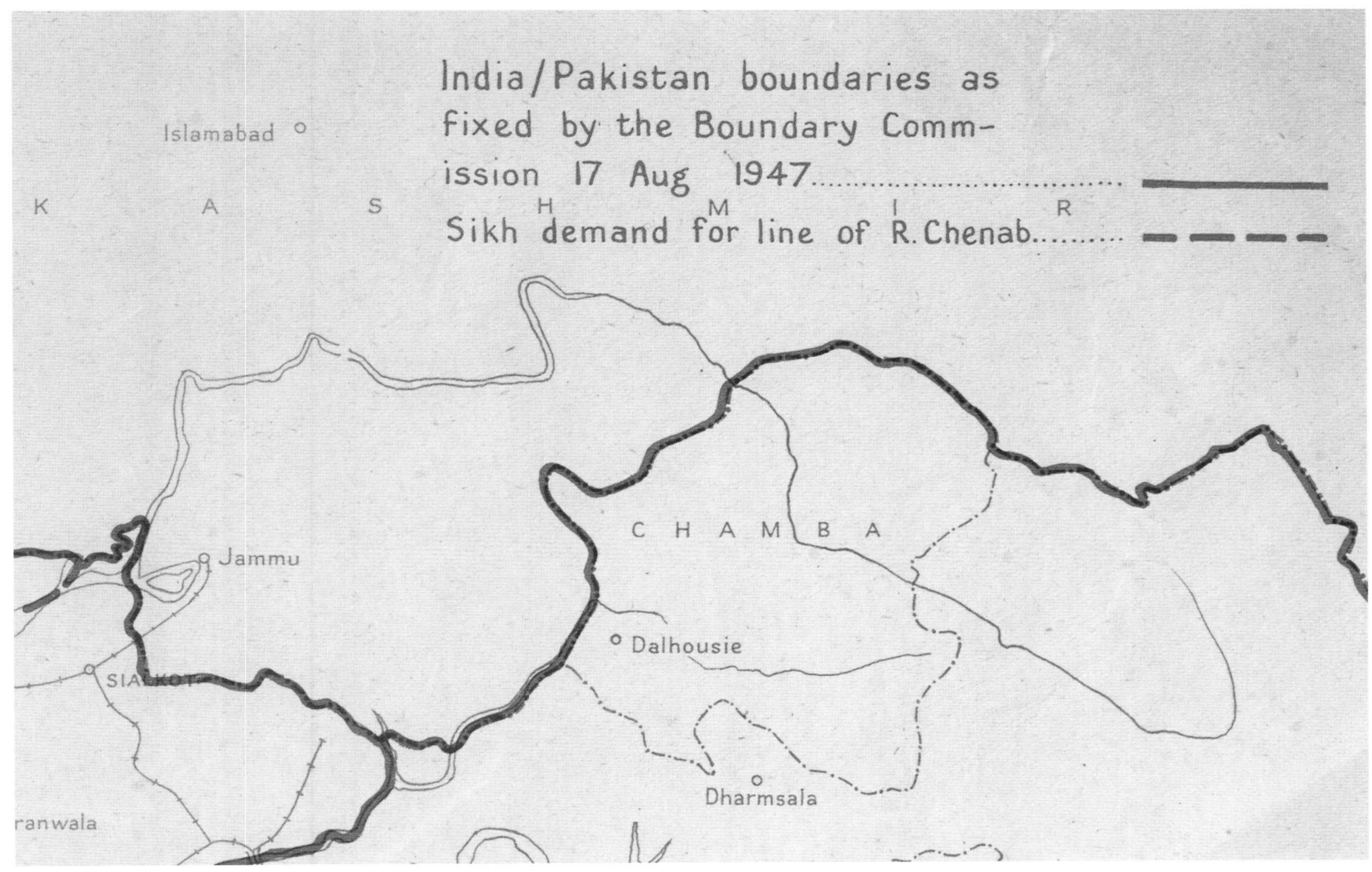

Map of Punjab showing just part of the boundary line defined by the Boundary Commission in 1947 as part of the process of the Partition of India. British Library, London, UK.

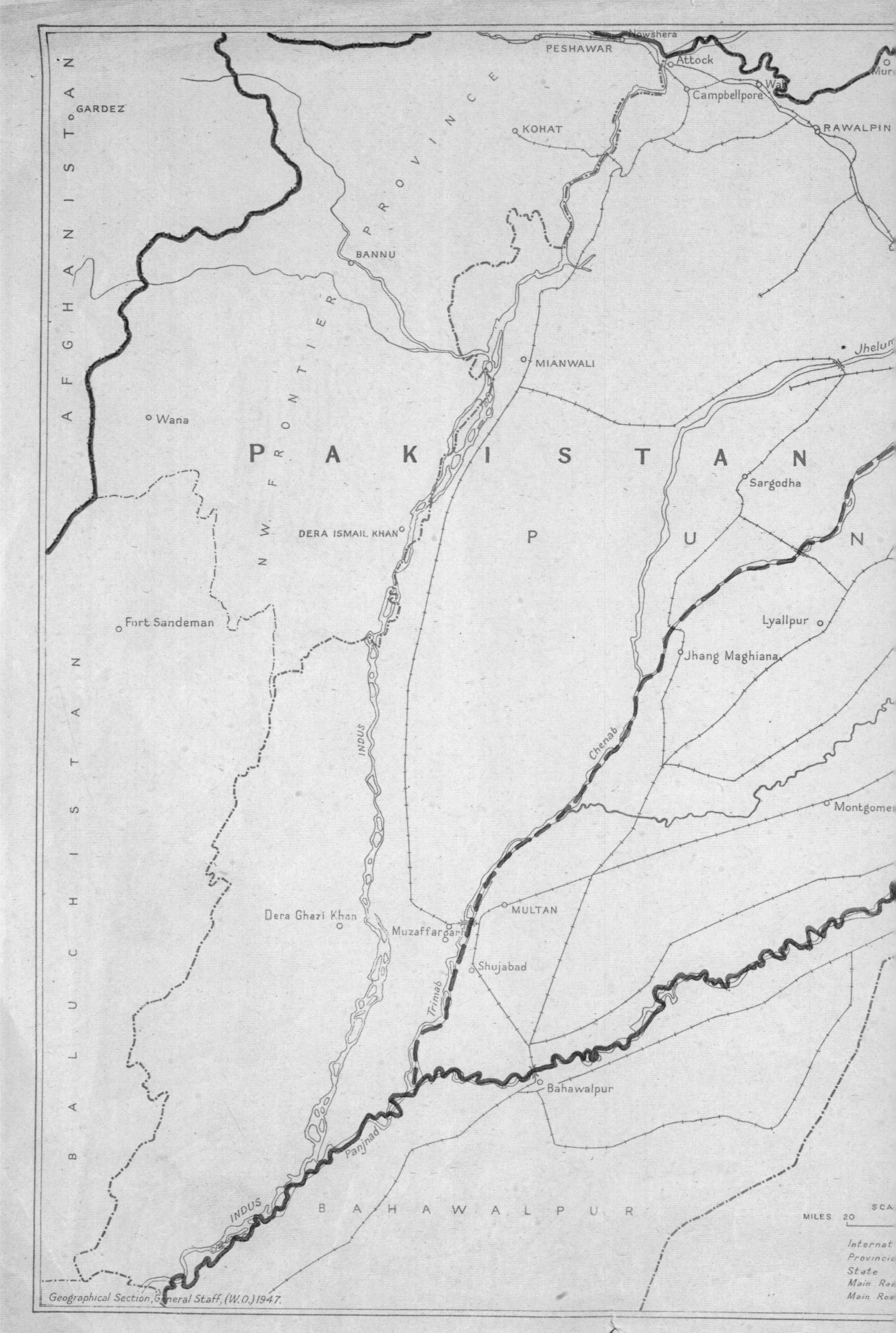

AFGHANISTAN
GARDEZ
PESHAWAR
Nowshera
Attock
Campbellpore
Wah
RAWALPIN
KOHAT
N.W. FRONTIER PROVINCE
BANNU
MIANWALI
Wana
PAKISTAN
Sargodha
P U N
DERA ISMAIL KHAN
Fort Sandeman
Lyallpur
Jhang Maghiana
BALUCHISTAN
INDUS
Chenab
Montgomer
MULTAN
Dera Ghazi Khan
Muzaffargarh
Shujabad
Trimab
Bahawalpur
Panjnad
INDUS
BAHAWALPUR
MILES 20
Geographical Section, General Staff, (W.O.) 1947.

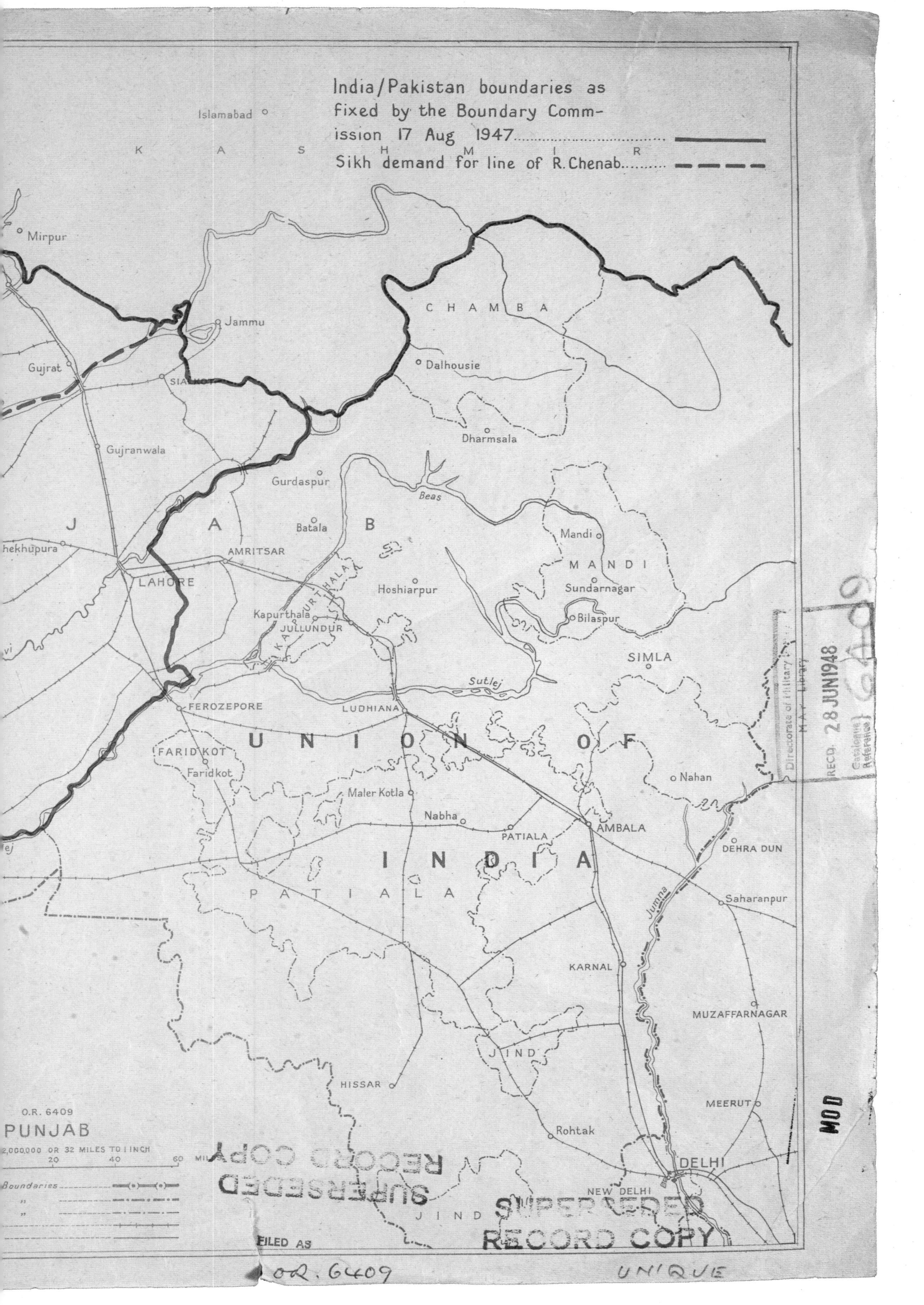
India/Pakistan boundaries as fixed by the Boundary Commission 17 Aug 1947
Sikh demand for line of R. Chenab
Islamabad
K A S H M I R
Mirpur
Jammu
Gujrat
CHAMBA
Dalhousie
Gujranwala
Dharmsala
Gurdaspur
Beas
J A B
Batala
Mandi
MANDI
AMRITSAR
LAHORE
Hoshiarpur
Sundarnagar
Kapurthala
KAPURTHALA
JULLUNDUR
Bilaspur
SIMLA
Sutlej
FEROZEPORE
LUDHIANA
UNION OF INDIA
FARID KOT
Faridkot
Nahan
Maler Kotla
Nabha
PATIALA
AMBALA
DEHRA DUN
PATIALA
Jumna
Saharanpur
KARNAL
MUZAFFARNAGAR
JIND
HISSAR
MEERUT
Rohtak
DELHI
NEW DELHI
JIND
O.R. 6409
PUNJAB
2,000,000 OR 32 MILES TO 1 INCH
20
40
60
Boundaries

Persian Gulf

Mapping has always been important to the energy industry, whether through geological surveys showing potential areas for exploration, or mapping the routes of oil and gas pipelines. By the mid-1950s the Middle East was emerging as the world's key source of supply of petroleum resources. Significantly, the colour coding on this map distinguishes between areas where oil discoveries had been made or were likely to be made on the one hand, and areas where likelihood of finding oil was small; as well as areas where finds were very unlikely. Such distinctions were fundamental to the understanding of the evolving economics and geopolitics of the industry.

From its beginnings in the second half of the nineteenth century, oil was a global industry with a global market. Between the 1870s and 1930s, apart from replacement of whale oil by kerosene for illumination, demand was especially driven by naval interests. Oil was twice as economical as coal for raising steam, and four times as economical when used in internal combustion engines. To this demand was added that of the growing motoring and aviation markets. Overall production was relatively small by present standards: in mid-1918, for example, annual imports into Britain were 3.2 million tons (mt) for home consumption; 1 mt to build up stocks of naval fuel; and 1.5 mt for other purposes. An additional 4 mt served the needs of the British Empire. The main sources of supply included the United States, Mexico, Venezuela, the Caspian Sea, the Middle East, Burma and Indonesia. There was also competition between American and European oil companies, resulting in the 'Red Line' agreement of 1928 which limited competition between them in the former Ottoman Empire.

The Great Depression of the 1930s was associated with a glut of oil, and price-fixing by the oil companies, as well as restrictions on development, especially in Iraq. Also important were mergers and acquisitions to create the huge oil companies which came to dominate the industry by the 1950s. It was widely thought that oil resources would run out within a few decades. However, at this time there were also step-changes in the application of science and technology in exploration and production which acted as a springboard for expansion during the next stage of development.

This second stage began after the Second World War. Demand accelerated due to rapid growth of the global economy, especially in North America, Europe and Japan. As late as 1950 coal still met two-thirds of world energy demand, but oil was rapidly catching up under the influence of continued expansion of road transport, the replacement of steam by diesel at sea and on rail, and the beginnings of long haul jet travel (see page 220). Further influences were developments in the electrical power generation and petrochemical industries. Between 1948 and 1973 world oil consumption grew six-fold. The majority of fields developed by this time were oilfields, as illustrated by the map, although there were significant reserves of gas as well. Also clearly illustrated is the emerging scale and technology of production, including refining and transportation by overland pipeline and sea routes for oil tankers.

The development and exploitation of oil resources was complicated by geopolitics, including the drive by the key market countries to establish secure sources of supply. After the First World War, this was especially manifested in the leading British role in the establishment of the Anglo-Persian Oil Company (to become British Petroleum in 1954), and the establishment of Iraq in the 1920s. The concern with security emerged again after the Second World War in the special relationship which developed between the United States and Saudi Arabia, which was partly inspired by the perceived need to provide secure supplies for Europe.

At the regional level the situation was complicated by foreign interference, Arab nationalism, and anti-Zionism. The Paris Peace Conference of 1919 was the starting point for the dismemberment of the Ottoman Empire, where a primary concern was access to oil resources. The drive to improve the economic benefit of the industry for the countries of the Middle East themselves accelerated in the 1950s under local pressure and varying degrees of political unrest in Iran, Iraq and Saudi Arabia, as well as through the nationalization of the Suez Canal in 1956. However, outright nationalization of oil resources still lay in the future, as did the emergence of the Organisation of Petroleum Exporting Countries (OPEC) cartel, not least because the countries concerned still remained highly dependent upon technological, economic and political support from the principal market states.

OIL AND ENERGY

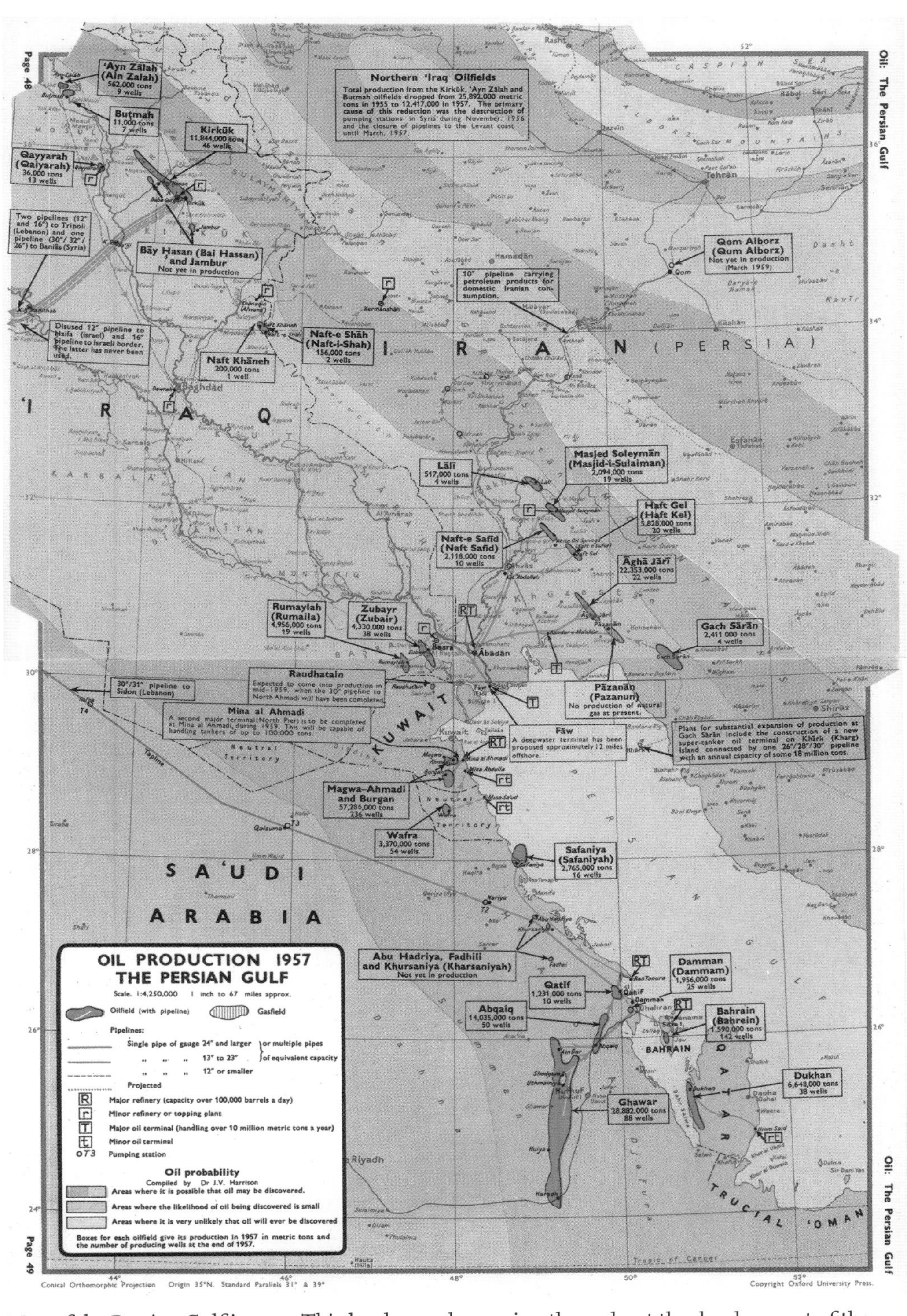

Map of the Persian Gulf in 1957. This has been a key region throughout the development of the global oil and energy industry. *Oxford Regional Economic Atlas The Middle East and North Africa*, 1960. Clarendon Press/Collins Bartholomew.

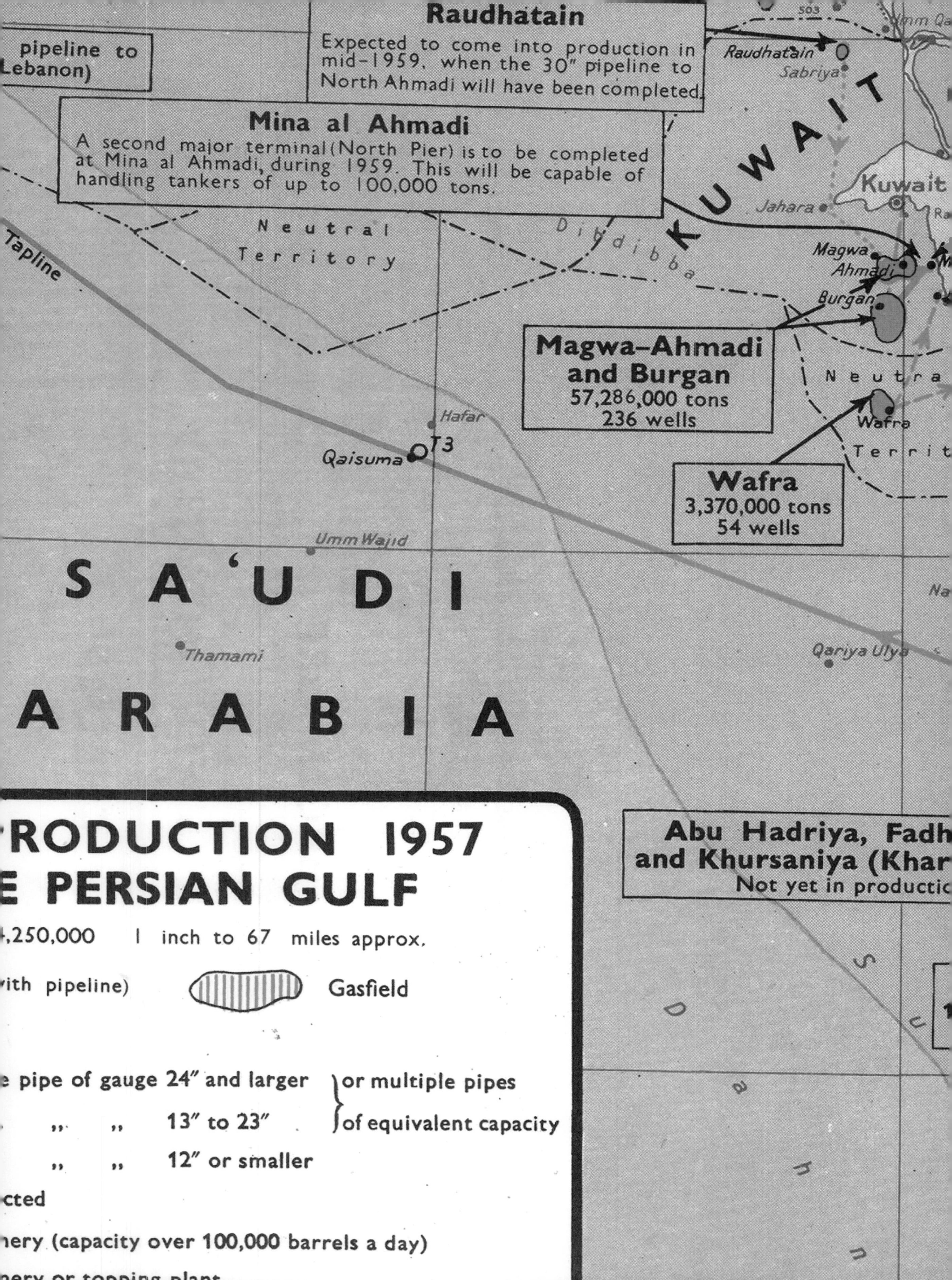
Raudhatain
Expected to come into production in mid-1959, when the 30" pipeline to North Ahmadi will have been completed.
pipeline to
Lebanon)
Mina al Ahmadi
A second major terminal (North Pier) is to be completed at Mina al Ahmadi, during 1959. This will be capable of handling tankers of up to 100,000 tons.
Raudhatain
Sabriya
KUWAIT
Kuwait
Jahara
Neutral Territory
Diddibba
Tapline
Magwa
Ahmadi
Burgan
Magwa-Ahmadi and Burgan
57,286,000 tons
236 wells
Neutra
Wafra
Territ
Wafra
3,370,000 tons
54 wells
Hafar
Qaisuma
T3
Umm Wajid
SA'UDI
Thamami
ARABIA
Qariya Ulya
Abu Hadriya, Fadh
and Khursaniya (Khar
Not yet in productic
RODUCTION 1957
E PERSIAN GULF
4,250,000 1 inch to 67 miles approx.
vith pipeline)
Gasfield
e pipe of gauge 24" and larger
,, ,, 13" to 23"
,, ,, 12" or smaller
or multiple pipes
of equivalent capacity
cted
nery (capacity over 100,000 barrels a day)
nery or topping plant

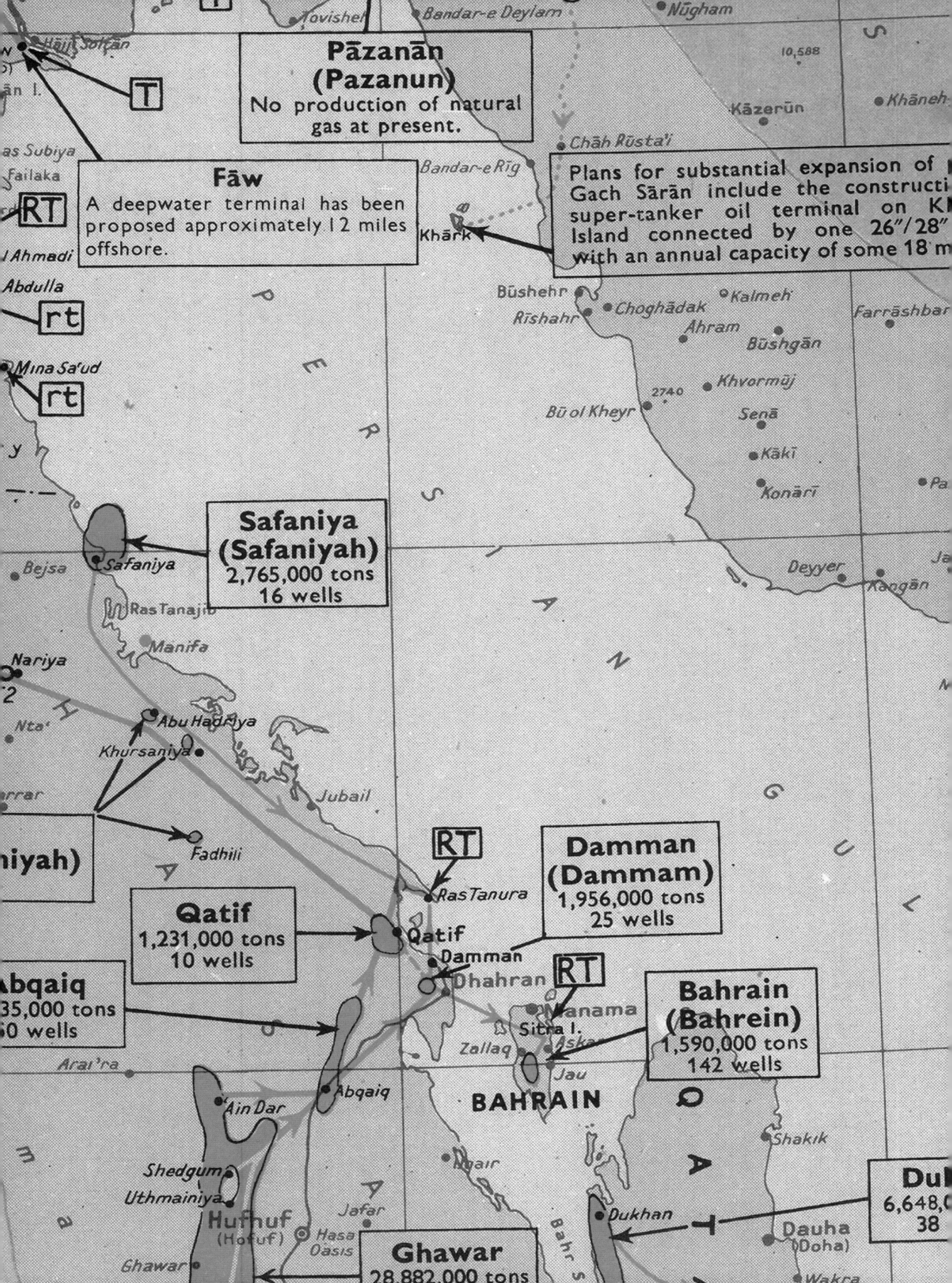

Pāzanān
(Pazanun)
No production of natural gas at present.
Fāw
A deepwater terminal has been proposed approximately 12 miles offshore.
Plans for substantial expansion of
Gach Sārān include the constructi
super-tanker oil terminal on K
Island connected by one 26"/28"
with an annual capacity of some 18 m
Safaniya
(Safaniyah)
2,765,000 tons
16 wells
Qatif
1,231,000 tons
10 wells
Damman
(Dammam)
1,956,000 tons
25 wells
Bahrain
(Bahrein)
1,590,000 tons
142 wells
Abqaiq
35,000 tons
50 wells
iyah)
Ghawar
28,882,000 tons
Du
6,648,
38
T
RT
rt
rt
RT
RT
Tovishel
Bandar-e Deylam
Nūgham
Hājī Soltān
Kāzerūn
Khāneh
Chāh Rūsta'i
Bandar-e Rīg
Khārk
as Subiya
Failaka
Ahmadi
Abdulla
Mina Sa'ud
Būshehr
Rīshahr
Choghādak
Kalmeh
Ahram
Būshgān
Farrāshbar
Khvormūj
Bū ol Kheyr
Senā
Kākī
Konārī
Deyyer
Kangān
Safaniya
Bejsa
Ras Tanajib
Manifa
Nariya
Nta'
Abu Hadriya
Khursaniya
Jubail
Fadhili
Ras Tanura
Qatif
Damman
Dhahran
Manama
Sitra I.
Askar
Zallaq
Jau
BAHRAIN
Arai'ra
Abqaiq
'Ain Dar
Shedgum
Uthmainiya
Hufhuf
(Hofuf)
Hasa
Oasis
Jafar
Ghawar
Shakik
Dukhan
Dauha
(Doha)
Wakra
Bahr S
P E R S I A N G U L F
Q A T

WORLD POWERS

The decade following the end of the Second World War saw the breaking apart of the wartime alliances and the emergence of two geopolitical blocs grouped around the United States and the USSR. Political maps of the time show a stark division, with countries on the periphery of the two groups often the sites of proxy wars or attempts to absorb them into one or other alliance.

The most striking feature of this map is its unusual viewpoint, or map projection. Devised in 1948 by John Bartholomew of the famous Edinburgh cartographic dynasty, the Atlantis Projection abandons the common atlas convention of depicting maps with the Arctic at the top and Antarctica at the bottom, with the Atlantic Ocean in between. Here the projection is tilted and centred at a point 30°W, 45°N to allow a focus on the world's oceans, in particular the Atlantic, and to illustrate the relationship between the western and eastern hemispheres in an innovative way.

The use of the Atlantis Projection in this instance is particularly effective in conveying the combative nature of relations between the United States and the Soviet Union, the two superpowers which emerged to dominate the new world order following the Second World War. Like giant beasts poised to grapple, the major landmasses of the capitalist West and the communist East face each other across the Arctic and North Atlantic Oceans – this was an accurate reflection of the state of international politics at this time.

With the defeat of Fascism in 1945, any sense of unity and common purpose between the Soviet Union and the other Allied powers swiftly evaporated. This was soon replaced by a growing mutual distrust and outright hostility which was underpinned by the polarizing effects of their respective political ideologies. Both power blocs possessed nuclear arsenals but were reluctant to confront each other in either nuclear or conventional warfare. The uneasy armed truce that emerged became known as the Cold War.

In spite of these geopolitical tensions, the United Nations (UN) organization had been founded in 1947 as a vehicle for maintaining international peace and security and developing international economic and social cooperation. Ten years later, the UN had over eighty member nations, although their division into hostile armed camps did little to further the aims of the organization.

Both the USA and the Soviets sought to build economic, military and diplomatic alliances to support their particular strategic ambitions. In 1949, the USA established the North Atlantic Treaty Organization (NATO), formally allying itself to Western Europe in order to contain the spread of Communism there. Elsewhere in the world, in 1954–5, two similar alliances were formed to counter communist expansion – the South East Asian Treaty Organization (SEATO) in the Philippines and Indo-China, and the Baghdad Pact in the Middle East. The USSR and communist China retaliated by providing military and economic support to anti-colonial or nationalist struggles in Africa, Asia and Latin America, while in 1955, the Soviets established the Warsaw Pact, their own military alliance among the communist countries of Eastern Europe.

Within a few short years of this particular map being drawn, significant colour changes would be required for a number of countries: Alaska would become a full member state of the USA (1959); Fidel Castro would establish a Marxist government on America's doorstep in Cuba (1959 – see page 228), and the process by which many African nations would shake off the last remnants of European colonialism would begin in earnest (see page 216).

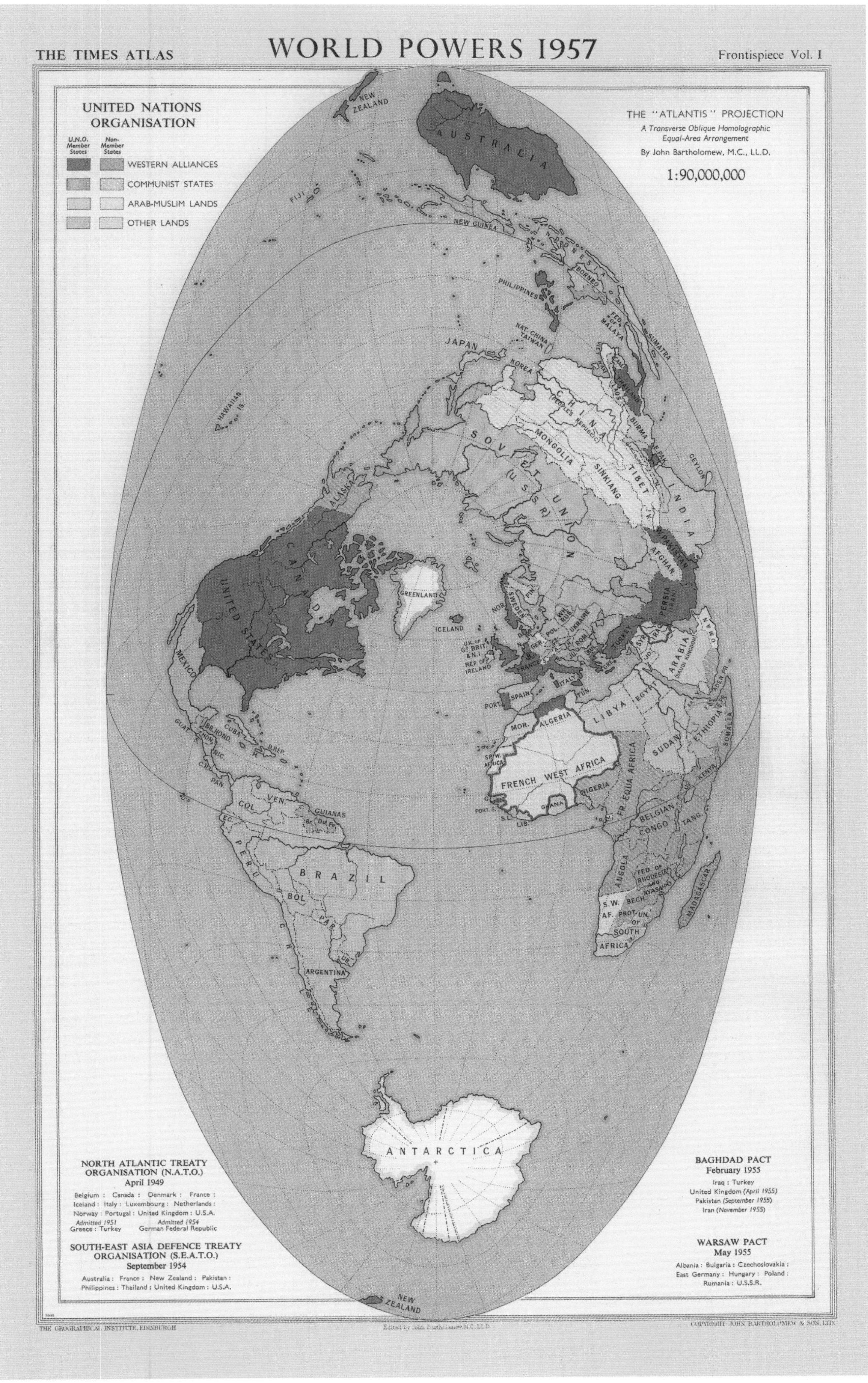

The proximity and zones of influence of the world's new superpowers is clear on this innovative 'Atlantis' map projection. From *The Times Atlas of the World, Mid-Century Edition*, 1958. Collins Bartholomew.

ALASKA
CANADA
GREENLA
UNITED STATES
MEXICO
CUBA
BR.HOND.
GUAT.
S.
HON.
NIC.
C.R.
PAN.
D.REP.
H.
VEN.
COL.
GUIANAS
Br.
Du.
Fr.
EC.
PERU
BRAZIL
BOL.

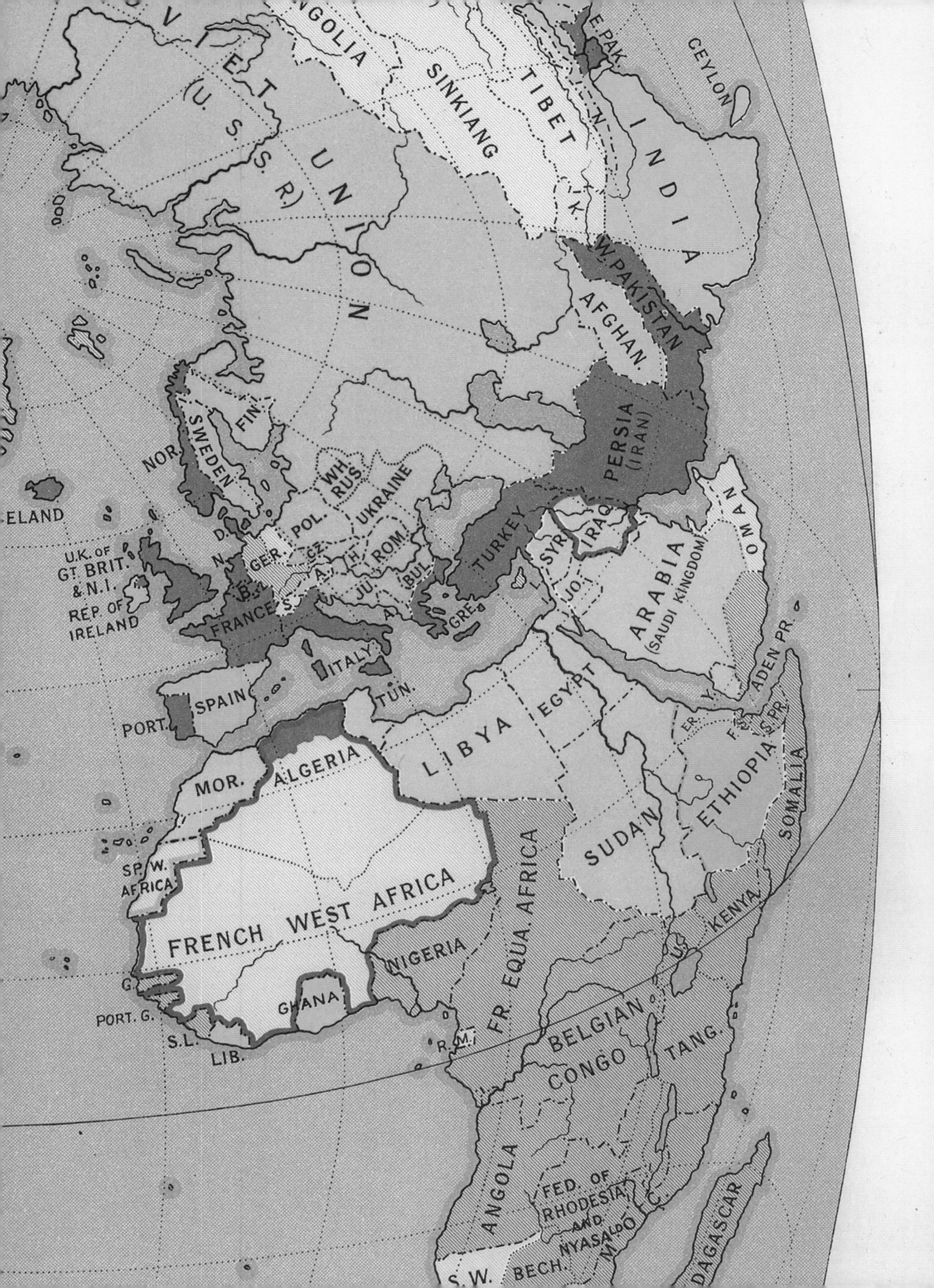
SOVIET UNION (U.S.S.R.)
MONGOLIA
SINKIANG
TIBET
E. PAK.
CEYLON
INDIA
N.
K.
W. PAKISTAN
AFGHAN.
PERSIA (IRAN)
OMAN
FIN.
SWEDEN
NOR.
ICELAND
WH. RUS.
UKRAINE
POL.
CZ.
H.
ROM.
BUL.
GER.
D.
N.
B.
U.K. OF GT. BRIT. & N.I.
REP. OF IRELAND
FRANCE
S.
A.
JU.
A.
GRE.
TURKEY
SYR.
IRAQ
L.
JO.
ARABIA (SAUDI KINGDOM)
ADEN PR.
Y.
ITALY
TUN.
SPAIN
PORT.
MOR.
ALGERIA
LIBYA
EGYPT
ER.
F.S.
S.PR.
ETHIOPIA
SOMALIA
SUDAN
SP. W. AFRICA
FRENCH WEST AFRICA
FR. EQUA. AFRICA
NIGERIA
GHANA
G.
PORT. G.
S.L.
LIB.
R.M.
KENYA
U.
BELGIAN CONGO
TANG.
ANGOLA
FED. OF RHODESIA AND NYASALD.
MOC.
S.W.
BECH.
DAGASCAR

Berlin

On 13 August 1961 the government of the German Democratic Republic began erecting the Berlin Wall – a barrier dividing the eastern part of Berlin from the sectors of the city controlled by the Western Allies – in a bid to prevent East German citizens escaping to the West. For the next three decades Berlin was a divided city, and maps showed the Berlin Wall snaking across the city, splitting it into apparently irreconcilable halves.

At the end of the Second World War in 1945, Germany was divided into four zones occupied by the victorious allies – the USSR, France, the United Kingdom, and the United States. The zones controlled by the latter three became gradually more integrated, leading ultimately to the creation of the pro-Western Federal Republic of Germany in May 1949. The USSR had tried to throttle the infant western state and in particular West Berlin, which lay surrounded by a sea of Soviet-controlled territory, accessible only by restricted air corridors, sealed trains and military convoys. But these attempts, notably the Berlin Blockade from June 1948 to May 1949, all failed.

By early 1960 the German Democratic Republic, which the USSR had established in its own occupation zone of Germany, was suffering an economic crisis with serious shortages of raw materials and a growing debt. Worst of all, many East Germans were escaping through the open border to West Berlin – an exodus which accelerated to 199,000 people in 1960, and to over 19,000 escapees in June 1961 alone.

The East German communist leader Walter Ulbricht acted to staunch this flow by developing a plan to seal off West Berlin entirely. He ordered the establishment of a 1,500-strong security unit and reinforced the local riot police who began assembling 150 tons of barbed wire and 18,000 concrete posts to create a temporary barrier along the whole of the 209 km (103 mile) perimeter of West Berlin. During the weekend of 6–7 August, 3,628 people fled East Germany and Ulbricht felt he could delay no longer. On the night of 13 August, thousands of troops entered Berlin. Tanks blocked access to the main streets, sentries were placed at two-metre intervals to prevent anyone crossing the border and soldiers rolled out barbed wire barriers to seal the frontier. By the time most East Berliners woke the next day, it was too late; only twenty-eight people reached the West that day and forty-one the next, some by jumping from the upper storeys of buildings in the Bernauer Strasse which directly overlooked the western sector.

As time went on, the initial makeshift barrier was reinforced by concrete, creating a true wall. In January 1962 a road was built behind this, and gradually additional obstacles were constructed, including anti-crash barriers, tank traps, ditches, observation towers, and barbed wire fences further back in East Berlin. Even sewer tunnels were sealed to deter those intrepid escapers who had begun to crawl through them. Some, though, still did try to escape and a few succeeded, but the risks were high – up to 227 people were killed while trying to reach the West between 1961 and 1989.

The maps of the time both reflect and seek to shape the reality of a city which now had two halves which rarely communicated with each other. This map from 1961 shows the wall as a vivid red gash partitioning Berlin, with a menacing barbed wire barrier enclosing its western half. The East German sector (the Soviet Zone) is simply labelled as 'Area IV' and 'Ostberlin'. Contemporary West German transport maps continued to show subway lines which passed through East German territory; the stations, at which no-one could alight or embark, were marked simply as 'out of service'. In East Germany, by contrast, West Berlin was sometimes simply indicated by a complete blank in the map, or by a small bubble about which little detail was given.

The Berlin Wall, including its extensions to cover the entire frontier between West and East Germany, seemed an immovable symbol of a city and country divided until 1989, when demonstrations throughout Eastern Europe threatened the hegemony of the long-entrenched communist governments. In East Germany, tens of thousands of people crossed into Hungary and Czechoslovakia, taking refuge in the West German embassies, while anti-government demonstrations in Leipzig grew to 300,000-strong by late October. Faced with unparalleled opposition, the communist regime buckled, and on 9 November announced it would open the border to limited traffic between East and West Berlin. The result was an uncontrollable surge of East Germans dashing to cross into the long-forbidden West. As border guards stood helplessly by, thousands of people rushed the crossings and the authorities were even forced to smash extra passages through the Wall to cope with the exodus.

The Brandenburg Gate – the symbolic heart of Berlin which had sat across the path of the Wall – reopened on 22 December. By the time East Germany itself disappeared on 3 October 1990, when it was reunited with West Germany, the wall had largely been dismantled. With it the maps of the wall and the divided city became historical relics of a Cold War wound which had long scarred the heart of Europe.

THE BERLIN WALL

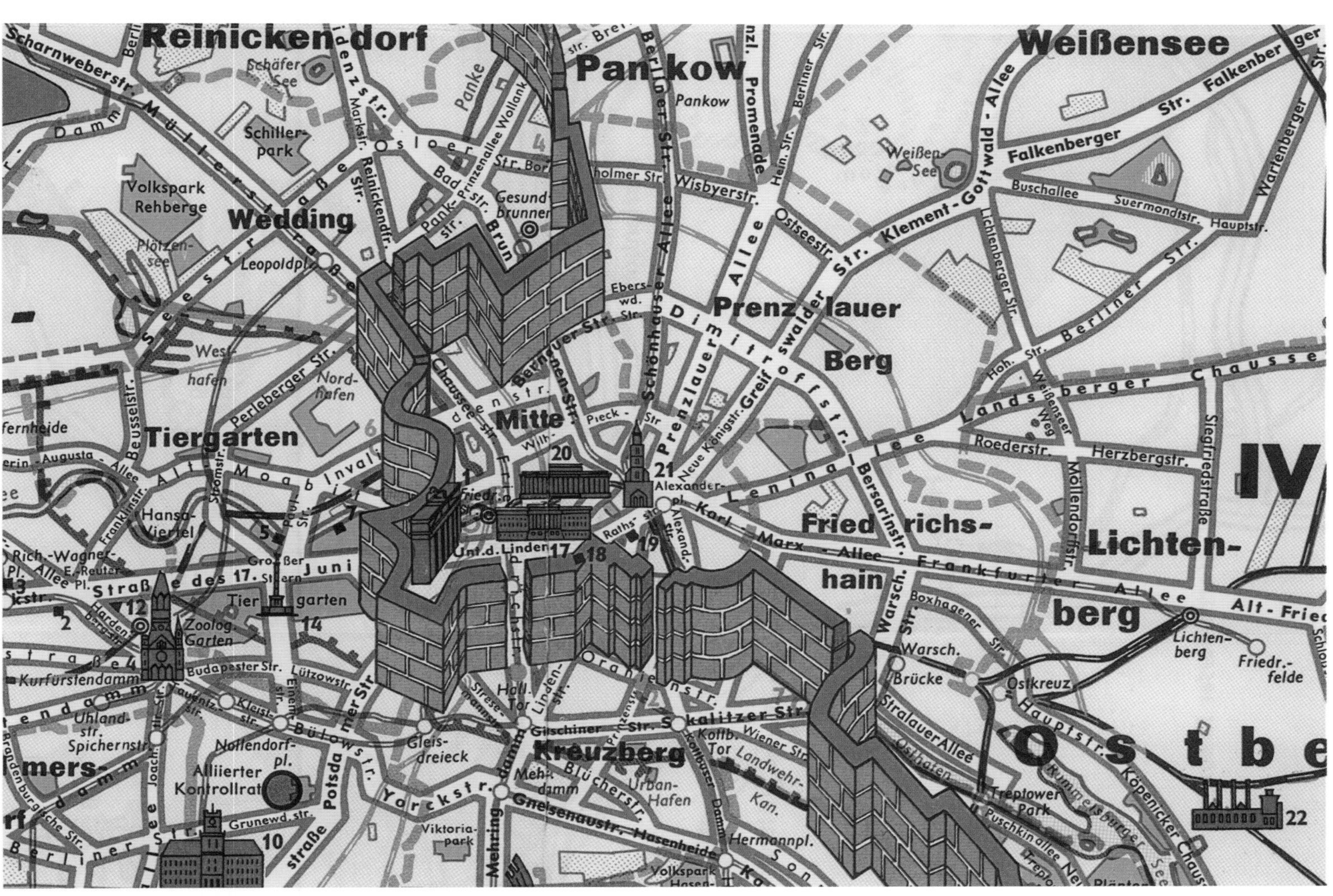

This illustrative map published in Berlin graphically emphasizes the dividing wall and details checkpoints and the different sectors of the city: I US sector; II British Sector; III French Sector; IV Soviet Sector. National Library of Scotland, Edinburgh, UK.

Erläuterung der Ziffern in der Karte
Numbers on the map
Signification des chiffres qui figurent sur la carte
Explicación de las cifras que se dan en el mapa
1 = Brandenburger Tor
2 = Schillertheater
3 = Deutsche Oper Berlin
4 = Gedächtniskirche
5 = Schloß Bellevue
6 = Haus des Rundfunks
7 = Kongreßhalle
8 = Olympia-Stadion
9 = Kraftwerk Reuter
10 = Rathaus Schöneberg
11 = Freie Universität
12 = Technische Universität
13 = Schloß Charlottenburg
14 = Siegessäule
15 = Luftbrückendenkmal
16 = Funkturm
17 = Humboldt-Universität
18 = Deutsche Staatsoper
19 = Neues Stadthaus
20 = Museums-Insel
21 = Marienkirche
22 = Kraftwerk Klingenberg
West-
Berlin
Spandau
Reinicken-dorf
Wedding
Tiergarten
Charlottenburg
Wilmers-dorf
Schöneberg
Steglitz
Zehlendorf
Potsdam
Nedlitz
Staaken
Heerstraße
Kladow
„Dreilinden“
Neubabelsberg
Steinstücken (Zehlendorf)
Nuthewiese (Zehlendorf)
Laßzins-Wiesen (Spandau)
Papenberger Wiesen (Spandau)
Eiskeller (Spandau)
Große Kuhlake (Spandau)
Frohnau
Tegel
Heiligensee
Grunewald
Glienicker Brücke
Düppel
Wannsee
Nikolassee
Allierter Kontrollrat
Alliierte Kommandantur
Hamburg, Hannover ← München
Autobahnen Helmstedt-Hannover, Hof-München
0 5 km
0 3 Statute Miles
= Grenze von Gesamt-Berlin
= Grenze der westlichen Besatzungssektoren bis zu ihrer Aufhebung als Einzelsektoren, zugleich Bezirksgrenze (I = US-Sektor, II = brit. Sektor, III = franz. Sektor, IV = sowjet. Sektor)
= Grenze der übrigen Bezirke
= West-Berliner Kontrollstelle (Heerstraße, Dreilinden, Schiffahrtskontrollstelle Kladow)
= Ostzonaler Kontrollpunkt (Staaken, Neubabelsberg, Schiffahrtskontrollstelle Nedlitz)
= Kontrollstelle nur für westalliierte Militärmissionen und Diplomaten
= Borderline of Berlin's total area
= Borderline of Western occupation sectors before their abolitio as individual sectors, at the same time district border (I = U sector, II = British sector, III = French sector, IV = Sovi sector)
= Borderline of remaining districts
= West Berlin checkpoint (Heerstrasse, Dreilinden, Navigatio checkpoint Kladow)
= Soviet zone checkpoint (Staaken, Neubabelsberg, Navigatio checkpoint Nedlitz)
= Checkpoint for control of Western Allied military missions an diplomats only

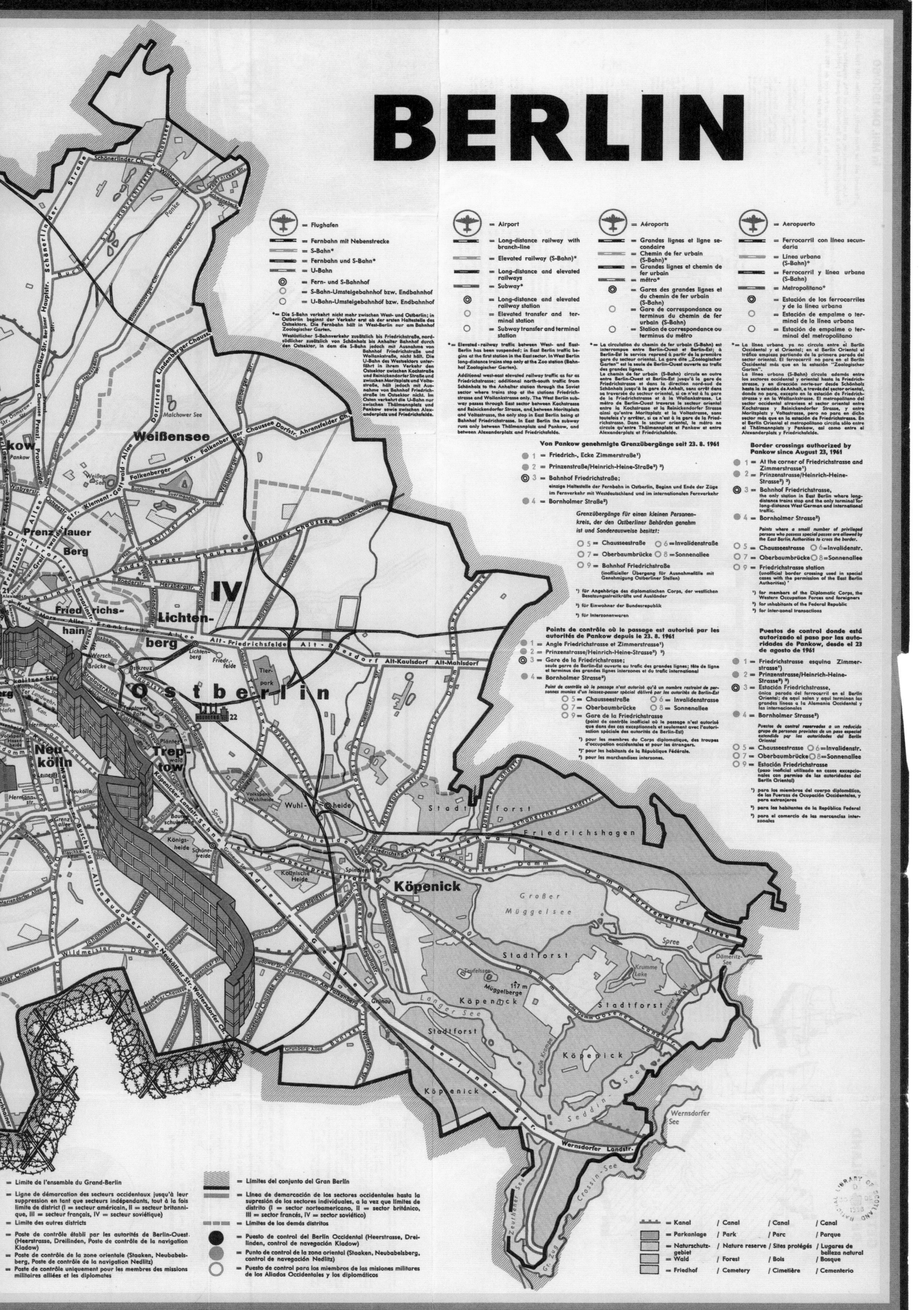

BERLIN
= Flughafen
= Fernbahn mit Nebenstrecke
= S-Bahn*
= Fernbahn und S-Bahn*
= U-Bahn
= Fern- und S-Bahnhof
= S-Bahn-Umsteigebahnhof bzw. Endbahnhof
= U-Bahn-Umsteigebahnhof bzw. Endbahnhof
*= Die S-Bahn verkehrt nicht mehr zwischen West- und Ostberlin; in Ostberlin beginnt der Verkehr erst ab der ersten Haltestelle des Ostsektors. Die Fernbahn hält in West-Berlin nur am Bahnhof Zoologischer Garten.
Westöstlicher S-Bahnverkehr zusätzlich bis Friedrichstraße, nord-südlicher zusätzlich von Schönholz bis Anhalter Bahnhof durch den Ostsektor, in dem die S-Bahn jedoch mit Ausnahme von Bahnhof Friedrichstraße und Wollankstraße, nicht hält. Die U-Bahn des Westsektors unterfährt in ihrem Verkehr den Ostsektor zwischen Kochstraße und Reinickendorfer Straße und zwischen Moritzplatz und Voltastraße, hält jedoch mit Ausnahme von Bahnhof Friedrichstraße im Ostsektor nicht. Im Osten verkehrt die U-Bahn nur zwischen Thälmannplatz und Pankow sowie zwischen Alexanderplatz und Friedrichsfelde.
= Airport
= Long-distance railway with branch-line
= Elevated railway (S-Bahn)*
= Long-distance and elevated railways
= Subway*
= Long-distance and elevated railway station
= Elevated transfer and terminal station
= Subway transfer and terminal station
*= Elevated-railway traffic between West- and East-Berlin has been suspended; in East Berlin traffic begins at the first station in the East sector. In West Berlin long-distance trains stop only at the Zoo station (Bahnhof Zoologischer Garten).
Additional west-east elevated railway traffic as far as Friedrichstrasse; additional north-south traffic from Schönholz to the Anhalter station through the Soviet sector where trains stop at the stations Friedrichstrasse and Wollankstrasse only. The West Berlin subway passes through East sector between Kochstrasse and Reinickendorfer Strasse, and between Moritzplatz and Voltastrasse, the only stop in East Berlin being at Bahnhof Friedrichstrasse. In East Berlin the subway runs only between Thälmannplatz and Pankow, and between Alexanderplatz and Friedrichsfelde.
= Aéroports
= Grandes lignes et ligne secondaire
= Chemin de fer urbain (S-Bahn)*
= Grandes lignes et chemin de fer urbain
= métro*
= Gares des grandes lignes et du chemin de fer urbain (S-Bahn)
= Gare de correspondance ou terminus du chemin de fer urbain (S-Bahn)
= Station de correspondance ou terminus du métro
*= La circulation du chemin de fer urbain (S-Bahn) est interrompue entre Berlin-Ouest et Berlin-Est; à Berlin-Est le service reprend à partir de la première gare du secteur oriental. La gare dite „Zoologischer Garten" est la seule de Berlin-Ouest ouverte au trafic des grandes lignes.
Le chemin de fer urbain (S-Bahn) circule en outre entre Berlin-Ouest et Berlin-Est jusqu'à la gare de Friedrichstrasse et dans la direction nord-sud de Schönholz jusqu'à la gare de Anhalt, sans arrêt dans sa traversée du secteur oriental, si ce n'est à la gare de la Friedrichstrasse et à la Wollankstrasse. Le métro de Berlin-Ouest traverse le secteur oriental entre la Kochstrasse et la Reinickendorfer Strasse ainsi qu'entre Moritzplatz et la Voltastrasse, sans toutefois s'y arrêter, si ce n'est à la gare de la Friedrichstrasse. Dans le secteur oriental, le métro ne circule qu'entre Thälmannplatz et Pankow et entre Alexanderplatz et Friedrichsfelde.
= Aeropuerto
= Ferrocarril con línea secundaria
= Línea urbana (S-Bahn)*
= Ferrocarril y línea urbana (S-Bahn)
= Metropolitano*
= Estación de los ferrocarriles y de la línea urbana
= Estación de empalme o terminal de la línea urbana
= Estación de empalme o terminal del metropolitano
*= La línea urbana ya no circula entre el Berlín Occidental y el Oriental; en el Berlín Oriental el tráfico empieza partiendo de la primera parada del sector oriental. El ferrocarril no para en el Berlín Occidental más que en la estación "Zoologischer Garten".
La línea urbana (S-Bahn) circula además entre los sectores occidental y oriental hasta la Friedrichstrasse, y en dirección norte-sur desde Schönholz hasta la estación de Anhalt, a través del sector oriental, donde no para, excepto en la estación de Friedrichstrasse y en la Wollankstrasse. El metropolitano del sector occidental atraviesa el sector oriental entre Kochstrasse y Reinickendorfer Strasse, y entre Moritzplatz y Voltastrasse, pero no para en dicho sector más que en la estación de Friedrichstrasse. En el Berlín Oriental el metropolitano circula sólo entre el Thälmannplatz y Pankow, así como entre el Alexanderplatz y Friedrichsfelde.
Von Pankow genehmigte Grenzübergänge seit 23. 8. 1961
1 = Friedrich-, Ecke Zimmerstraße¹)
2 = Prinzenstraße/Heinrich-Heine-Straße²) ³)
3 = Bahnhof Friedrichstraße;
einzige Haltestelle der Fernbahn in Ostberlin, Beginn und Ende der Züge im Fernverkehr mit Westdeutschland und im internationalen Fernverkehr
4 = Bornholmer Straße²)
Grenzübergänge für einen kleinen Personenkreis, der den Ostberliner Behörden genehm ist und Sonderausweise besitzt:
5 = Chausseestraße 6 = Invalidenstraße
7 = Oberbaumbrücke 8 = Sonnenallee
9 = Bahnhof Friedrichstraße
(inoffizieller Übergang für Ausnahmefälle mit Genehmigung Ostberliner Stellen)
¹) für Angehörige des diplomatischen Corps, der westlichen Besatzungsstreitkräfte und Ausländer
²) für Einwohner der Bundesrepublik
³) für Interzonenwaren
Border crossings authorized by Pankow since August 23, 1961
1 = At the corner of Friedrichstrasse and Zimmerstrasse¹)
2 = Prinzenstrasse/Heinrich-Heine-Strasse²) ³)
3 = Bahnhof Friedrichstrasse,
the only station in East Berlin where long-distance trains stop and the only terminal for long-distance West German and international traffic.
4 = Bornholmer Strasse²)
Points where a small number of privileged persons who possess special passes are allowed by the East Berlin Authorities to cross the border.
5 = Chausseestrasse 6 = Invalidenstr.
7 = Oberbaumbrücke 8 = Sonnenallee
9 = Friedrichstrasse station
(unofficial border crossing used in special cases with the permission of the East Berlin Authorities)
¹) for members of the Diplomatic Corps, the Western Occupation Forces and foreigners
²) for inhabitants of the Federal Republic
³) for inter-zonal transactions
Points de contrôle où le passage est autorisé par les autorités de Pankow depuis le 23. 8. 1961
1 = Angle Friedrichstrasse et Zimmerstrasse¹)
2 = Prinzenstrasse/Heinrich-Heine-Strasse²) ³)
3 = Gare de la Friedrichstrasse;
seule garre de Berlin-Est ouverte au trafic des grandes lignes; tête de ligne et terminus des grandes lignes interzones et du trafic international
4 = Bornholmer Strasse²)
Point de contrôle où le passage n'est autorisé qu'à un nombre restreint de personnes munies d'un laissez-passer spécial délivré par les autorités de Berlin-Est
5 = Chausseestraße 6 = Invalidenstrasse
7 = Oberbaumbrücke 8 = Sonnenallee
9 = Gare de la Friedrichstrasse
(point de contrôle inofficiel où le passage n'est autorisé que dans des cas exceptionnels et seulement avec l'autorisation spéciale des autorités de Berlin-Est)
¹) pour les membres du Corps diplomatique, des troupes d'occupation occidentales et pour les étrangers.
²) pour les habitants de la République Fédérale.
³) pour les marchandises interzones.
Puestos de control donde está autorizado el paso por las autoridades de Pankow, desde el 23 de agosto de 1961
1 = Friedrichstrasse esquina Zimmerstrasse¹)
2 = Prinzenstrasse/Heinrich-Heine-Strasse²) ³)
3 = Estación Friedrichstrasse,
única parada del ferrocarril en el Berlín Oriental; de aquí salen y aquí terminan las grandes líneas a la Alemania Occidental y las internacionales
4 = Bornholmer Strasse²)
Puestos de control reservados a un reducido grupo de personas provistas de un pase especial extendido por las autoridades del Berlín Oriental
5 = Chausseestrasse 6 = Invalidenstr.
7 = Oberbaumbrücke 8 = Sonnenallee
9 = Estación Friedrichstrasse
(paso inoficial utilizado en casos excepcionales con permiso de las autoridades del Berlín Oriental)
¹) para los miembros del cuerpo diplomático, de las Fuerzas de Ocupación Occidentales, y para extranjeros
²) para los habitantes de la República Federal
³) para el comercio de las mercancías interzonales
Weißensee
Prenzlauer Berg
Friedrichshain
Lichtenberg
IV
Ostberlin
Neukölln
Treptow
Köpenick
Friedrichshagen
Stadtforst
Großer Müggelsee
Müggelberge
Langer See
Seddin-See
Wernsdorfer See
Zeuthener See
Crossin-See
Dämeritz-See
= Limite de l'ensemble du Grand-Berlin
= Ligne de démarcation des secteurs occidentaux jusqu'à leur suppression en tant que secteurs indépendants, tout à la fois limite de district (I = secteur américain, II = secteur britannique, III = secteur français, IV = secteur soviétique)
= Limite des autres districts
= Poste de contrôle établi par les autorités de Berlin-Ouest. (Heerstrasse, Dreilinden, Poste de contrôle de la navigation Kladow)
= Poste de contrôle de la zone orientale (Staaken, Neubabelsberg, Poste de contrôle de la navigation Nedlitz)
= Poste de contrôle uniquement pour les membres des missions militaires alliées et les diplomates
= Límites del conjunto del Gran Berlín
= Línea de demarcación de los sectores occidentales hasta la supresión de los sectores individuales, a la vez que límites de distrito (I = sector norteamericano, II = sector británico, III = sector francés, IV = sector soviético)
= Límites de los demás distritos
= Puesto de control del Berlín Occidental (Heerstrasse, Dreilinden, control de navegación Kladow)
= Punto de control de la zona oriental (Staaken, Neubabelsberg, control de navegación Nedlitz)
= Puesto de control para los miembros de las misiones militares de los Aliados Occidentales y los diplomáticos
= Kanal / Canal / Canal / Canal
= Parkanlage / Park / Parc / Parque
= Naturschutzgebiet / Nature reserve / Sites protégés / Lugares de belleza natural
= Wald / Forest / Bois / Bosque
= Friedhof / Cemetery / Cimetière / Cementerio

Antarctica

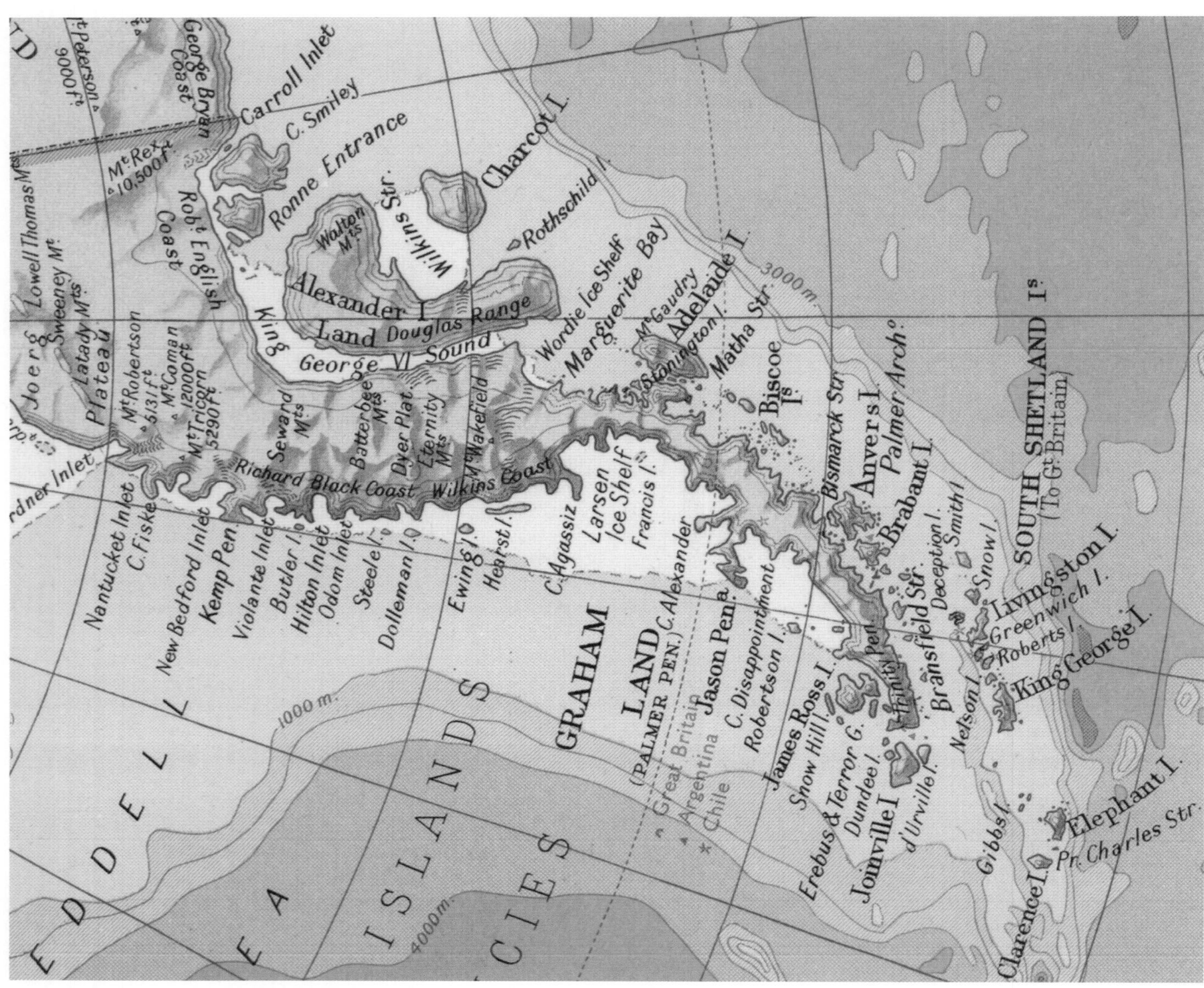

Map of Antarctica from *The Times Atlas of the World Mid-Century Edition*, 1958, showing territorial claims and research bases as they were just before the Antarctic Treaty was signed in 1959. The map reflects the state of mapping and knowledge of the continent at the time, before major new surveys were carried out in the 1950s. Collins Bartholomew.

PROTECTION OF A CONTINENT

Many countries have been involved in the exploration of Antarctica since the 1820s and seven have made territorial claims; Argentina, Australia, Chile, France, New Zealand, Norway and the UK. During the period after the Second World War, the competing claims and the presence in Antarctica of the US and USSR caused international tension, particularly in the Antarctic Peninsula where the claims of the UK, Argentina and Chile overlap. The need to manage these tensions eventually led to the Antarctic Treaty – signed in 1959 by its twelve original signatories and ratified in 1961. This was a unique system of international governance which established Antarctica as a continent for scientific collaboration, and set territorial claims to one side.

The Antarctic Peninsula in particular has been a significant region in both the geopolitics of the continent and the history of mapping in Antarctica. The UK has had a long involvement in this region and its legal claim for the British Antarctic Territory is the oldest in Antarctica and dates from 1908. Argentina has had a small, but long-term presence, operating a station in the South Orkney Islands established by a Scottish expedition from 1904. During the Second World War, Argentina and Chile increased their interest in Antarctica and the south Atlantic, opening bases and claiming overlapping sectors of the Antarctic Peninsula in 1942 and 1940.

From the beginning, mapping was seen as a way of projecting national presence and sovereignty, with, for example, Argentina and Chile publishing maps showing the Peninsula as national territory. Immediately after the war, the USA began an ambitious programme of establishing bases and carrying out airborne and overland survey in Antarctica, but without making a specific territorial claim. Britain established the Falkland Islands Dependencies Survey (FIDS) after the war, with mapping and related activities seen by the UK Foreign Office as an important part of 'effective occupation' to reinforce British territorial claims.

By the late 1950s, increasing international awareness of the geopolitical tensions in Antarctica led to a number of initiatives to encourage cooperation, rather than competition. The International Geophysical Year (IGY) in 1957–58 involved new international scientific and logistical collaborations. One of its legacies was the creation in 1958 of the Scientific Committee on Antarctic Research (SCAR) which promotes international scientific collaboration, including in the field of mapping. The cooperation during IGY and in SCAR eased tensions and led to the drafting of the Antarctic Treaty, ratified in 1961. Two of the Treaty's key achievements are to put territorial claims to one side and to encourage collaboration and sharing of data.

Early survey work by FIDS through the 1940s, which resulted in overview maps at 1:500,000 published by the British Directorate of Colonial Surveys, involved pioneering survey traverses, often hundreds of kilometres long and lasting months at a time, travelling into the unknown by dog-sledge and carrying out survey work in very difficult conditions. The resultant maps were inevitably approximate, with positions sometimes accurate to only several hundred metres and detail between the traverse routes sparse.

A subsequent survey programme in 1955–57 involved aerial photography and supporting ground control survey – believed to offer a faster and more accurate method for compiling maps. During this programme many major topographic features such as mountains and large glaciers were seen for the first time. It soon became apparent, however, that such mapping techniques would be difficult in the rugged mountains and across large areas of featureless snowy terrain. Government priorities were also changing at that time and the planned series of regional maps was never completed.

Interestingly, the aerial photographs and other material collected for this mapping programme form a unique record of the region sixty years ago and are now a valuable resource for measuring glacial and environmental change in the region.

In the last half century, Antarctic mapping has developed from the pioneering mapping work by FIDS and other contemporary organisations, to continent-wide coverage derived from satellite imagery and aerial survey. Antarctica is a key part of the global environmental system, and international scientists continue to seek to understand its role and venture into rarely visited areas to answer new science questions. There has been a revolution in the source information available for mapping, and the final product may have changed from paper maps to digital geographic data, but scientific fieldwork and the associated logistics continue to need accurate maps.

ANTA

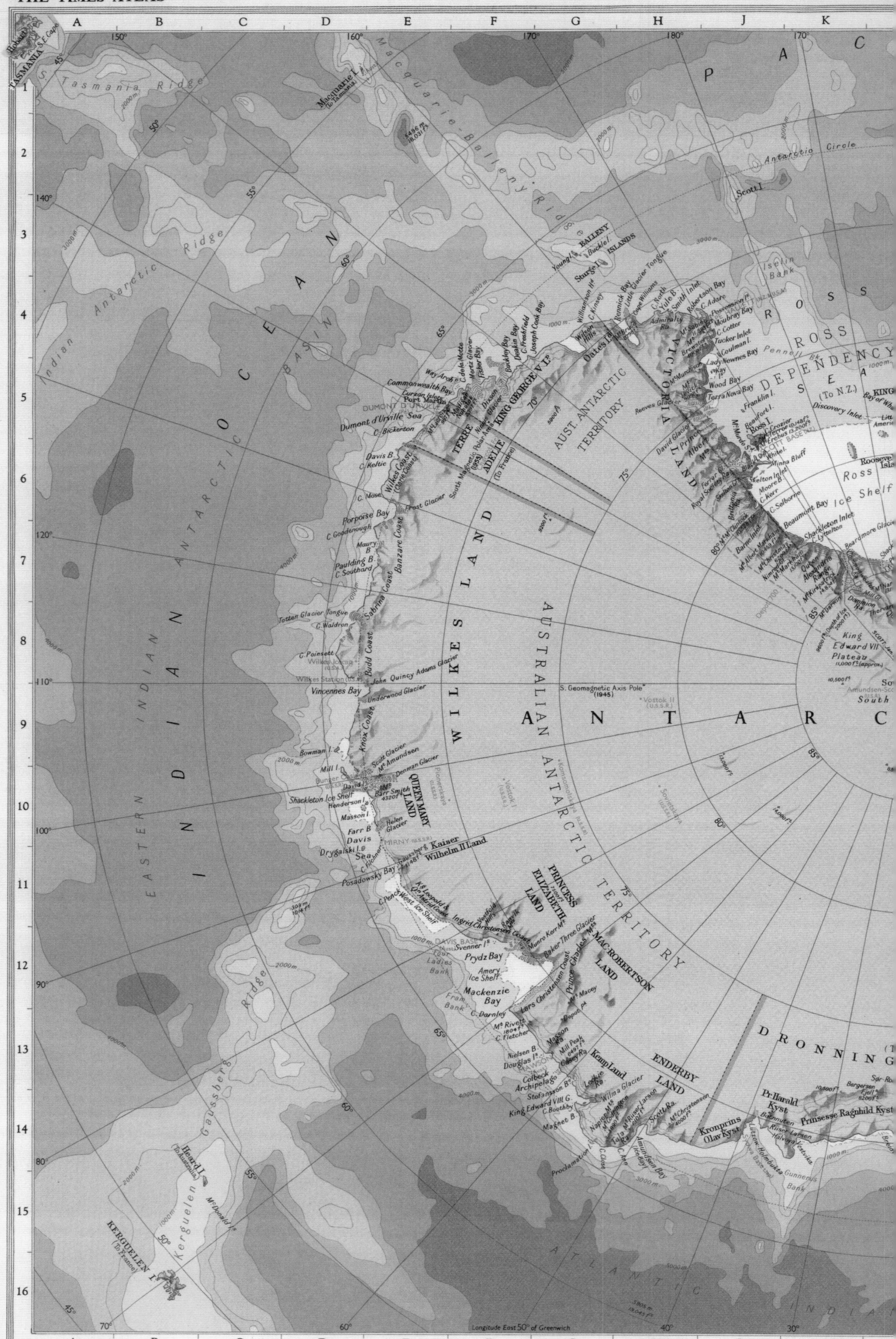

INDIAN OCEAN
EASTERN INDIAN ANTARCTIC BASIN
Indian Antarctic Ridge
Tasmania Ridge
Macquarie-Balleny Ridge
Macquarie I. (To Tasmania)
TASMANIA
S.E. Cape
PACIFIC
Antarctic Circle
Scott I.
BALLENY ISLANDS
Young I.
Buckle I.
Sturge I.
Iselin Bank
ROSS SEA
ROSS DEPENDENCY (To N.Z.)
Pennell Bk.
VICTORIA LAND
Oates Land
C. Adare
Robertson Bay
Possession Is.
Moubray Bay
C. Cotter
Tucker Inlet
Coulman I.
Lady Newnes Bay
Wood Bay
Terra Nova Bay
Franklin I.
Beaufort I.
Ross I.
C. Crozier
Discovery Inlet
Ross Ice Shelf
Roosevelt Island
Minna Bluff
Skelton Inlet
C. Selborne
Beaumont Bay
Shackleton Inlet
Beardmore Glacier
Queen Alexandra Range
David Glacier
Prince Albert Mts.
Reeves Glacier
Royal Society Ra.
AUST. ANTARCTIC TERRITORY
KING GEORGE V LAND
TERRE ADÉLIE (To France)
DUMONT D'URVILLE
Dumont d'Urville Sea
Commonwealth Bay
Port Martin
C. Bickerton
Davis B.
C. Keltie
Wilkes Coast
C. Mose
Frost Glacier
South Magnetic Pole
Porpoise Bay
C. Goodenough
Maury B.
Paulding B.
C. Southard
Banzare Coast
Sabrina Coast
Totten Glacier Tongue
C. Waldron
Budd Coast
C. Poinsett
Wilkes Station (U.S.A.)
Vincennes Bay
John Quincy Adams Glacier
Underwood Glacier
Knox Coast
WILKES LAND
AUSTRALIAN ANTARCTIC TERRITORY
ANTARCTICA
S. Geomagnetic Axis Pole (1945)
Vostok II (U.S.S.R.)
Vostok I (U.S.S.R.)
Komsomolskaya (U.S.S.R.)
Sovietskaya (U.S.S.R.)
Pionerskaya (U.S.S.R.)
King Edward VII Plateau
Amundsen-Scott (U.S.A.)
South Pole
Bowman I.
Scott Glacier
Mt. Amundsen
Denman Glacier
Mill I.
Bunger
David I.
Shackleton Ice Shelf
Henderson I.
Masson I.
QUEEN MARY LAND
Helen Glacier
MIRNY (U.S.S.R.)
Farr B.
Davis Sea
Drygalski I.
Kaiser Wilhelm II Land
Gaussberg
Posadowsky Bay
West Ice Shelf
Ingrid Christensen Coast
PRINCESS ELIZABETH LAND
DAVIS BASE
Svenner Is.
Prydz Bay
Amery Ice Shelf
Four Ladies Bank
Fram Bank
Mackenzie Bay
C. Darnley
Lars Christensen Coast
Prince Charles Mts.
Baker Three Glacier
MAC.ROBERTSON LAND
Mt. Rivett
C. Fletcher
Nielsen B.
Douglas Is.
MAWSON
Masson Ra.
Casey Ra.
Mill Peak
Kemp Land
Colbeck Archipelago
Stefansson B.
King Edward VIII G.
C. Boothby
Magnet B.
Wilma Glacier
Scott Ra.
ENDERBY LAND
Proclamation I.
C. Close
C. Ann
Amundsen Bay
Kronprins Olav Kyst
Pr. Harald Kyst
Prinsesse Ragnhild Kyst
DRONNING
Gunnerus Bank
Gaussberg Ridge
Kerguelen Gaussberg Ridge
Heard I. (To Australia)
McDonald I.
KERGUELEN I. (To France)
ATLANTIC
INDIAN
Longitude East 50° of Greenwich

THE GEOGRAPHICAL INSTITUTE, EDINBURGH

Boundaries of International Dependency Claims

Bases and Stations, International Geophysical Year (1957-8), shown in red

Feet	19686	16409	13124	9843	3281	Feet
Metres	6000	5000	4000	3000	1000	Metres

Edited by John

1:15,

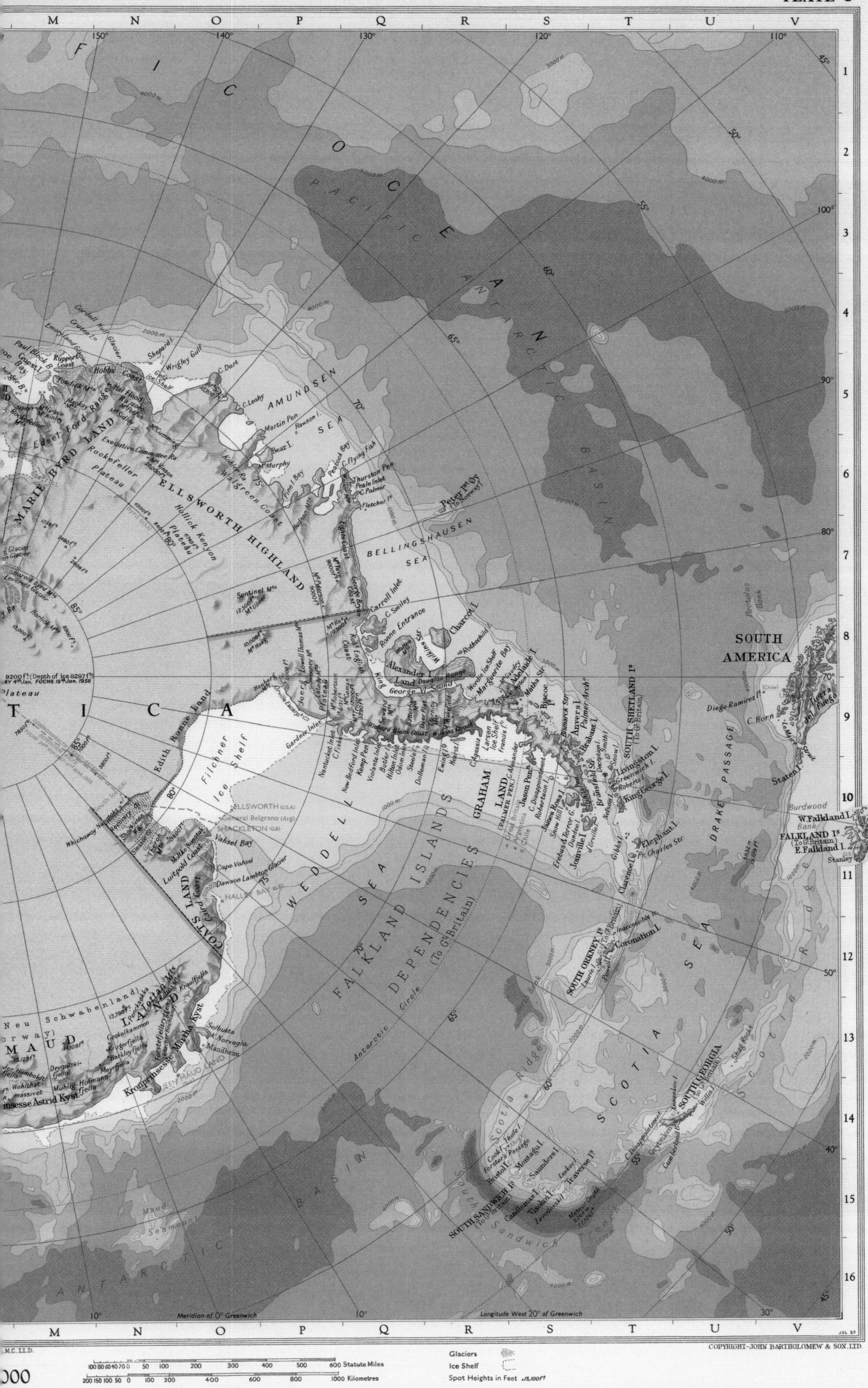
PACIFIC OCEAN
ANTARCTIC BASIN
AMUNDSEN SEA
BELLINGSHAUSEN SEA
MARIE BYRD LAND
ELLSWORTH HIGHLAND
Rockefeller Plateau
Hollick Kenyon Plateau
Sentinel Mts
Peter Ist Oy (To Norway)
Thurston Pen
Charcot I.
Alexander I. Land
George VI Sound
Ronne Entrance
Carroll Inlet
Marguerite Bay
GRAHAM LAND
(PALMER PEN.)
Edith Ronne Land
Filchner Ice Shelf
ELLSWORTH (U.S.A.)
General Belgrano (Arg.)
SHACKLETON (G.B.)
Vahsel Bay
HALLEY BAY (G.B.)
COATS LAND
Luitpold Coast
Caird Coast
WEDDELL SEA
FALKLAND ISLANDS DEPENDENCIES (To Gt Britain)
Antarctic Circle
SOUTH SHETLAND Is (To Gt Britain)
King George I.
Livingston I.
Elephant I.
Clarence I.
SOUTH ORKNEY Is (To Gt Britain)
Coronation I.
DRAKE PASSAGE
SCOTIA SEA
Scotia Ridge
SOUTH GEORGIA (To Gt Britain)
SOUTH SANDWICH Is (To Gt Britain)
South Sandwich Trench
SOUTH AMERICA
C. Horn
Diego Ramirez Is
Staten I.
Burdwood Bank
W. Falkland I.
FALKLAND Is (To Gt Britain)
E. Falkland I.
Stanley
Falkland Sound
Neu Schwabenland
MAUD LAND
Prinsesse Astrid Kyst
Prinsesse Martha Kyst
Kronprinsesse Märtha Kyst
Maudheim
Meridian of 0° Greenwich
Longitude West 20° of Greenwich
M N O P Q R S T U V

J.M.C. LTD.

COPYRIGHT–JOHN BARTHOLOMEW & SON, LTD.

000

100 80 60 40 20 0 50 100 200 300 400 500 600 Statute Miles
200 150 100 50 0 100 200 400 600 800 1000 Kilometres

Glaciers
Ice Shelf
Spot Heights in Feet 15,100 ft

1945/1977 LATE COLONIALISM

The three decades following the Second World War saw dramatic changes in political boundaries. As European colonial control over vast areas of Asia, Africa and the Pacific and Caribbean islands unwound, scores of new nations emerged. This process of decolonization is highlighted by the differences between the world map in 1945, vividly splashed with patches of French green and British pink, and that of 1977 where those nations' empires had shrunk to almost nothing.

European colonial empires had come under increasing pressure from nationalist movements demanding greater autonomy or independence, particularly in India, where Mahatma Gandhi and the Indian National Congress ran non-violent campaigns of non-cooperation with the British authorities from the 1920s. But the Second World War proved to be a catalyst. It had weakened Britain, France and the other colonial powers economically and politically, and it gave hope to those nationalist movements which had suspended their campaigns during the War, that they would be rewarded afterwards.

The first post-war decolonizations were rushed and chaotic, as the British colonial authorities in particular struggled to respond to a rapidly changing environment. In 1947 in Palestine – which had been acquired by Britain under a League of Nations Mandate in 1923 – a UN plan for partition into Jewish and Arab states foundered amidst violent disagreements between the two sides (see page 172). War ensued, which resulted in the acquisition by the Jewish State of Israel of significantly more territory, and the flight of over 700,000 Palestinian Arab refugees. The matter of borders also blighted the partition of British India into a predominantly Hindu India and an overwhelmingly Muslim Pakistan. The Commission under Sir Cyril Radcliffe which decided exactly where the boundary should fall, had to work very quickly over a short period in summer 1947 and its decisions often left large religious minority communities stranded on the wrong side of the line (see page 196). This resulted in a humanitarian crisis after independence was declared in August 1947 in which over 200,000 people died and a massive exodus of up to ten million refugees took place. Violence also accompanied the independence of French Indo-China. In Vietnam, a communist-nationalist insurgency led by Ho Chi Minh fought a prolonged war against France until the colonial power admitted defeat after a searing defeat by the nationalists at Dien Bien Phu in March 1954. This led to the partition of the territory between a communist North and a western-leaning South Vietnam.

A second wave of decolonization took place between 1956 and 1968, notably in Africa, where thirty-seven British, French, Belgian and Spanish colonies received their independence, with the process beginning in the French North African colonies of Morocco and Tunisia (1956) and the British West African territory of Ghana (1957). Where possible, the colonial authorities handed over to nationalist parties to which they had ceded a level of self-rule beforehand, but labour unrest, political dissent and violence broke out in those colonies where it was deemed the European authorities were moving too slowly. In 1952 the Mau Mau uprising, centred on Kenya's ethnic Kikuyu, led to the deaths of over 20,000 civilians, insurgents and British security forces, before the revolt was put down. Bloodier still was the Algerian War of Independence which erupted in 1954; the French largely defeated the nationalists by 1959, but a counter-revolution by French settlers, spear-headed by their military wing, the OAS, re-ignited the conflict, which only ended with the granting of independence in 1962, after 153,000 had died.

The new nations inherited the boundaries which their colonial administrators had sketched out for them. In many cases (and especially in Africa) these were straight lines which had little connection either with the topography of the land or the ethnographic composition of local peoples, who now found themselves split by apparently arbitrary decisions made in far-away metropolitan capitals. States were left with a complex mix of populations which often meant democracy degenerated into a squabbling cabal of parties representing ethnic and tribal interests. There seemed, though, little choice but to accept the status quo. The alternative would have been prolonged civil wars (such as erupted in Nigeria, when the state of Biafra tried to break away in 1967–70) or even international conflicts. To avoid this the Organization of African Unity (OAU), set up in 1963 to provide a collective voice for the newly independent countries, made clear in its founding acts that the boundaries of the continent were to be immutable.

There were a few unfinished acts of decolonization – Rhodesia's white settlers had made a Unilateral Declaration of Independence in 1965 and formal independence did not take place until 1979 (in the face of a concerted campaign by black African nationalist guerrillas), while a scattering of island territories remained under European rule even in the twenty-first century. In a few cases the post-independence cocktail of ethnicities proved too potent – Eritrea split off from Ethiopia in 1991, while South Sudan gained independence from the north in 2011, after a twenty-year civil war. In the main, however, the political boundaries of the post-colonial world have remained as they were in 1977.

DECOLONIZATION

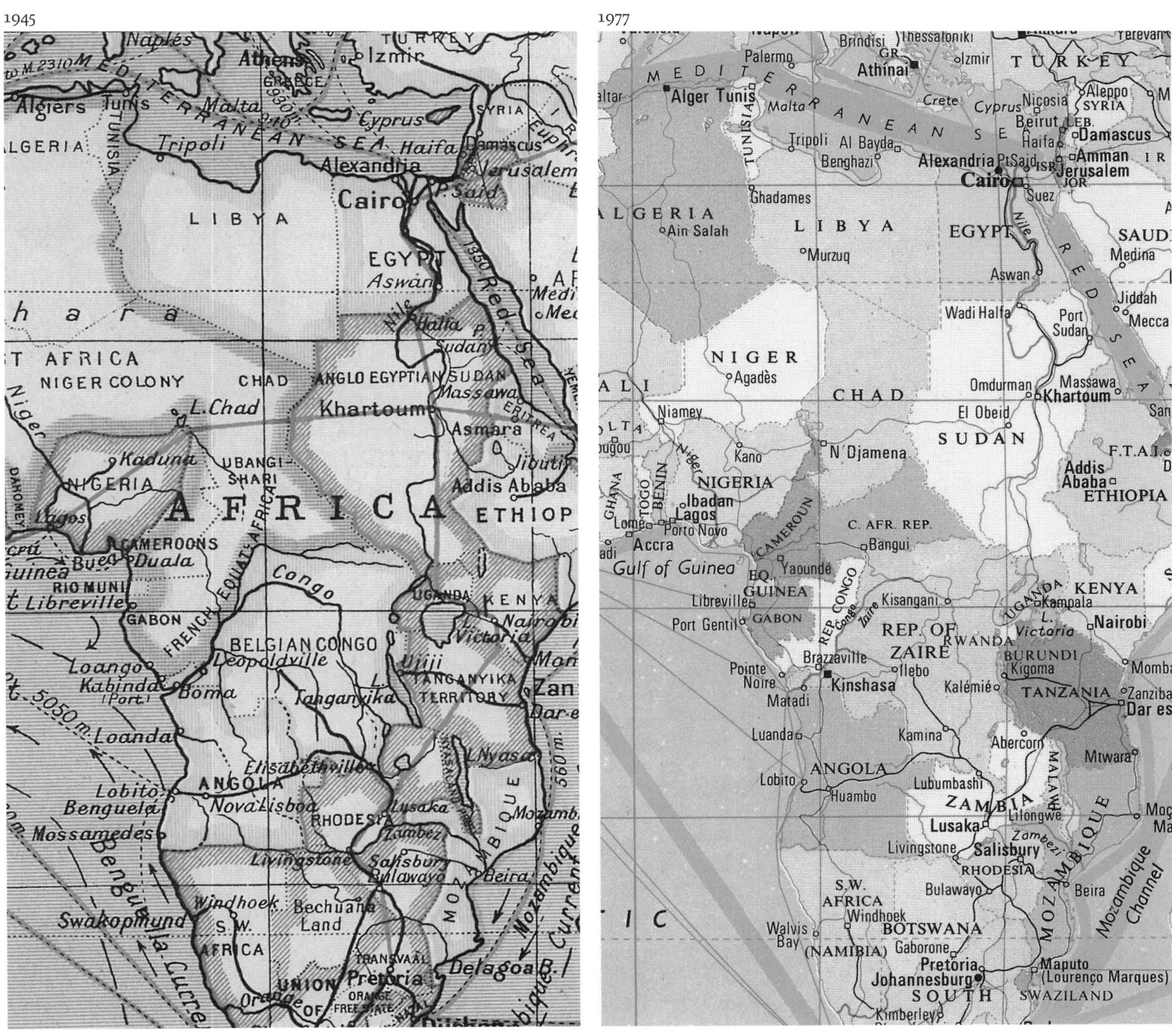

Comparative world political maps illustrating the changes of ownership of countries during the main period of decolonization. From *Meiklejohn's Intermediate School Atlas*, 1945, John Bartholomew & Son Ltd, Edinburgh and *The Times Concise Atlas of the World*, 1977. Collins Bartholomew.

UNITED STATES
Salt Lake City
Chicago
Detroit
Boston
New York
Philadelphia
Pittsburgh
Baltimore
Washington
Cincinnati
Denver
Kansas C.
St Louis
San Francisco
Los Angeles
Colorado
Oklahoma City
Mississippi
El Paso
New Orleans
Charleston
Newport News
Bermudas
Azores
Gulf Stream
California Current
Honolulu to Vancouver 2430 m.
Honolulu to S.F. 2100 m.
Honolulu to L.A. 2228 m.
Honolulu to Panama 4710 m.
San Francisco to Panama 3280 m.
York to Colon 1970
Colon to Liverpl 4530 m.
Guaymas
Galveston
Gulf of Mexico
Miami
Bahama
Havana
Cuba
MEXICO
Tropic of Cancer
Hawaiian Is.
WEST INDIES
ATLANTIC
OCEAN
Mexico
Vera Cruz
Jamaica
BR. HONDURAS
St Thomas
Hispaniola
Puerto Rico
Caribbean Sea
Acapulco
GUATEMALA
HONDURAS
SALVADOR
NICARAGUA
CENTRAL AMERICA
COSTA RICA
Colon
Panama
Balboa
Curaçao
La Guayra
Caracas
VENEZUELA
Barbados
Trinidad
North Equatorial Current
Georgetown
Paramaribo
Cayenne
GUIANA
New York to C. Town 6800 m.
Madeira
C. Verde Is.
Equatorial Counter Current
Equator
Christmas I.
Fanning I.
Jarvis I.
Malden I.
Starbuck I.
Bogota
COLOMBIA
Quito
ECUADOR
Galapagos
Guayaquil
Amazon
Manáos
Para
Natal
C. S. Roque
Pernambuco
South Equatorial Current
Marquesas Is.
Caroline I.
Manihiki Is.
Auckland to Panama 6590 m.
Peru Current
Lima
Callao
PERU
BRAZIL
Bahia
SOUTH AMERICA
BOLIVIA
La Paz
Sucre
Mollendo
Iquique
Society Is.
Cook Is.
Tuamotu or Low Arch.
Tropic of Capricorn
Austral Is.
Pitcairn I.
Ducie I.
Easter I.
Antofagasta
Caldera
PARAGUAY
Parana
Asuncion
S. Paulo
Rio de Janeiro
Santos
Buenos Aires to Liverpool
Brazil Current
URUGUAY
Valparaiso
Santiago
Juan Fernandez
Rosario
Buenos Aires
Montevideo
Concepcion
ARGENTINA
Rio de la Plata
Milwaukee
Detroit
Buffalo
Boston
Omaha
Chicago
Cleveland
Paterson
Newark
New York
Salt Lake City
UNITED STATES OF AMERICA
Denver
Kansas City
Indianapolis
Pittsbg.
Baltimore
Philadelphia
Washington
Norfolk
St. Louis
Cincinnati
San Francisco
San Jose
Azores (Port.)
NORTH ATLANTIC OCEAN
Oklahoma City
Memphis
Atlanta
Birmingham
Los Angeles
San Bernadino
San Diego
Ft. Worth
Dallas
PACIFIC OCEAN
El Paso
San Antonio
Jacksonville
Bermuda (UK)
R. Grande
Houston
New Orleans
Tampa
MEXICO
Monterrey
GULF OF MEXICO
Miami
The Bahamas
La Habana
Tampico
CUBA
Hawaiian Is. (USA)
Hawaii
Guadalajara
Mexico City
Veracruz
Hispaniola
HAITI
DOM. REP.
Port au Prince
Santo Domingo
San Juan
PUERTO RICO
JAMAICA
Kingston
Belmopan
Acapulco
Pto. Barrios
Leeward Is.
Cape Verde Is.
GUAT.
HOND.
Guatemala
San Salvador
Tegucigalpa
NIC.
Managua
COSTA RICA
San Jose
PANAMA
CARIBBEAN SEA
Barbados
Windward Is.
Barranquilla
Maracaibo
La Guaira
Caracas
Colón
Panama
Trinidad
VENEZUELA
Clipperton I. (Fr.)
Georgetown
Paramaribo
Cayenne
GUYANA
SUR.
FR. GUI.
Medellín
Bogotá
Orinoco
Buenaventura
COLOMBIA
Equator
Islas Galápagos (Ecuador)
Quito
ECUAD.
Guayaquil
Manaus
Amazon
Belém
Sao Luiz
Fortaleza
Fernando de Noronha (Braz.)
BRAZIL
Marquesas (Fr.)
PERU
Lima
Pto. Velho
Recife
Callao
Cusco
Salvador
Society Is. (Fr.)
Tahiti
Tuamotu (Fr.)
Cook Is.
La Paz
BOLIVIA
Brasília
Mollendo
Arica
Sucre
Iquique
Belo Horizonte
Trinidade (Braz.)
PARAGUAY
Tropic of Capricorn
São Paulo
Rio de Janeiro
SOUTH
Santos
Antofagasta
Pitcairn I. (UK)
Asuncion
Sala y Gomez (Chile)
Easter I.
Tucumán
Parana
Coquimbo
Córdoba
Pôrto Alegre
Juan Fernandez (Chile)
Valparaíso
Rosario
URUG.
Santiago
Buenos Aires
Montevideo

Madrid
Lisbon
SPAIN
Rome
Naples
Athens
Istanbul
Ankara
TURKEY
Black Sea
Izmir
Algiers
Tunis
Malta
Tripoli
ALGERIA
MOROCCO
Marrakesh
Casablanca
Gibraltar
Fez
Cyprus
Haifa
Damascus
Alexandria
Jerusalem
Cairo
Baghdad
Tehran
PERSIA
Basra
Bushire
Kabul
AFGHANISTAN
Kandahar
TIBET
Lhasa
Lahore
Delhi
Lucknow
Benares
INDIA
Calcutta
Karachi
Bombay
Goa
Madras
Colombo
CEYLON
Chungking
Kunming
CHINA
Hankow
Lanchow
The Gobi
Peking
Tientsin
Canton
Hong Kong
Hanoi
Hainan
Mandalay
BURMA
Rangoon
SIAM
Bangkok
Saigon
G. of Siam
Penang
Kuala Lumpur
Singapore
Sumatra
Borneo
NETHERLANDS
Batavia
Java
Christmas I.
Cocos Is.
LIBYA
EGYPT
Aswan
Sahara
MAURITANIA
RIO DE ORO
FRENCH WEST AFRICA
NIGER COLONY
Timbuktu
FRENCH SUDAN
Dakar
FRENCH GUINEA
IVORY COAST
LIBERIA
Monrovia
NIGERIA
Lagos
Kaduna
L. Chad
CHAD
ANGLO EGYPTIAN SUDAN
Khartoum
Massawa
Aden
Perim I.
Asmara
Jibuti
Addis Ababa
ETHIOPIA
Berbera
ARABIA
Medina
Mecca
Riyadh
Bahrein I.
OMAN
Muscat
AFRICA
Red Sea
Arabian Sea
Socotra
Laccadive Is.
Maldive Is.
Andaman Is.
Nicobar Is.
Bay of Bengal
Gulf of Guinea
Guinea Current
Equatorial Current
Ascension
St. Helena
Libreville
GABON
CAMEROONS
Duala
Congo
BELGIAN CONGO
Leopoldville
Boma
Kabinda
Loango
Loanda
Lobito
Benguela
ANGOLA
Nova Lisboa
Mossamedes
Benguela Current
Swakopmund
Windhoek
S.W. AFRICA
Bechuana Land
UGANDA
KENYA
Nairobi
Mombasa
Zanzibar
Dar-es-Salaam
TANGANYIKA TERRITORY
Ujiji
Tanganyika
L. Nyasa
Elisabethville
RHODESIA
Lusaka
Livingstone
Salisbury
Bulawayo
Beira
Mozambique
Mozambique Channel
Mozambique Current
MADAGASCAR
Antananarivo
Reunion
Mauritius
Seychelles
Amirantes
Chagos Archo.
TRANSVAAL
Pretoria
Delagoa B.
UNION OF SOUTH AFRICA
Bloemfontein
CAPE PROV.
Durban
Capetown
C. of Good Hope
Port Elizabeth
Agulhas Curnt.
Tristan da Cunha
South West Monsoon Drift
North East Monsoon Drift
Indian Counter Current
Equatorial Current
INDIAN OCEAN
Tropic of Capricorn
West Australian Current
N.W. Cape
Fremantle
Durban to Fremantle 4248m.
Colombo to Fremantle 3120m.
Aden to Fremantle 4930m.
C. Town to Colombo 4360m.
Madeira to C. Town 4860m.
Marseille
Barcelona
Valencia
Lisboa
PORT.
Roma
Napoli
Palermo
Brindisi
Tirane
Istanbul
Thessaloniki
Uskudar
Ankara
Batumi
Tbilisi
Yerevan
Baku
Krasnovodsk
Bukhara
Tashkent
Samarkand
Dushanbe
Ashkhabad
Mashhad
Tabriz
Athinai
Izmir
TURKEY
Crete
Nicosia
Cyprus
Aleppo
SYRIA
Mosul
Beirut
Damascus
Tehran
IRAN
Baghdad
Haifa
Amman
Jerusalem
IRAQ
Esfahan
Yazd
Alexandria
Cairo
Suez
Basra
Abadan
Al Kuwayt
Persian Gulf
Zahedan
Quetta
Herat
Kabul
AFGHANISTAN
Peshawar
Islamabad
Rawalpindi
Lahore
PAKISTAN
Bandar Abbas
Karachi
Ahmadabad
Baroda
Delhi
Agra
Kanpur
Lucknow
Allahabad
Kathmandu
NEPAL
Thimbu
Lhasa
TIBET
Yarkand
Patna
BANGLA DESH
Calcutta
Chittagong
INDIA
Nagpur
Bombay
Hyderabad
Vishakhapatnam
Panjim
Bangalore
Mangalore
Madras
Cochin
Colombo
SRI LANKA (CEYLON)
MALDIVES
Bay of Bengal
ARABIAN SEA
Madeira (Port.)
Gibraltar
Cádiz
Tanger
Rabat
Casablanca
Marrakech
MOROCCO
Canarias (Sp.)
Aaiun
Alger
Tunis
Malta
TUNISIA
MEDITERRANEAN SEA
Tripoli
Al Bayda
Benghazi
Ghadames
ALGERIA
Ain Salah
LIBYA
Murzuq
EGYPT
Aswan
SAUDI ARABIA
Medina
Riyadh
Jiddah
Mecca
UN. ARAB EMIRATES
Muscat
OMAN
RED SEA
Wadi Halfa
Port Sudan
Massawa
YEM.
SOUTH YEMEN
San'a
Madinat ash Sha'ab
Aden
Socotra (S.Y.)
MAURITANIA
Nouakchott
St. Louis
SENEGAL
Dakar
Banjul
Bissau
G.B.
GUINEA
Conakry
Freetown
S.L.
Monrovia
LIB.
Tombouctou
MALI
Bamako
VOLTA
Ouagadougou
Niamey
NIGER
Agadès
Kano
NIGERIA
Ibadan
Lagos
Porto Novo
BENIN
TOGO
GHANA
Lome
Accra
IVORY COAST
Abidjan
Takoradi
Gulf of Guinea
CHAD
N'Djamena
SUDAN
Omdurman
Khartoum
El Obeid
F.T.A.I.
Djibouti
Addis Ababa
ETHIOPIA
SOMALIA
Mogadishu
CAMEROUN
C. AFR. REP.
Bangui
Yaounde
EQ. GUINEA
Libreville
GABON
Port Gentil
REP. CONGO
Brazzaville
Pointe Noire
Kinshasa
Matadi
Kisangani
REP. OF ZAIRE
Ilebo
Kalémié
UGANDA
Kampala
L. Victoria
KENYA
Nairobi
RWANDA
BURUNDI
Kigoma
TANZANIA
Zanzibar
Dar es Salaam
Mombasa
Mtwara
Ascension (UK)
Luanda
Lobito
Huambo
ANGOLA
Kamina
Lubumbashi
ZAMBIA
Lusaka
Abercorn
MALAWI
Lilongwe
Livingstone
Salisbury
RHODESIA
Bulawayo
Beira
MOZAMBIQUE
Moçambique
Majunga
Mozambique Channel
Madagascar
MALAGASY REP.
Tamatave
Tananarive
Mauritius
Réunion (Fr.)
St. Helena (UK)
ATLANTIC OCEAN
S.W. AFRICA (NAMIBIA)
Windhoek
Walvis Bay
BOTSWANA
Gaborone
Pretoria
Johannesburg
Maputo (Lourenço Marques)
SWAZILAND
SOUTH AFRICA
Kimberley
Bloemfontein
LESOTHO
Durban
Cape Town
East London
Port Elizabeth
SEYCHELLES
INDIAN OCEAN
Tropic of Capricorn
Equator
Chiuch'üan
Peking
Tienching
Shihchiachuang
Taiyüan
CHINA
Lanchow
Hsian
Nanch'ing
Wuhan
Ch'engtu
Ch'ungch'ing
Ch'angsha
Kunming
Canton
Hanoi
Haiphong
Hainan
BURMA
Mandalay
Chiengmai
Vientiane
LAOS
Rangoon
THAILAND
Hue
VIETNAM
Bangkok
CAMBODIA
Phnom Penh
Saigon
G. of Thailand
SOUTH CHINA SEA
Georgetown
Kuala Lumpur
MALAYSIA
Singapore
Kuching
Borneo
Balikpapan
Sumatra
Padang
Palembang
JAVA SEA
Jakarta
Bandung
Surabaya
Java
Cocos (Aust.)
Perth
Fremantle

World Air Routes

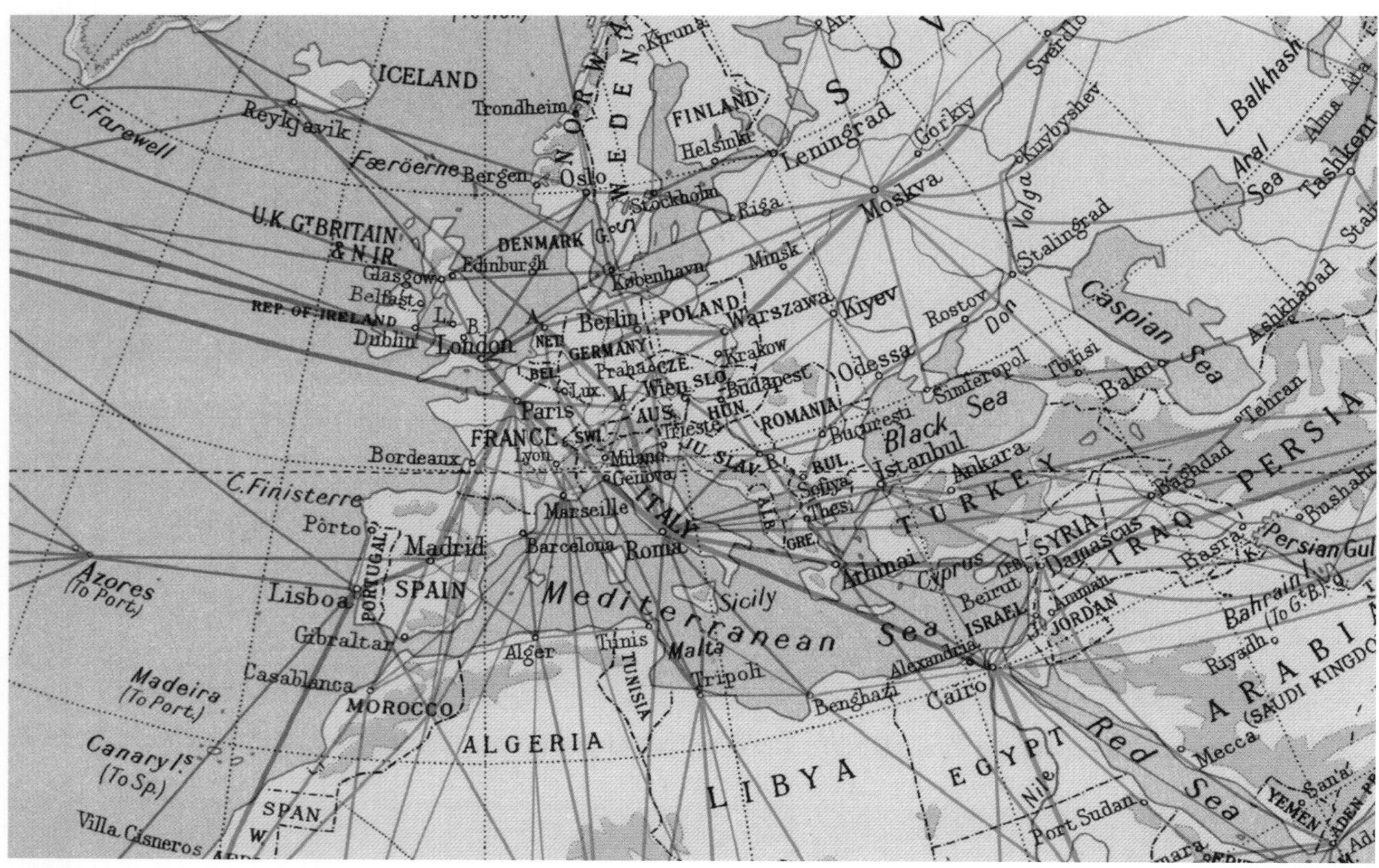

World Air Routes, from *The Times Atlas of the World Mid-Century Edition*, 1958. Collins Bartholomew.

AIR TRAVEL

The *World Air Routes* map, published in 1958, records the dawn of the jet age and passenger jet travel. The world's first jet airliner, the Comet, began a London to Johannesburg service in 1952. The first transatlantic jet service (London to New York) followed in 1958, heralding a golden age of air travel.

Passenger air travel grew quickly after the Second World War, as lighter aircraft were developed for mass production, and as thousands of trained pilots finished military service and entered commercial service. In the 1950s, new turbo-propeller aircraft were superseded by transatlantic jets, as demand for air travel took off. The *World Air Routes* map from this time reflects the mass movement of air travellers, air cargo, and airmail – especially between Europe and North America.

British technology initially led the world in aircraft engineering with the development of the De Havilland Comet 1, the first jet airliner. The Comet could carry thirty-six passengers, had a top speed of 810 km/hr (503 miles/hr) and a range of 2,400 km (1,500 miles). In 1952, BOAC (British Overseas Airways Corporation) celebrated the Comet's maiden flight, which went some 11,200 km (7,000 miles) from London to Johannesburg, with stops at Rome, Beirut, Khartoum, Entebbe, and Livingstone. The next year BOAC started a jet service to Tokyo, reducing the London–Tokyo flight time from eighty-six to thirty-three hours. However, after a series of crashes and mid-air explosions, the Comet was grounded, and airlines began purchasing American aircraft, including the Boeing 707 and the Douglas DC-8.

The *World Air Routes* map shows that the 'Principal World Air Routes' lines (the thickest red lines) converge on the cities of London, Paris, and Rome in Europe, and New York and Chicago in North America. The air routes south of Europe still connected Europeans to their colonial possessions in Africa. In the 1950s, exotic places such as French West Africa and the Belgian Congo beckoned the adventurous traveller. The London to Sydney 'Kangaroo' route took about four days, travelling via six cities, including Cairo and Singapore. Limited range required planes to island-hop across the oceans, using the Azores and Newfoundland to get over the Atlantic. The North Pacific Ocean map (lower centre) traces the air routes over the ocean, revealing the importance of Hawaii, Wake Island, and Guam to planes flying over the vast Pacific.

As air traffic surged, so did the need to keep track of aeroplanes. Americans increasingly favoured the airport over the railway station. Air traffic safety became a major issue when two passenger planes crashed over the Grand Canyon in 1956, killing 128 people. Such mid-air collisions brought about passage of the Federal Aviation Act of 1958. This act created the Federal Aviation Agency (FAA), charged with developing and maintaining air navigation and air traffic control.

New and expanded airports transformed the cultural geography of cities as more people took to the skies. During the Second World War, the British Royal Air Force (RAF) took over land in and around Heath Row, an ancient agricultural village. Heathrow became a civil airport in 1946, recording 63,000 passengers in its first year, but seeing passenger numbers soar to 796,000 by 1951. In fact, Heathrow was growing so fast that a new London airport was opened in 1958 at Gatwick. All over the world, airports became the rapidly growing cathedrals of the jet age. The events of the 1950s set in motion the prospect of routine air travel, making today's airports the primary gateways to the world.

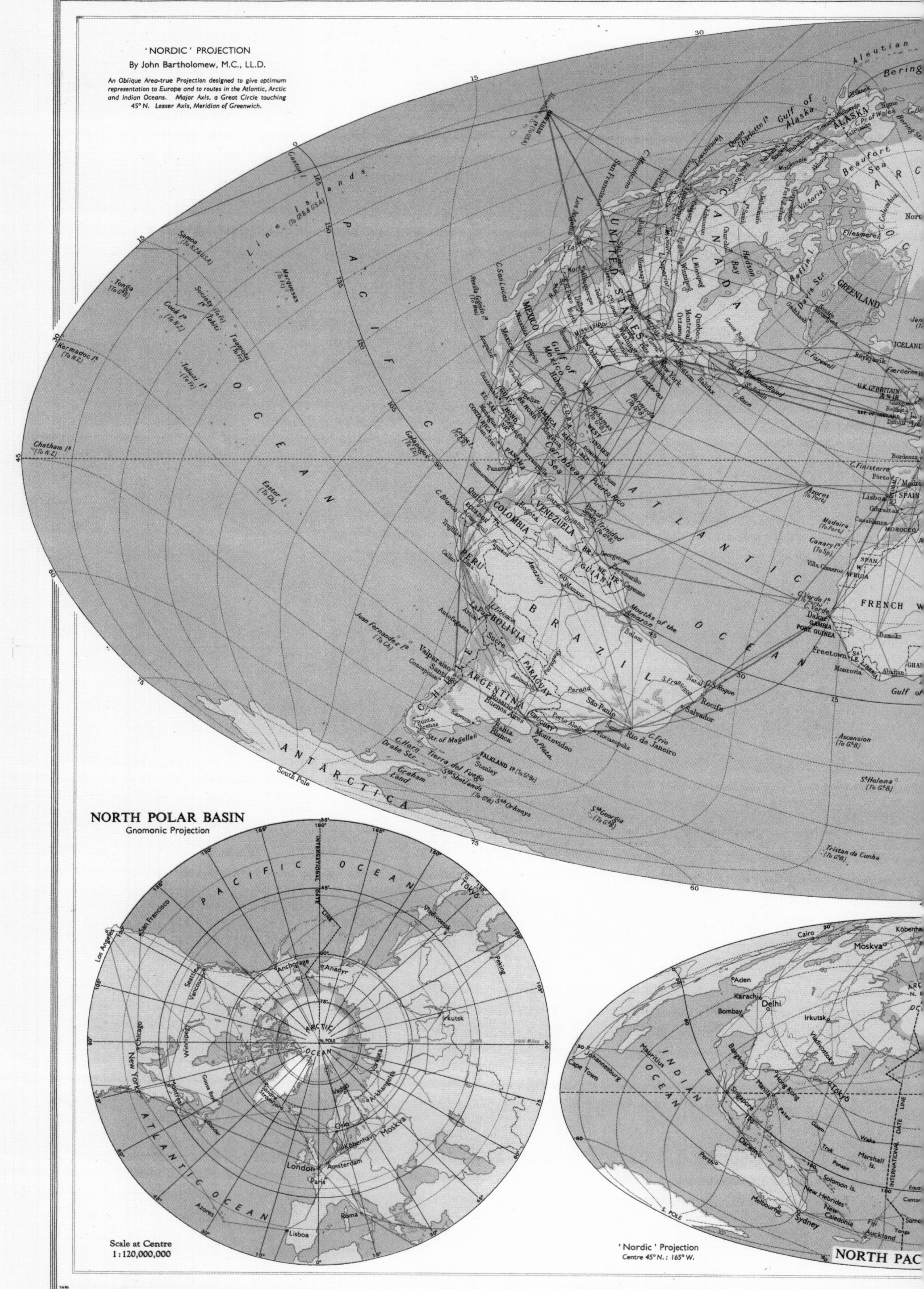
'NORDIC' PROJECTION
By John Bartholomew, M.C., LL.D.
An Oblique Area-true Projection designed to give optimum representation to Europe and to routes in the Atlantic, Arctic and Indian Oceans. Major Axis, a Great Circle touching 45° N. Lesser Axis, Meridian of Greenwich.
PACIFIC OCEAN
ATLANTIC OCEAN
UNITED STATES
CANADA
MEXICO
BRAZIL
ARGENTINA
ANTARCTICA
GREENLAND
NORTH POLAR BASIN
Gnomonic Projection
PACIFIC OCEAN
ARCTIC OCEAN
ATLANTIC OCEAN
Scale at Centre
1:120,000,000
INDIAN OCEAN
'Nordic' Projection
Centre 45° N.: 165° W.
NORTH PAC

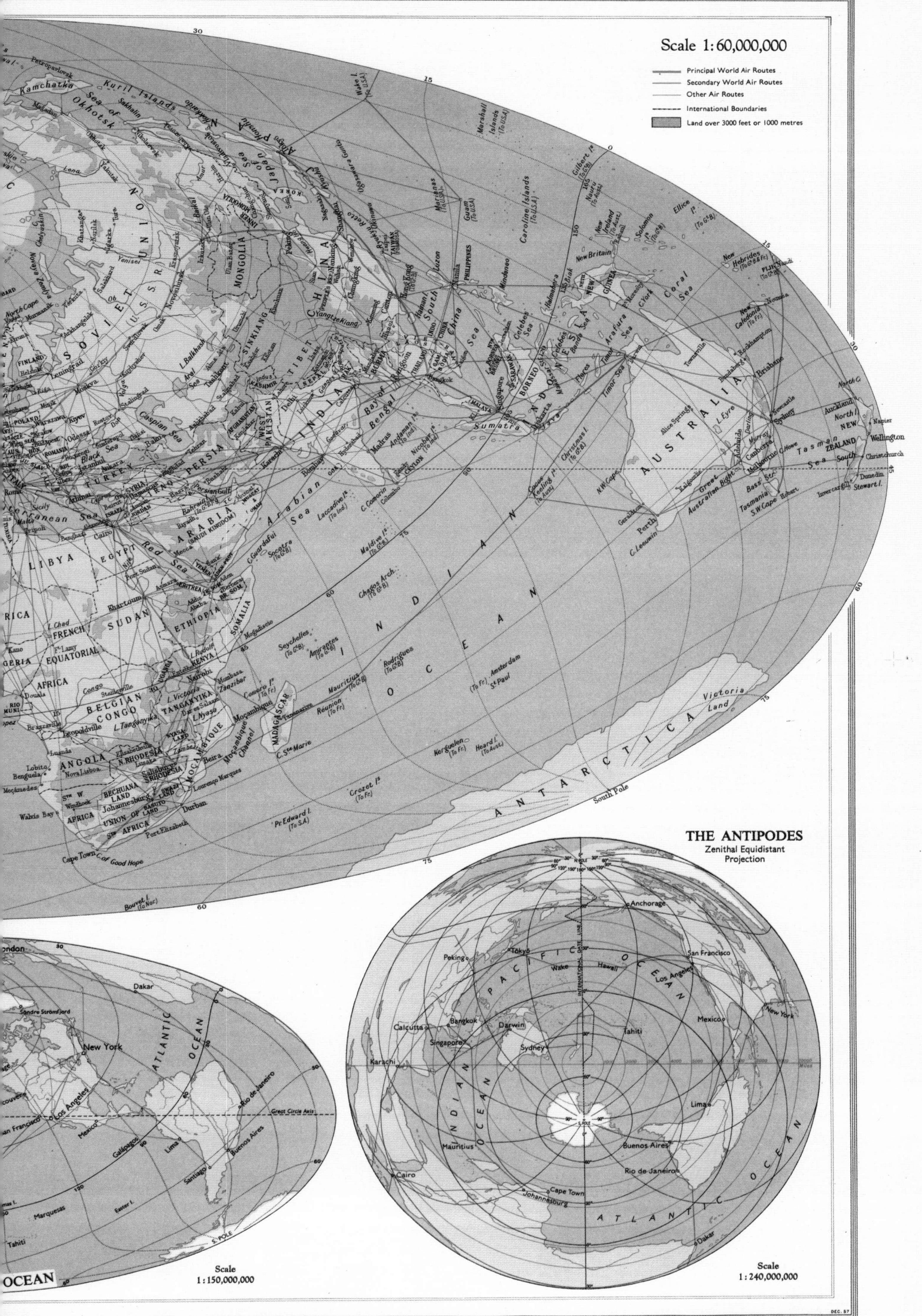
Scale 1:60,000,000
Principal World Air Routes
Secondary World Air Routes
Other Air Routes
International Boundaries
Land over 3000 feet or 1000 metres
INDIAN OCEAN
AUSTRALIA
ANTARCTICA
THE ANTIPODES
Zenithal Equidistant Projection
Scale
1:150,000,000
Scale
1:240,000,000
OCEAN
DEC. 57

MAPPING THE MOON

The US–USSR space race started in 1957, when the Soviet Union launched Sputnik I, the world's first artificial satellite. Americans saw Soviet spaceflight as a threat. The National Aeronautics and Space Administration (NASA), created in 1958, achieved its first lunar flyby in 1959, responding to a Soviet moon mission. At the same time, work began at the United States Geological Survey (USGS) to compile a map of lunar landforms, which would result in the first *Generalized Photogeologic Map of the Moon.*

Soviet pride rose on 12 April 1961, when Vostok, 'the world's first spaceship,' orbited the planet, carrying cosmonaut Yuri Gagarin. Weeks later on 5 May, American patriotism soared when astronaut Alan Shepard achieved orbit in the Freedom 7 capsule. On 21 May 1961, President John F. Kennedy told Americans, 'I believe that this nation should commit itself to achieving the goal, before this decade is out, of landing a man on the moon and returning him safely to the earth.' Space exploration captured the American imagination, and on 20 July 1969, Apollo 11 landed the first two men on the moon, Neil Armstrong and Edwin 'Buzz' Aldrin. However, getting to the moon required an intense mapping effort to identify potential landing sites.

The first map of the moon produced by the US Government after Kennedy's speech was published later in 1961 by the USGS and was titled, *Generalized Photogeologic Map of the Moon.* This was the first map to illustrate the landform and surface history of the moon. The photogeologic map was based on photographic plates from large telescopes, such as that at the McDonald Observatory in Texas.

The first drawings of the moon based on telescopic observation were made in 1609 and are attributed to Englishman Thomas Harriot and Italian Galileo Galilei. The US effort to make a modern map of the moon started in 1959 and was led by Dr Arnold Mason of the Military Geology Branch at the USGS. The Sputnik launch sparked Mason's interest in lunar geology, and he started analyzing the moon's terrain to estimate site suitability for such things as spacecraft landings and base construction. The photogeologic moon map, a product of his research, was published in the *Engineer Special Study of the Surface of the Moon.*

This colourful map classifies the moon's nearside surface into three categories: Pre-Maria (orange); Maria (yellow); and Post-Maria (green). Pre-Maria regions, the oldest rocks, consist of cratered highlands formed by the accumulation of thousands of meteorites crashing into the surface. The Maria (Latin for 'seas') encompasses vast lava fields, which cover lowlands and appear as dark plains. These layers of basalt were deposited when the moon experienced most of its volcanic activity some three billion years ago. The Post-Maria surface mostly identifies meteoric impacts occurring after the extensive lava flows; there are also some features of relatively recent volcanic origin, perhaps one billion years old.

The map represents a key moment for US space exploration and the Apollo program. Apollo 11 landed on Mare Tranquillitatis (Sea of Tranquility), a relatively level region (yellow area) which is labelled near the centre of the map. When Neil Armstrong took his first steps on the moon an estimated 530 million people were watching his televised image. In 1969, Apollo 11 accomplished a national goal set by a visionary US President.

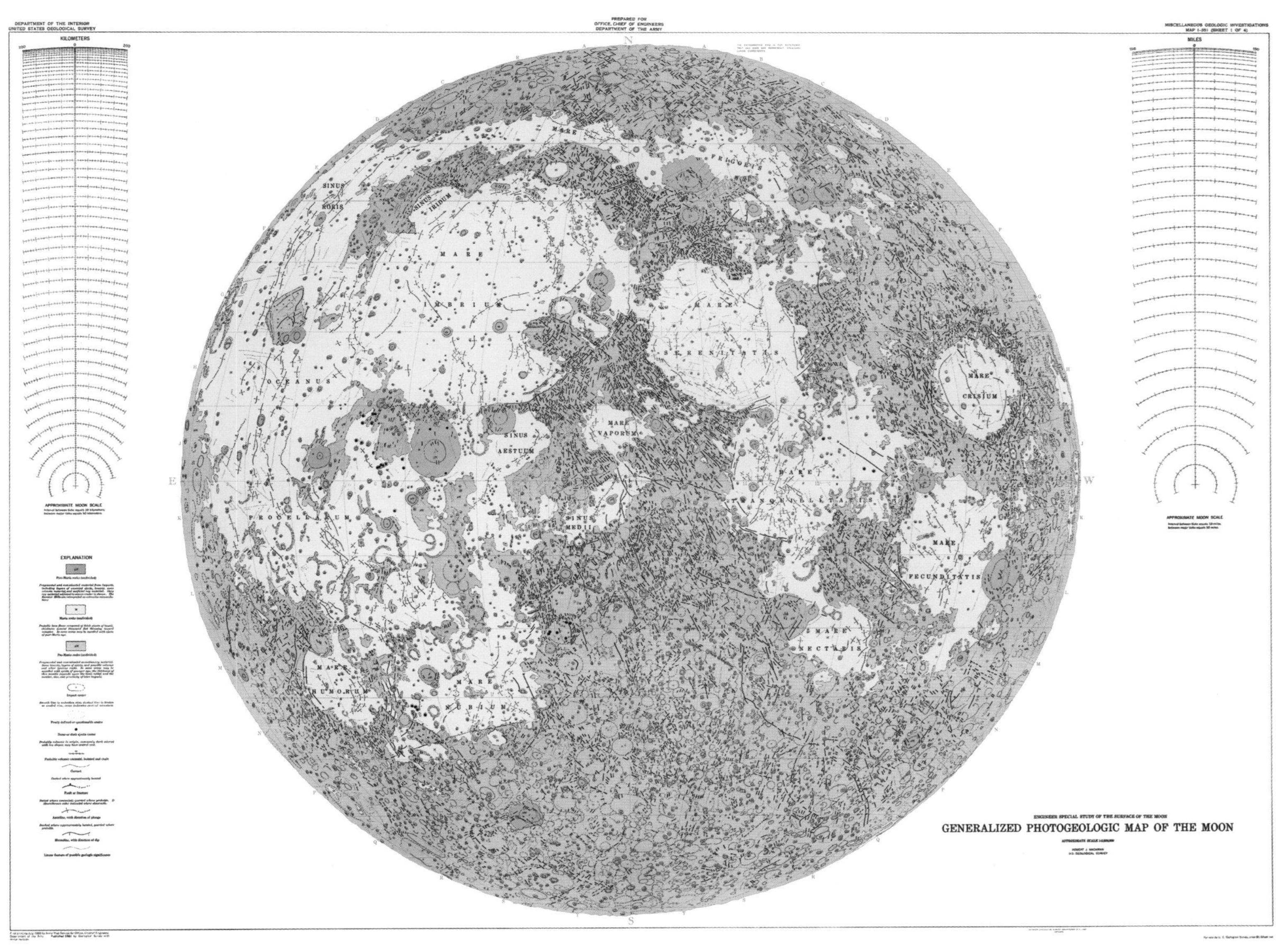

Generalized Photogeologic Map of the Moon, 1961 – the first official US map of the moon to be published after President Kennedy vowed to land a man on the moon by the end of the 1960s. Lunar and Planetary Institute, Houston, TX, USA.

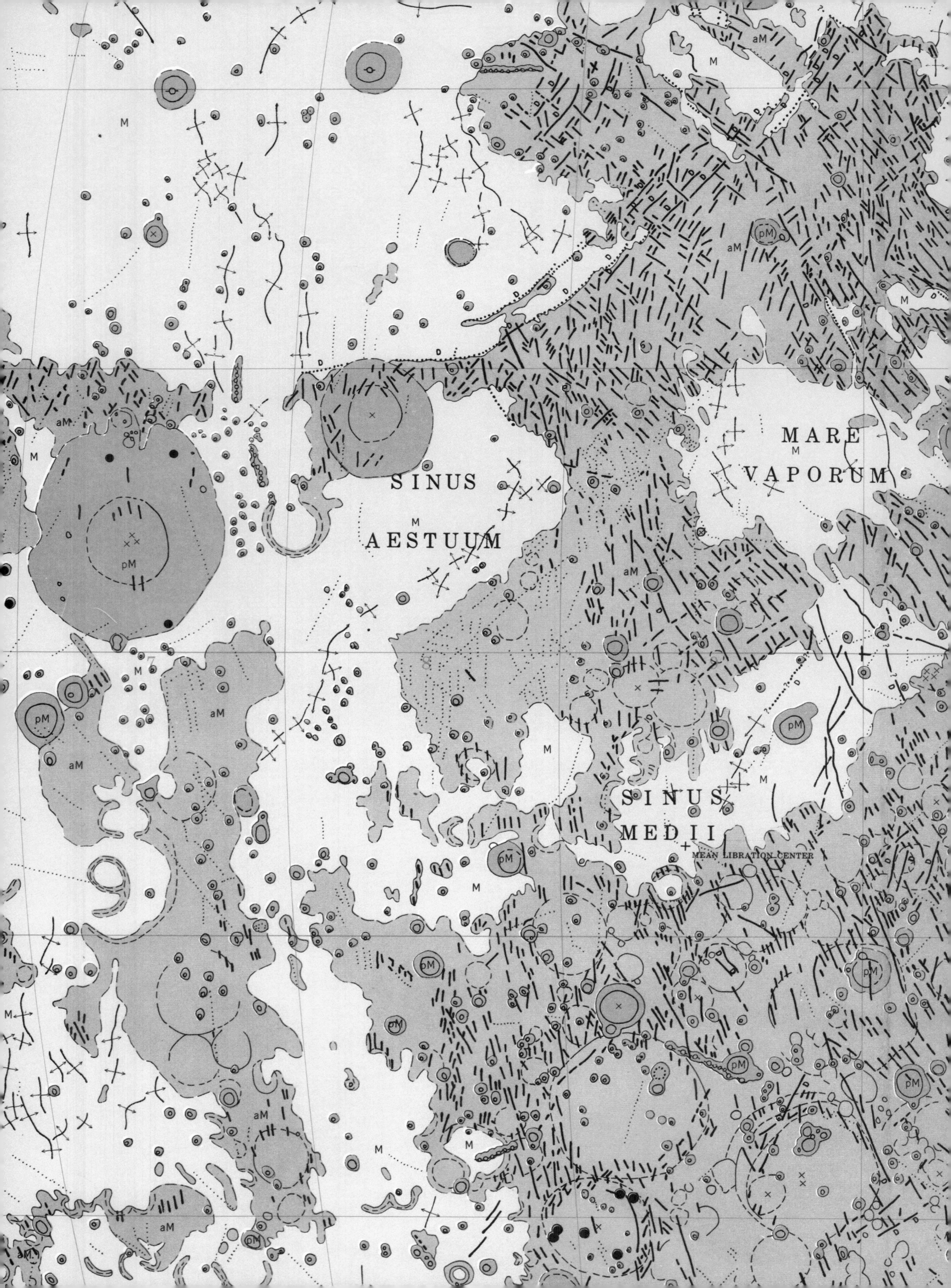

MARE
VAPORUM
SINUS
AESTUUM
SINUS
MEDII
MEAN LIBRATION CENTER

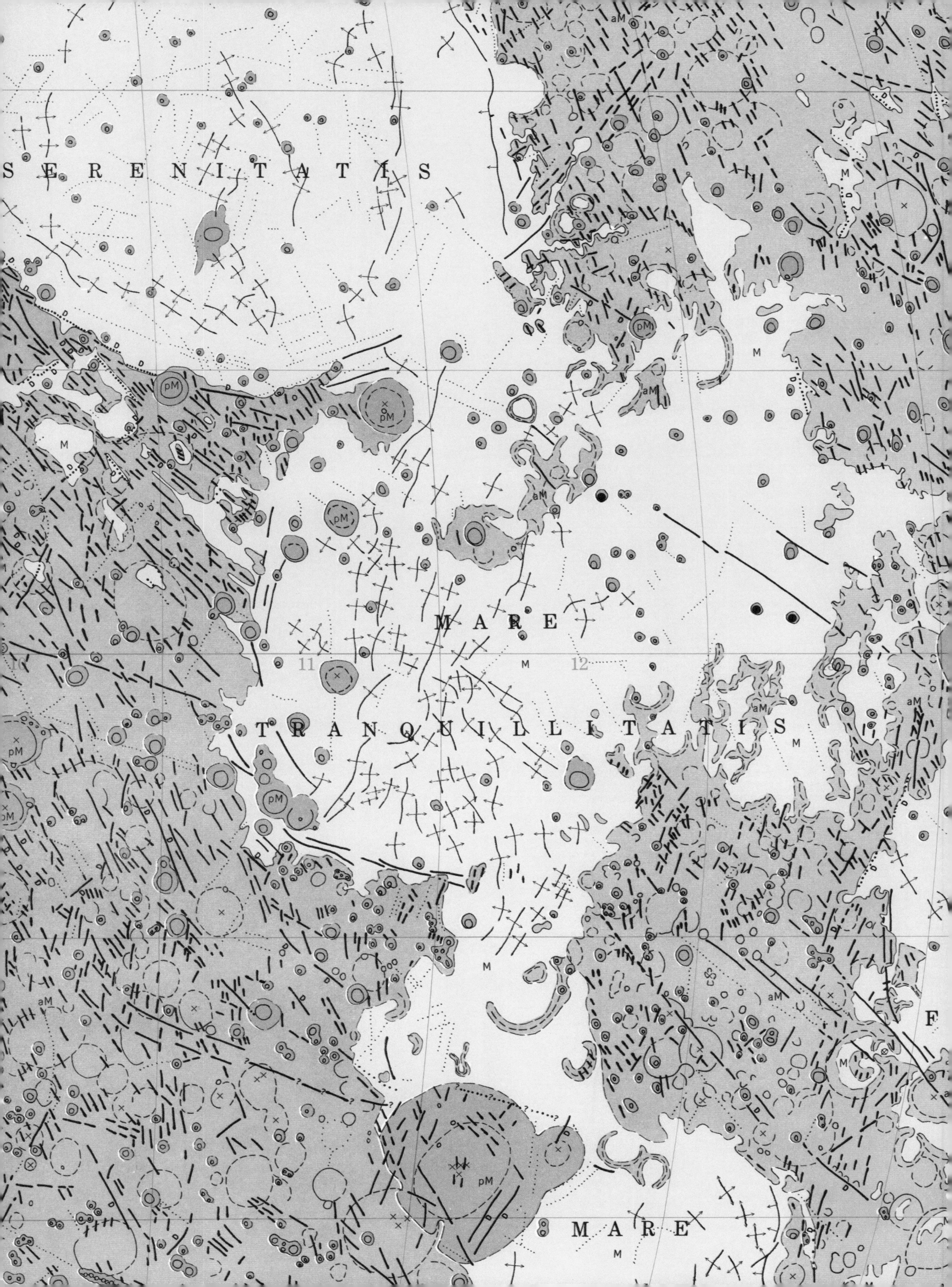

SERENITATIS
MARE
TRANQUILLITATIS
MARE
F
10
11
12
M
pM
aM

CUBAN MISSILE CRISIS

The geographical location of Cuba, just ninety miles from the coastline of Florida, became a hot geopolitical issue in October 1962 as maps appeared in American newspapers showing how strategic nuclear missiles, which the Soviet Union had been found to have deployed on the island, could target almost all parts of the United States.

The thirteen days of the Cuban Missile Crisis were the closest the world came to nuclear conflict during the Cold War. Tensions were already high between the United States and the USSR. The building of the Berlin Wall in August 1961 had cemented Soviet control over the German capital and threatened to isolate the Western-controlled sectors of the city economically and politically (see page 208). And the Cuban Revolution which overthrew the pro-American dictator Fulgencio Batista in January 1959 had installed Fidel Castro, a Marxist firebrand, as president of a communist, Soviet-inspired regime situated dangerously close, as the Americans saw it, to their territory.

Although the US military and political establishment had made much play in the late 1950s of the missile gap – the perceived Soviet advantage in strategic nuclear missiles – in truth, the nuclear balance was very much in Washington's favour (at 27,300 warheads to Moscow's 3,300). Even so, Soviet President Nikita Khrushchev considered nuclear weapons to be a comparatively cheap way of countering the United States's vastly superior military budgets. So, following a decision in May 1962, the Soviet Union decided to deploy nuclear missiles to Cuba. It was a secret operation, intended to ship to the island twenty-four R-12 missiles, each with a 1,100-mile range and sixteen R-14 rockets, which could reach targets almost twice that distance. Once the missiles were fully operational, Khruschev would have the equivalent of an unsinkable aircraft carrier which could strike at any part of the United States.

The Soviets had repeatedly assured the Americans that they had no intention of deploying offensive weapons on Cuba, but a U-2 spy plane took photographs on 14 October showing indisputably that they had been lying and that missile installations were already being built there. On 16 October, the CIA passed this intelligence to President John F. Kennedy. He marked the map used during that briefing with the annotation 'missile sites', a chilling testament to the impending crisis. After intensive debate, Excomm, the president's crisis committee, decided to ignore the hawks who advocated a pre-emptive strike against Cuba, which it was feared would provoke Khrushchev into seizing West Berlin. Instead it was decided to impose a quarantine, or naval blockade, on Cuba to prevent any military supplies getting through, and in the hope that this would prevent the missiles from being fully armed. On Monday 22 October Kennedy went public, calling on Khrushchev to halt and eliminate this clandestine, reckless and provocative threat to world peace.

American newspapers were suddenly full of maps graphically showing how the Soviet missiles could hit targets in any part of the United States (and Mexico for that matter), although the contention of the map shown here that Seattle would be in range is an exaggeration. The blockade was successful and prevented the longer-range warheads from arriving, and only a limited number of tactical missiles with ranges up to fifty miles were ever actually operational. For a while, though, brinkmanship on both sides looked likely to drag the world into a nuclear war. The Soviets raised their state of military readiness on Cuba, and American ships aggressively enforced the quarantine zone. Khrushchev, however, seems to have been having second thoughts, in part out of alarm at the bellicose statements being made by Castro, who was urging a first-strike against the United States. A Soviet proposal on 27 October that the Americans withdraw their Jupiter nuclear missiles from Turkey (seen as the USSR's analogy to Cuba) was initially received negatively, but the next day Kennedy put forward the idea that if Khrushchev agreed to pull out the missiles from Cuba, then four or five months later the Jupiters – which were in any case obsolete – would also be removed.

Khrushchev quickly agreed, believing that an American attack on Cuba was imminent, and the missiles were withdrawn. Inspections to ensure that the withdrawal had in fact been carried out were made by US jets which flew low over Soviet ships and witnessed the ships' crews pulling back tarpaulins to reveal their cargoes. The two superpowers had pulled away from the brink, but the maps the crisis had generated showed how close the world had been to catastrophe.

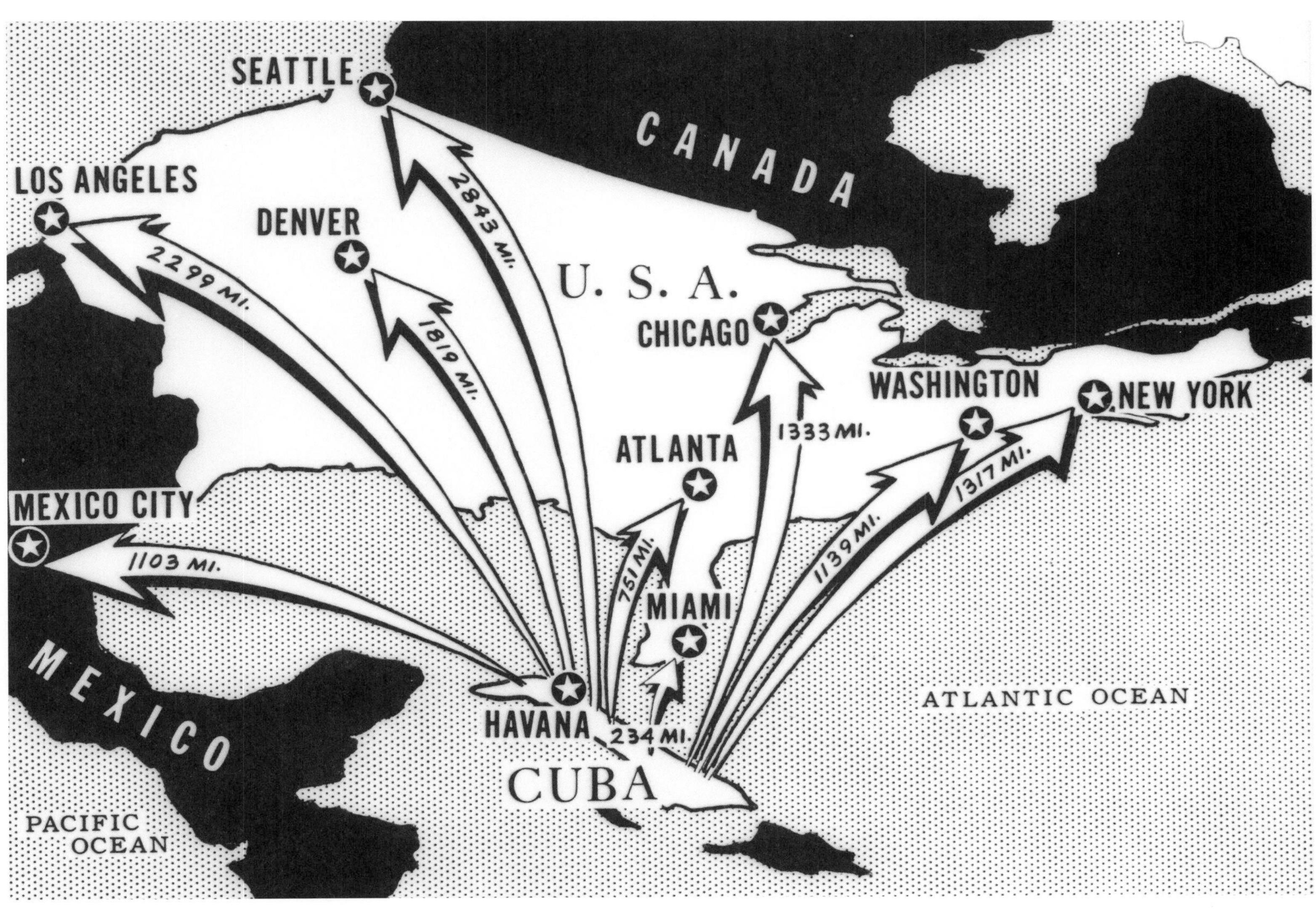

Newspaper map illustrating the threat posed by nuclear weapons in Cuba, and a CIA briefing map of Cuba annotated by President Kennedy during the missile crisis. Corbis; John F. Kennedy Presidential Library and Museum, Boston, USA.

CUBA
GULF OF MEXICO
CARIBBEAN SEA
missle sites
HAVANA
Guanabacoa
Santa Cruz del Norte
Jaruco
MATANZAS
Varadero
Cárdenas
CAYO CRUZ DEL PADRE
CAYO BLANQUIZAL
CAY SAL
Corralillo
Isabela
Quemado de Güines
Sagua la Grande
Encrucijada
Esperanza
SANTA CLARA
Marti
Máximo Gómez
Jovellanos
Unión de Reyes
Perico
Colón
Los Arabos
Pedro Betancourt
Jagüey Grande
Aguada de Pasajeros
Rodas
Lajas
Cruces
Palmira
Cienfuegos
Trinidad
Mariel
Bauta
Guanajay
Artemisa
San José de las Lajas
Madruga
Güines
San Nicolás
Güira de Melena
Surgidero de Batabanó
Bahía Honda
San Cristóbal
Candelaria
Los Palacios
Cantón
La Esperanza
Santa Lucía
Viñales
Consolación del Sur
PINAR DEL RIO
Los Arroyos
Mantua
Guane
La Fe
ARCHIPIELAGO DE LOS COLORADOS
GOLFO DE GUANAHACABIBES
CABO SAN ANTONIO
CABO CORRIENTES
GOLFO DE BATABANÓ
ENSENADA DE LA BROA
Ciénaga de Zapata
MATANZAS
BANCO DE LOS JARDINES
Nueva Gerona
Santa Barbara
Santa Fe
ISLE OF PINES
(LA HABANA)
CABO PEPE
GRAND CAYMAN (Jamaica)
LESSER CAYMANS (Jamaica)
85
84
83
82
81
80
23
22
21
20
Provincia boundary
National capital
CAMAGÜEY Provincia capital
2000 Spot height (in feet)
Railroad
Road
Swamp or marsh
Reef
0 10 20 40 60 80 100
Statute Miles
0 10 20 40 60 80 100
Kilometers

29053.1 6-60

79
78
77
76
75
74
ANDROS ISLAND
GREAT EXUMA
RUM CAY
LONG ISLAND
B A H A M A
(U. K.)
I S L A N D S
JUMENTOS CAYS
CROOKED ISLAND
ACKLINS ISLAND
23
22
21
20
CAYO SANTA MARIA
CAYO COCO
BAHÍA DE BUENA VISTA
CAYO ROMANO
BAHÍA JIGÜEY
Yaguajay
Morón
Jatibonico
Ciego de Ávila
Esmeralda
Júcaro
Florida
PUNTA MATERNILLOS
Nuevitas
CAMAGÜEY
Río Saramaguacán
Río Durán
Río Altamira
Río Najasa
Marti
Guáimaro
Victoria de las Tunas
Puerto Padre
Gibara
Santa Lucia
CABO LUCRECIA
Banes
Holguin
Antilla
JARDINES DE LA REINA
Santa Cruz del Sur
Guayabal
Río Cauto
Río Salado
Mayarí
Sagua de Tánamo
PUNTA GUARICO
Baracoa
PUNTA MAISI
Manzanillo
Bayamo
Jiguani
Campechuela
Media Luna
Niquero
CABO CRUZ
Palma Soriano
San Luis
SANTIAGO DE CUBA
La Maya
Guantánamo
Caimanera
U.S. NAVAL BASE
GREAT INAGUA
W I N D W A R D   P A S S A G E

EARTH OBSERVATION

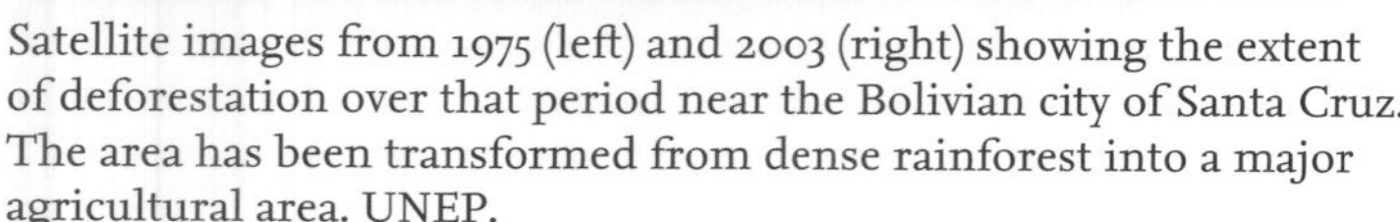

Satellite images from 1975 (left) and 2003 (right) showing the extent of deforestation over that period near the Bolivian city of Santa Cruz. The area has been transformed from dense rainforest into a major agricultural area. UNEP.

Since the first photographs looking down at the earth were taken from balloons in the mid-nineteenth century and from early aeroplanes a few decades later, the ability to observe the earth from above has been an invaluable tool for mapping and analysing the planet. The advent of satellites and the ability to capture images of the earth from space revolutionized this science. It greatly increased the potential to observe the earth over time and space and, with the development of new sensors, in new ways.

The earth is a dynamic and fragile place, and our awareness of the need to protect it has been steadily increasing. In order to direct these efforts, we need to observe and monitor change and we therefore need to gather information on as wide a scale, in as much detail, and as regularly as possible. Traditional means of collecting and recording information – ground survey, scientific measurements, terrestrial photography and mapping – are still as valid today as they have been for centuries. However, from the earliest days of flight the ability to observe and photograph the earth's surface from above has provided perhaps the greatest insight into how the planet looks, works and changes.

The value of aerial photography as an intelligence-gathering tool was proved during the First World War, and since then, as techniques and technologies have developed, air photos and photogrammetry – the science of taking detailed measurements from such images – have been widely used for creating detailed topographic maps and for many other purposes. But perhaps the most significant development in terms of capturing images of the earth from above has been the emergence of earth-observing satellites. The first satellite dedicated to this purpose was the Earth Resources Technology Satellite (ERTS), later to be known as Landsat 1, which was launched in 1972 and completely changed the way we view the earth and gather information about it. The Landsat programme continues to this day – Landsat 7 still operates and Landsat 8 was launched in 2013 – making it the longest-running programme for acquiring satellite imagery of the earth.

Satellite imagery, and the related science of satellite remote sensing – the acquisition, processing and interpretation of images captured by satellite-borne sensors – is now an invaluable tool in observing and monitoring the earth at a global level. Such satellites can carry sensors which capture data at different spectral wavelengths – 'seeing' things which would not be visible to the eye or to conventional photographic cameras. These data can be processed to allow the detailed interpretation of landscapes, vegetation, environmental phenomena, and meteorological and atmospheric conditions. Images derived from these data are now vital tools in such fields as agriculture, forestry, environmental monitoring, geology and the analysis of climate change. The level of detail discernible in such images has steadily increased – Landsat 1 images had resolutions of 80 m (262 feet), while the latest commercial satellites now capture images with resolutions of less than 1 m (3.3 feet).

One crucial aspect of remote sensing is that satellites regularly revisit the same point above the earth and so can gather time-sequence images of exactly the same area. Earth-observing satellites follow one of two types of orbit: geostationary or polar. Satellites in geostationary satellites (most commonly used for meteorological applications) effectively sit above the same point of the earth's surface, allowing the constant collection of images of the same area. Satellites in polar orbits travel around the earth in a north-south-north direction. As the earth rotates they progressively capture images of adjacent (or partially overlapping) areas. Such satellites typically revisit the same point every 16–26 days, but this period can be decreased dramatically if a satellite is able to capture oblique images – the French SPOT satellite can in fact revisit the same point every 1–4 days. Such capabilities allow enormous amounts of change data to be gathered, and for dramatic images to be produced, such as those of deforestation in South America shown here.

Satellites and sensors continue to be developed and the resolution of the images they capture continues to improve. Sensors recording very detailed environmental and atmospheric data will remain a critical tool in observing the earth and monitoring the way it is changing, and the images they provide will continue to form an emotive and compelling view of its beauty and vulnerabilities.

South Africa

From 1948, successive South African governments implemented a policy of apartheid – the exclusion of black South Africans from full economic, social and political rights. This reached its culmination with the establishment of independent 'Homelands' (or Bantustans) in the 1970s, a development displayed in this 1973 map which shows the boundaries of the ten Homelands in which millions of black South Africans were forced to live.

South Africa came into being in 1910, when the British colonies of Natal and the Cape were united with the Afrikaaner colonies of Transvaal and the Orange Free State. Right from the start the majority black African population and smaller numbers of Indians and 'coloured' or mixed-race people, were subject to the political dominance of the descendants of white European settlers.

The introduction of apartheid legislation began with the victory in the 1948 General Election of the predominantly Afrikaaner National Party. By this time blacks accounted for nearly 80 per cent of the population and the National Party pushed forward with a policy of formal separation of races, with measures such as the 1950 Population Registration Act. This established formal racial classifications and introduced identity cards which bore a person's race. The Group Areas Act of the same year divided the country up by racial areas, and determined in which districts each could live.

Even so, the growth of the black urban population began to concern the apartheid authorities, as it reached 3.4 million in 1960 and threatened to reach a position of demographic dominance in the cities. Pass laws were introduced to limit the movement of black workers (on whom the cities and factories depended) and 384,000 Africans were convicted of offences under these measures in 1962 alone.

International pressure on the apartheid regime grew, with economic and cultural boycotts beginning to bite and internal dissent spreading after the Sharpeville Massacre in 1960 in which sixty-nine Africans demonstrating against the Pass Laws were shot dead by police. The South African government pressed ahead with the most extreme form of apartheid, the political separation of black Africans, by creating 'Homelands' (Bantustans) to which self-governing powers, and ultimately independence, would be granted. The Bantu Authorities Act of 1951 had begun the development of Tribal, Regional and Territorial Authorities for blacks, often based on traditional tribal chieftainships. In 1959 this went one stage further with the Promotion of Bantu Self-Government Act which established eight (and later ten) 'national units' in which it was planned the vast majority of black South Africans would ultimately find themselves.

The map shows the boundaries of the Bantustans in 1973, just before the first, Transkei, was granted its notional independence in 1976, and shows in graphic form the apartheid ambition of separate developments for South Africa's racial groups. The ten Bantustans, Transkei, Ciskei, Venda, Bophuthatswana, Basotho Qwaqwa, Gazankulu, KwaZulu, Lebowa, South Ndebele and Swazi form a great, broken arc in the northeast and southeast of South Africa. Their fragmented nature – Bophuthatswana had no fewer than seven land-locked enclaves – rendered them politically weak and unlikely ever to assert themselves against the country which had spawned them. In reality, they played a part in depriving millions of blacks of South African citizenship.

In the event, only four of the Bantustans reached independence (Transkei in 1976, Bophutatswana in 1977, Venda in 1979 and Ciskei in 1981) with the other six achieving lesser degrees of autonomy between 1971 and 1981. Only around half of South Africa's black population lived in the Bantustans and, as pressure for reforms grew both within the country and from outside, the policy was pursued with less vigour. Following the election of F.W. de Klerk as president in 1990, the National Party opened negotiations with black nationalist organizations, principally the African National Congress (ANC) whose leader Nelson Mandela had been released from prison in February that year.

After lengthy negotiations, elections were held in April 1994 on a universal franchise, which the ANC won by a landslide, leading to Mandela's election as president. The Bantustans were dismantled – despite some last minute resistance by the authorities in Bophuthatswana – and reincorporated into South Africa. Apartheid, and the Bantustans, were at an end.

HOMELANDS MAP

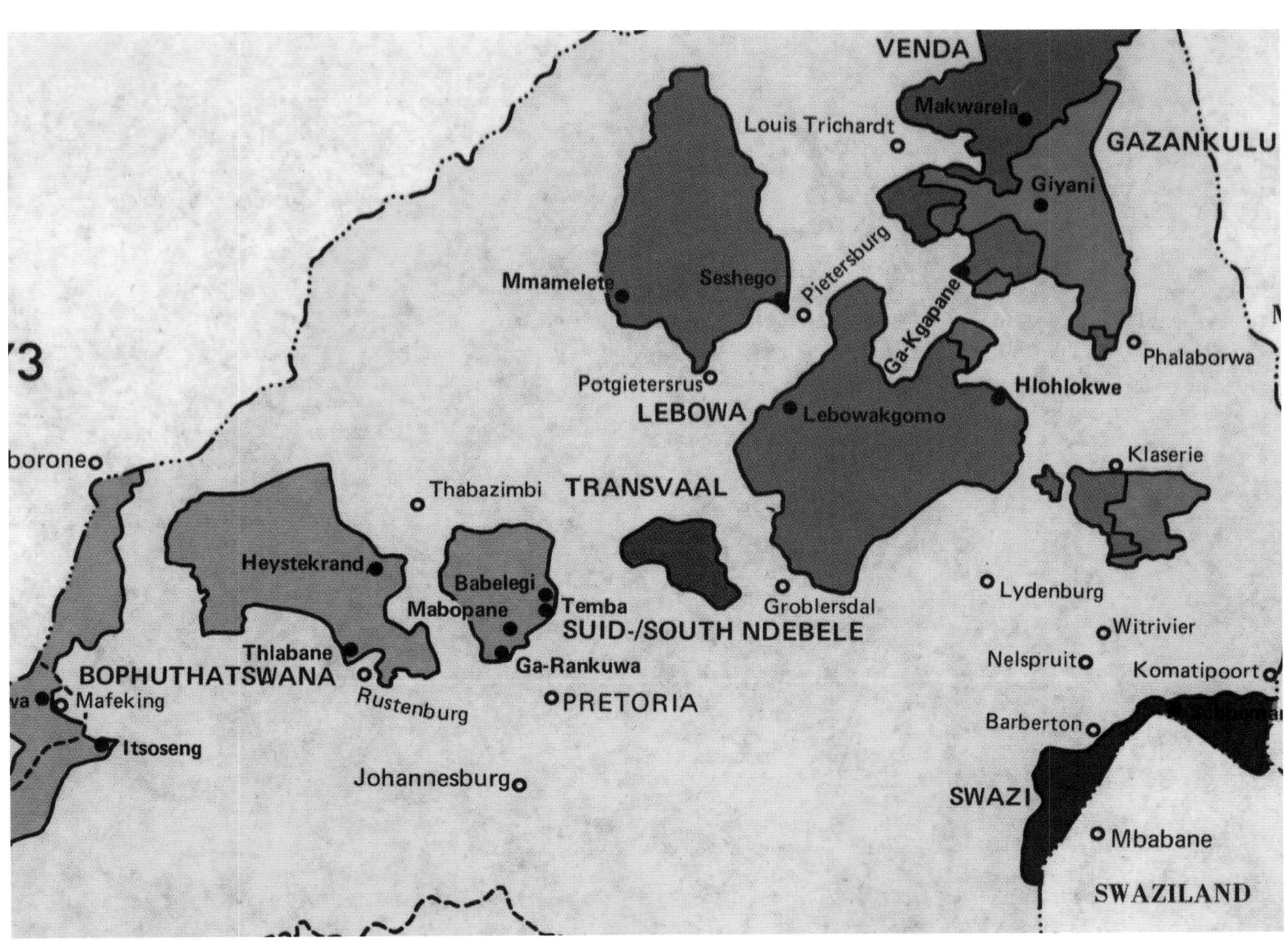

Map of the black 'Homelands', or Bantustans, proposed by the apartheid regime as separate political entities for South Africa's black population. Library of Congress, Washington D.C., USA.

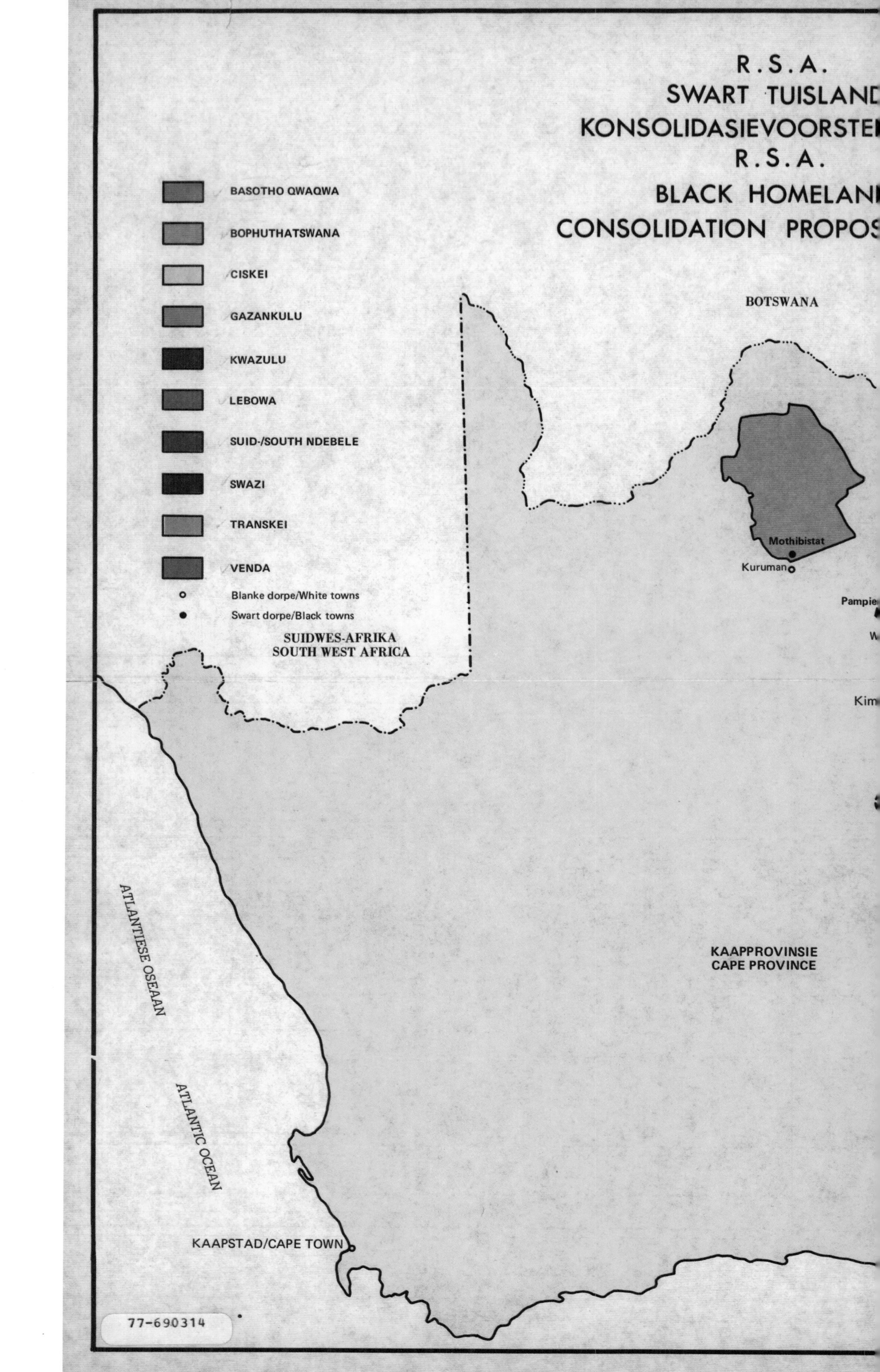
R.S.A.
SWART
R.S.A.
BLACK
CONSOLIDATION
BASOTHO QWAQWA
BOPHUTHATSWANA
CISKEI
GAZANKULU
KWAZULU
LEBOWA
SUID-/SOUTH NDEBELE
SWAZI
TRANSKEI
VENDA
Blanke dorpe/White towns
Swart dorpe/Black towns
SUIDWES-AFRIKA
SOUTH WEST AFRICA
BOTSWANA
Mothibistat
Kuruman
KAAPPROVINSIE
CAPE PROVINCE
ATLANTIESE OSEAAN
ATLANTIC OCEAN
KAAPSTAD/CAPE TOWN

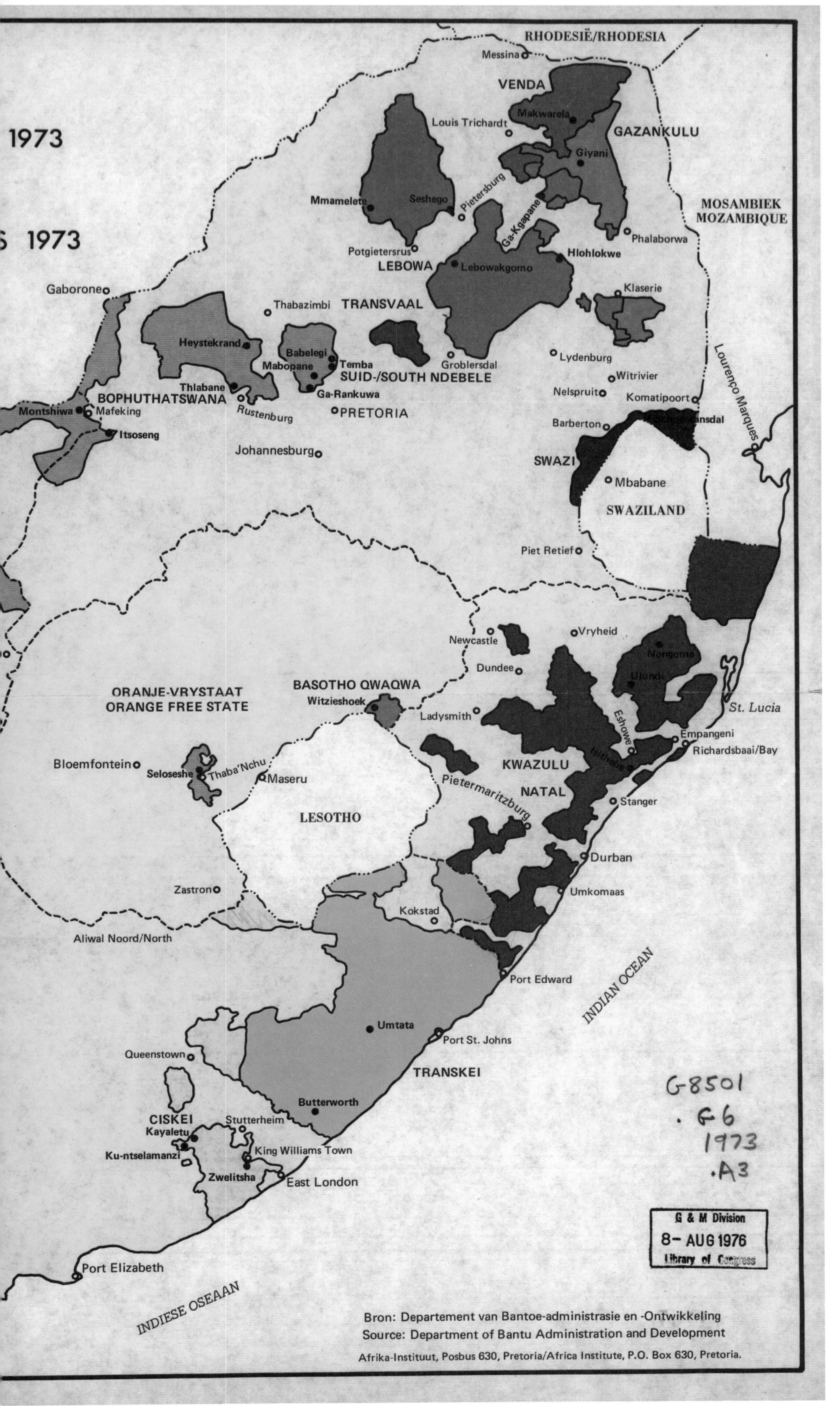

1973
1973
RHODESIË/RHODESIA
Messina
VENDA
Makwarela
Louis Trichardt
GAZANKULU
Giyani
Mmamelete
Seshego
Pietersburg
Ga-Kgapane
MOSAMBIEK
MOZAMBIQUE
Phalaborwa
Potgietersrus
LEBOWA
Lebowakgomo
Hlohlokwe
Klaserie
Gaborone
Thabazimbi
TRANSVAAL
Heystekrand
Babelegi
Mabopane
Temba
SUID-/SOUTH NDEBELE
Groblersdal
Lydenburg
Witrivier
Thlabane
BOPHUTHATSWANA
Ga-Rankuwa
Nelspruit
Komatipoort
Lourenço Marques
Montshiwa
Mafeking
Rustenburg
PRETORIA
Barberton
Itsoseng
Johannesburg
SWAZI
Mbabane
SWAZILAND
Piet Retief
Newcastle
Vryheid
Dundee
Nongoma
Ulundi
BASOTHO QWAQWA
ORANJE-VRYSTAAT
ORANGE FREE STATE
Witzieshoek
Ladysmith
Eshowe
St. Lucia
Empangeni
Richardsbaai/Bay
Bloemfontein
Seloseshe
Thaba'Nchu
Maseru
KWAZULU
NATAL
Pietermaritzburg
Stanger
LESOTHO
Durban
Zastron
Umkomaas
Kokstad
Aliwal Noord/North
Port Edward
INDIAN OCEAN
Umtata
Port St. Johns
Queenstown
TRANSKEI
Butterworth
CISKEI
Stutterheim
Kayaletu
Ku-ntselamanzi
King Williams Town
Zwelitsha
East London
Port Elizabeth
INDIESE OSEAAN
Bron: Departement van Bantoe-administrasie en -Ontwikkeling
Source: Department of Bantu Administration and Development
Afrika-Instituut, Posbus 630, Pretoria/Africa Institute, P.O. Box 630, Pretoria.

The Balkans

The bitter civil war which erupted in Bosnia-Herzegovina in 1992 – as part of a wider conflict after the collapse of Yugoslavia – became a protracted fight for territory between its Bosnian Croat, Muslim and Serb communities. The end of the war, after three years in which around 100,000 people died, came in 1995 after a conference at Dayton, Ohio, which involved protracted negotiations over the precise lines on the map which would form the boundaries of territory each community was to occupy.

Deep historical divisions among the complex ethnic mix of communities which made up Yugoslavia came to the fore after the death in 1980 of Josip Broz Tito, who had ruled the country for thirty-five years. A new Serb nationalism emerged, fomented by the Serb nationalist leader Slobodan Milošević, who sought to replace Yugoslavia's federal structure with a more centralized state dominated by Serbia. The other constituent republics of Yugoslavia responded by declaring their independence, beginning with Slovenia and Croatia in June 1991.

The Yugoslav army (JNA) fought an unsuccessful ten-day campaign to bring Slovenia back into the fold. A much bloodier conflict erupted in Croatia, where Bosnian Croat militias and the JNA seized around a quarter of the country, flattening the eastern Slavonian towns of Vukovar and Vinkovci and bombarding the ancient city of Dubrovnik, before a ceasefire in January 1992.

As fighting in Croatia subsided, ethnic tensions in Bosnia-Herzegovina, particularly after its declaration of independence in March 1992, ignited a civil war which lasted over three years and cost 100,000 lives (some 40 per cent of them civilians). The ethnic patchwork which made up pre-war Bosnia – with 43.6 per cent of the population Bosnian Muslim, 31.3 per cent Serb and 17.3 per cent Croat – led to a savage, brutal conflict. Serb forces in particular (backed by the JNA) engaged in campaigns of ethnic cleansing to carve out a Serbian homeland, the Republika Srpska. The Bosnian capital Sarajevo underwent a forty-four-month siege and indiscriminate shelling, while the world was shocked by a succession of atrocities. Most notable among these occurred in the Bosnian-Muslim enclave of Srebrenica, where a force led by the Bosnian Serb military commander Ratko Mladić massacred 8,000 men in July 1995.

A Croatian government offensive in summer 1995, which recaptured most of the land held by ethnic Serbs there since 1992, spilled over into Bosnia where Muslims and Croats retook a swathe of territory in the west. This, and a NATO bombing offensive in September against Bosnian Serb military positions, persuaded Milošević and the Bosnian Serb leader Radovan Karadzić to the negotiating table.

For three weeks in November 1995 the fate of Bosnia hung in the balance at Wright-Patterson Air Force base near Dayton, Ohio as the teams representing Serbs, Croats and Muslims haggled over the terms of a settlement. Agreement on the federal nature of a new Bosnia, divided between a joint Bosnian Muslim-Croat entity and the Republika Srpska, but sharing a central government (with rotating presidency and central bank), came early on. But as successive drafts of the map marking the dividing lines between them circulated, bitter disputes erupted. At first Milošević was not willing to cede Sarajevo, and disagreement on this and other thorny territorial issues such as how Sarajevo might be linked to Gorazde, the last Muslim enclave in eastern Bosnia, threatened to derail the whole peace process.

The Americans piled on the pressure, the tensions eased by a virtual reality programme called PowerScene which allowed the negotiators to go on virtual flights around Bosnia to inspect the ground over which they were arguing. Just when agreement seemed finally to have been reached on 18 November, Milošević, having given way on Sarajevo, then dug his heels in, angered that the Serbs had been given less than the 49 per cent of the territory they had been promised. Only the concession of a wedge of infertile, mountainous land south of Kljuc which was nicknamed the 'Egg' brought the proportions back into a balance the Serbs felt they could accept. The horse-trading went on until the final hours. The chief US negotiator Richard Holbrooke literally drew a line around the Posavina town of Samac, awarding it to the Croats, to push them away from a last-minute objection.

The Bosnian Muslims ended up with 30 per cent of the area of Bosnia-Herzegovina, and the Croats 21 per cent. Although the federal state which emerged from the agreement struggled to achieve cohesion and remained resolutely divided into ethnic cantons, the fighting did not re-erupt. Careful drawing of lines on the map and the use of cartography as a negotiating tool had secured an agreement where none seemed likely. Maps, in short, had brought peace.

THE DAYTON AGREEMENT

British military map produced for use by the NATO-led multinational peacekeeping Implementation Force (IFOR) following the Dayton Agreement and the definition of the Dayton ceasefire line in Bosnia. Defence Geographic Centre, Feltham, UK.

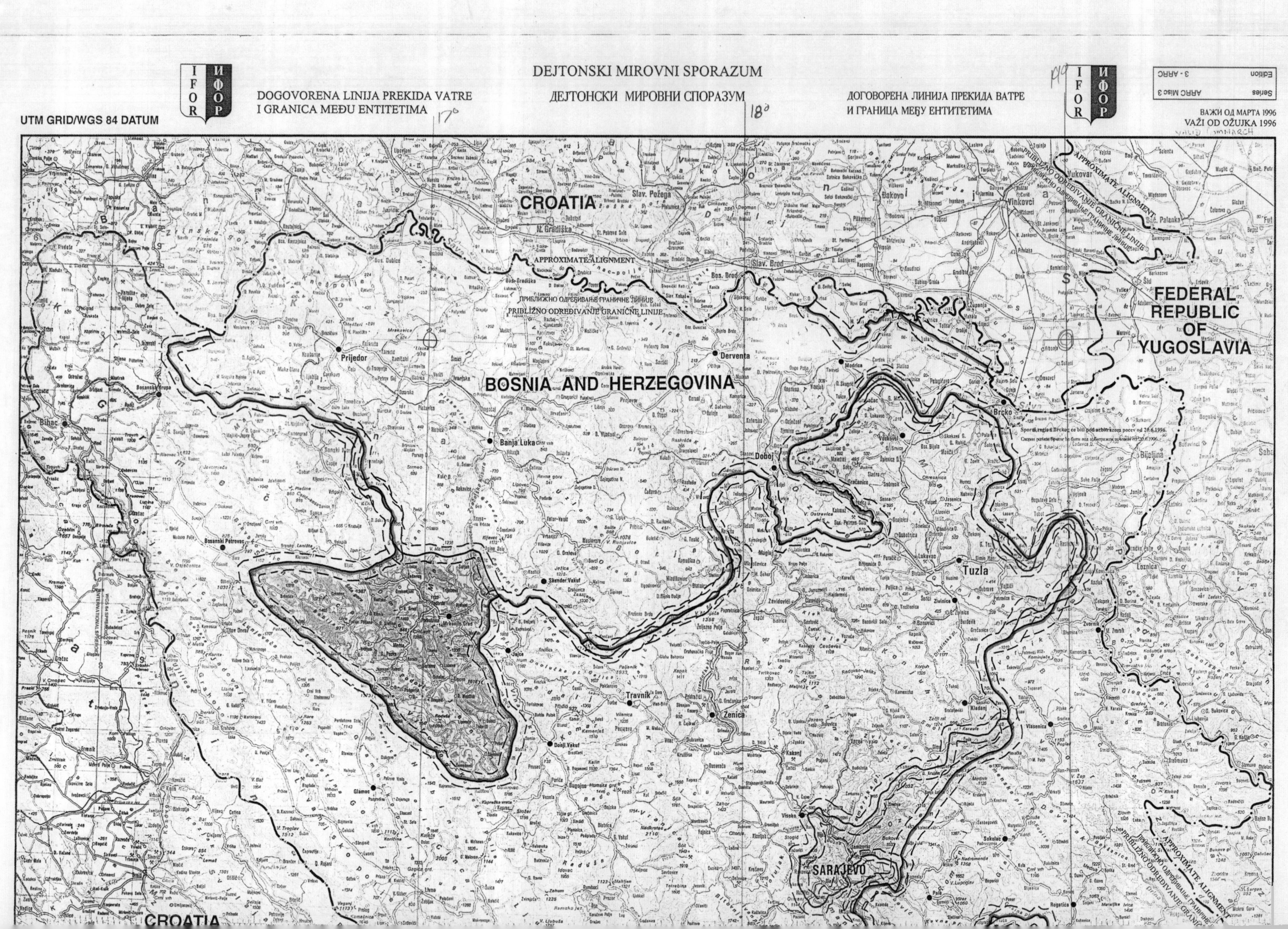
UTM GRID/WGS 84 DATUM
IFOR ИФОР
DOGOVORENA LINIJA PREKIDA VATRE
I GRANICA MEĐU ENTITETIMA
DEJTONSKI MIROVNI SPORAZUM
ДЕЈТОНСКИ МИРОВНИ СПОРАЗУМ
ДОГОВОРЕНА ЛИНИЈА ПРЕКИДА ВАТРЕ
И ГРАНИЦА МЕЂУ ЕНТИТЕТИМА
Series ARRC Misc 3
Edition 3 - ARRC
ВАЖИ ОД МАРТА 1996
VAŽI OD OŽUJKA 1996
CROATIA
BOSNIA AND HERZEGOVINA
FEDERAL REPUBLIC OF YUGOSLAVIA
APPROXIMATE ALIGNMENT
ПРИБЛИЖНО ОДРЕЂИВАЊЕ ГРАНИЧНЕ ЛИНИЈЕ
PRIBLIŽNO ODREĐIVANJE GRANIČNE LINIJE
Prijedor
Banja Luka
Derventa
Doboj
Tuzla
Brcko
Zenica
Travnik
Sarajevo
Bihać
Jajce
Vinkovci
Vukovar
Bijeljina
Zvornik
Slav. Brod
Slav. Požega
Đakovo
Glamoč
Bosanski Petrovac
Skender Vakuf
Mrkonjić Grad
Donji Vakuf
Kladanj
Vlasenica
Sokolac
Visoko
Kakanj
Modrica
Bosanska Krupa
N. Gradiška
Bos. Gradiška
Bos. Brod
Spornį region Brckog će biti pod arbitražom počev od 26.6.1996.
WGS 84 SPHEROID
INTERNATIONAL SPHEROID

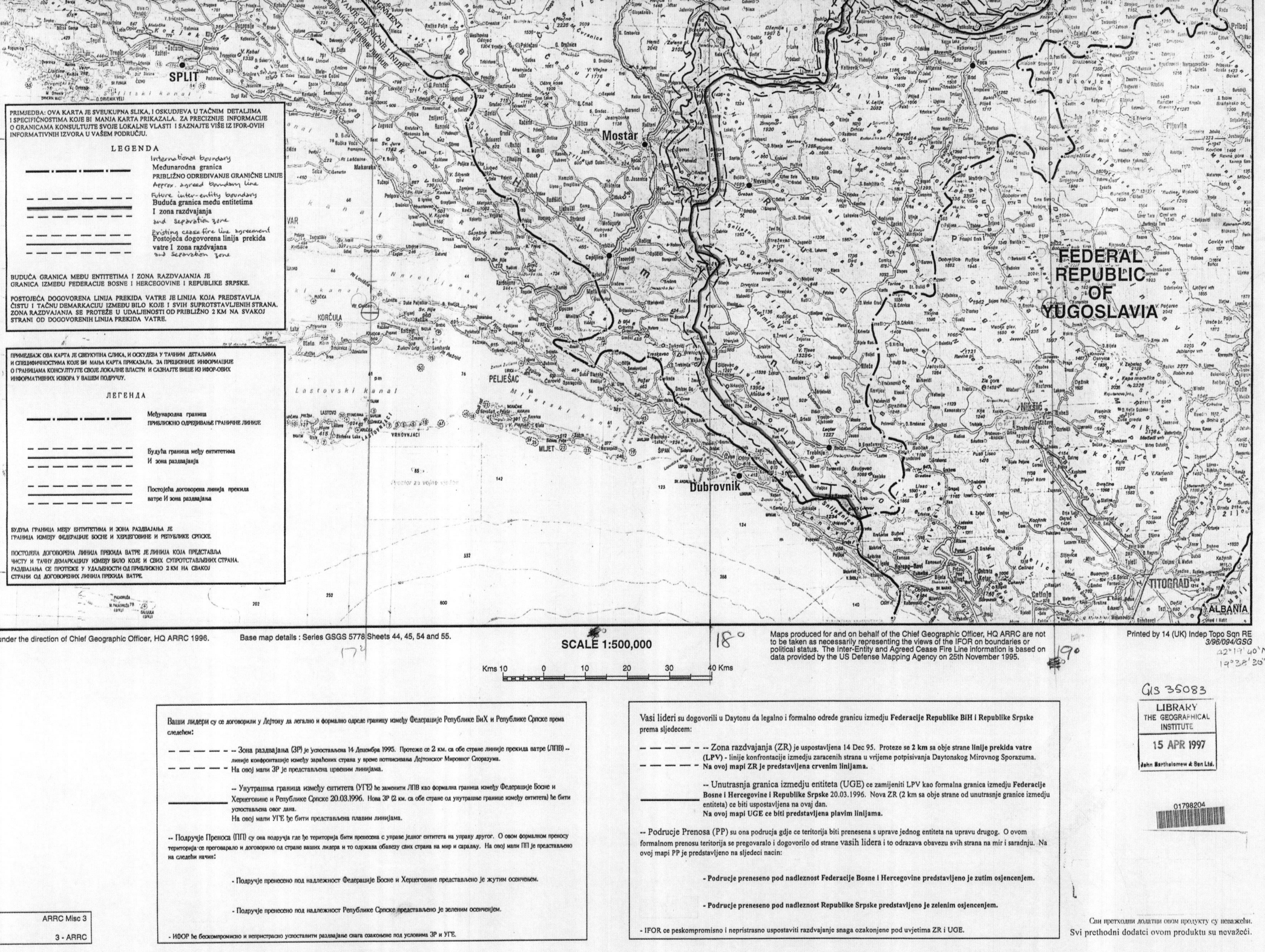

Produced under the direction of Chief Geographic Officer, HQ ARRC 1996.

Base map details : Series GSGS 5778 Sheets 44, 45, 54 and 55.

SCALE 1:500,000

Kms 10 0 10 20 30 40 Kms

Maps produced for and on behalf of the Chief Geographic Officer, HQ ARRC are not to be taken as necessarily representing the views of the IFOR on boundaries or political status. The Inter-Entity and Agreed Cease Fire Line information is based on data provided by the US Defense Mapping Agency on 25th November 1995.

Printed by 14 (UK) Indep Topo Sqn RE
3/96/094/GSG

17° 18° 19°

42°19'40"N
19°38'30"E

Ваши лидери су се договорили у Дејтону да легално и формално одреде границу између Федерације Републике БиХ и Републике Српске према следећем:

-- Зона раздвајања (ЗР) је успостављена 14 Децембра 1995. Протеже се 2 км. са обе стране линије прекида ватре (ЛПВ) -- линије конфронтације између зараћених страна у време потписивања Дејтонског Мировног Споразума.
На овој мапи ЗР је представљена црвеним линијама.

-- Унутрашња граница између ентитета (УГЕ) ће заменити ЛПВ као формална граница између Федерације Босне и Херцеговине и Републике Српске 20.03.1996. Нова ЗР (2 км. са обе стране од унутрашње границе између ентитета) ће бити успостављена овог дана.
На овој мапи УГЕ ће бити представљена плавим линијама.

-- Подручје Преноса (ПП) су она подручја где ће територија бити пренесена с управе једног ентитета на управу другог. О овом формалном преносу територија се преговарало и договорило од стране ваших лидера и то одржава обавезу свих страна на мир и сарадњу. На овој мапи ПП је представљено на следећи начин:

- Подручје пренесено под надлежност Федерације Босне и Херцеговине представљено је жутим осенчењем.

- Подручје пренесено под надлежност Републике Српске представљено је зеленим осенчењем.

- ИФОР ће бескомпромисно и непристрасно успоставити раздвајање снага озаконење под условима ЗР и УГЕ.

Vasi lideri su dogovorili u Daytonu da legalno i formalno odrede granicu izmedju **Federacije Republike BiH i Republike Srpske** prema sljedecem:

-- Zona razdvajanja (ZR) je uspostavljena 14 Dec 95. Proteze se **2 km** sa obje strane **linije prekida vatre (LPV)** - linije konfrontacije izmedju zaracenih strana u vrijeme potpisivanja Daytonskog Mirovnog Sporazuma.
Na ovoj mapi ZR je predstavljena crvenim linijama.

-- Unutrasnja granica izmedju entiteta (UGE) ce zamijeniti LPV kao formalna granica izmedju **Federacije Bosne i Hercegovine i Republike Srpske** 20.03.1996. Nova ZR (2 km sa obje strane od unutrasnje granice izmedju entiteta) ce biti uspostavljena na ovaj dan.
Na ovoj mapi UGE ce biti predstavljena plavim linijama.

-- Podrucje Prenosa (PP) su ona podrucja gdje ce teritorija biti prenesena s uprave jednog entiteta na upravu drugog. O ovom formalnom prenosu teritorija se pregovaralo i dogovorilo od strane vasih lidera i to odrazava obavezu svih strana na mir i saradnju. Na ovoj mapi PP je predstavljeno na sljedeci nacin:

- Podrucje preneseno pod nadleznost Federacije Bosne i Hercegovine predstavljeno je zutim osjencenjem.

- Podrucje preneseno pod nadleznost Republike Srpske predstavljeno je zelenim osjencenjem.

- IFOR ce peskompromisno i nepristrasno uspostaviti razdvajanje snaga ozakonjene pod uvjetima ZR i UGE.

Series	ARRC Misc 3
Edition	3 - ARRC

Сви претходни додатци овом продукту су неважећи.
Svi prethodni dodatci ovom produktu su nevažeći.

Middle East

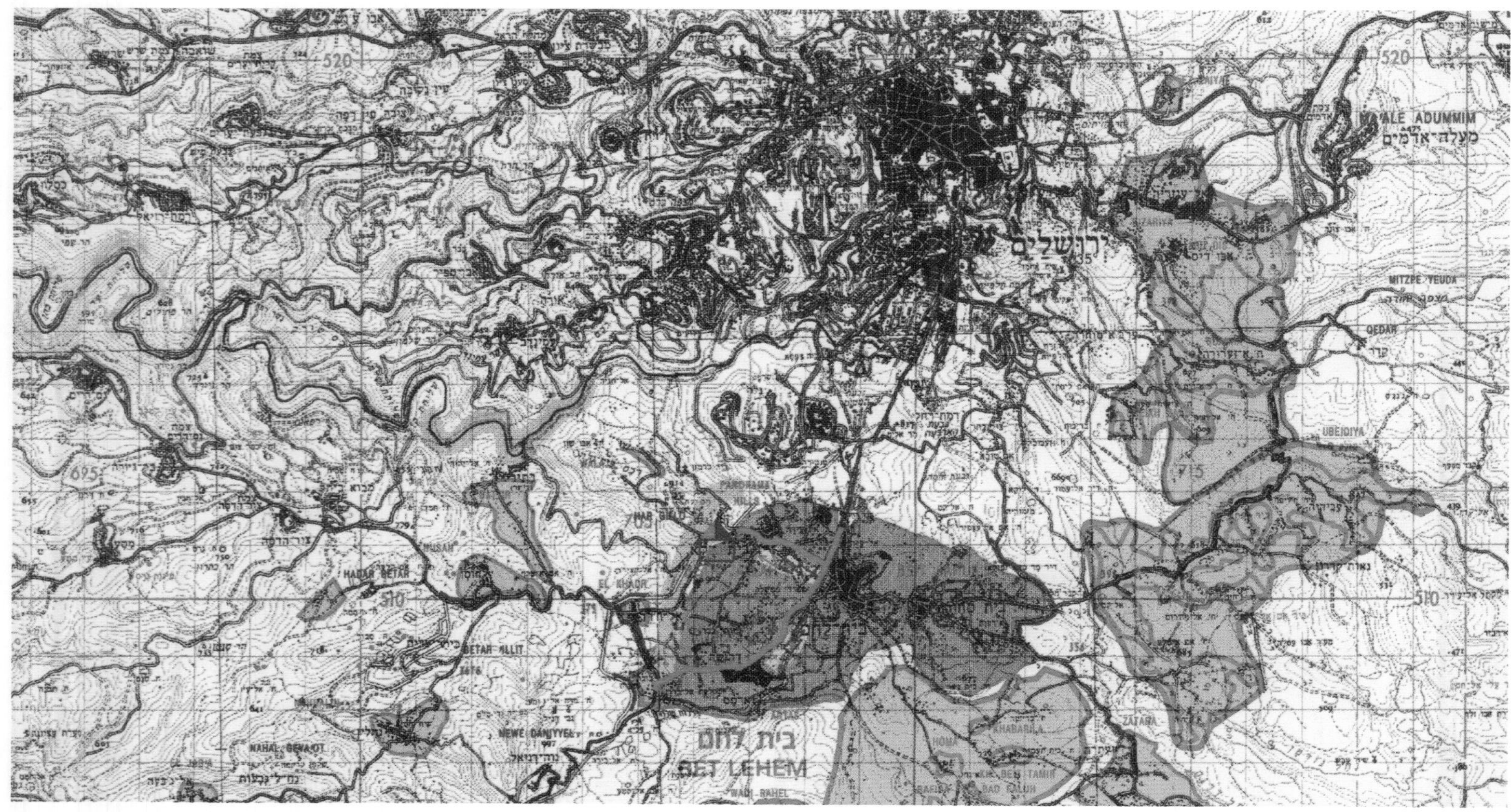

Map showing the levels of Palestinian control to be exercised in different parts of the West Bank (Areas A and B), the deployment of Palestinian police, joint activities and zoning arrangements. Published as part of the *Israeli-Palestinian Interim Agreement on the West Bank and the Gaza Strip*, or 'Oslo II'. University of Glasgow Library, Glasgow, UK.

THE OSLO ACCORDS

The search for a final solution to the problem of where the borders of Israel should lie has bedevilled the Middle East ever since the country's independence in 1948 (see page 192). Negotiations between Israelis and Palestinians in 1993–5 seemed to produce a blueprint for the establishment of an independent Palestinian state in Gaza and the West Bank, but successive attempts to agree and map the extent of its borders collapsed without any final agreement.

By the late 1980s, there seemed to be no resolution in sight to the Israeli-Arab conflict. The descendants of the 720,000 Palestinians who had fled their homes in 1948 remained unable to return. Israel's military successes in the 1967 Six-Day War and 1973 Yom Kippur War made it clear that its Arab neighbours were unlikely to be able to impose a military solution to the crisis. Terrorist attacks on Israel and its overseas interests by the Palestine Liberation Organization (PLO) and other armed groups had similarly failed to achieve any concessions from Israel.

The one glimmer of hope was the Camp David Accords, the US-brokered negotiations which achieved a peace treaty between Egypt and Israel in 1979. The agreement also included a 'Framework for Peace in the Middle East' which aimed at the establishment of an autonomous Palestinian authority in Gaza and the West Bank and the withdrawal of Israeli troops under the terms of UN Resolution 242 passed at the time of the Six-Day War. The hope that this would be achieved within five years was not realized and only after the outbreak in 1987 of the First Intifada – an uprising by Palestinians in the occupied territories against Israeli rule (which ultimately cost the lives of around 160 Israelis and over 2,000 Palestinians), did the peace process begin to gather momentum once more.

A conference held in Madrid in October 1991 affirmed the principle that Israel give up land to the Palestinians in exchange for firm security guarantees ('land for peace'). Bilateral negotiations between Israel and the Arabs then began (but not directly with the PLO). However, more progress was made in secret talks in Oslo which began in January 1993 and did involve the PLO. Israeli Prime Minister Yitzhak Rabin and PLO Chairman Yasser Arafat finally signed a 'Declaration of Principles on Interim Self-Government' (known as the Oslo Accord) in September 1993, by which Israel implicitly recognized the PLO as a representative of the Palestinian people. The agreement set down a timetable for further negotiations, establishing that an agreement on Israeli withdrawal from Gaza and the West Bank town of Jericho should be reached within two months. It also stated that Israel was to transfer to the Palestinians power over education, health, social welfare, taxation and tourism. The Palestinians in turn were to elect a Palestinian Council within nine months. A final settlement, including contentious issues such as the right of return of Palestinian refugees and the status of Jerusalem was to be concluded within five years.

Opposition from more radical Palestinian factions and from conservative Israelis (and the security forces there) slowed implementation of the agreement. The Cairo Agreement of February 1994 slightly watered down the Israeli commitment to reduce its armed forces in the West Bank and it took a further Oslo Accord (in September 1995) to define the borders of the areas from which it would withdraw. The map which accompanied the document (shown here) defines three areas: Area A, made up of large urban centres such as Ramallah, Jericho and Nablus, which constituted 3 per cent of the West Bank, in which the Palestinians would have complete control; Area B, consisting of smaller villages and rural areas, where the Israelis would retain military control, which constituted around 25 per cent of the West Bank; and the rest of the West Bank (referred to, but not labelled on the map, as Area C), from which Israel would not, at this stage, withdraw. The Palestinians were to receive the whole of the Gaza Strip, apart from Israeli settlement blocs and areas deemed essential for Israeli security.

On the map, the Palestinian area makes up a number of separate cantons, with little contiguous territory on which to build the solid foundations of a state. Although the initial withdrawals were implemented and a Palestinian Authority took control in those areas, severe opposition in Israel (and the assassination of Rabin in November 1995) hampered further progress. The 1998 Wye River Memorandum would have given the Palestinians a further 13 per cent of the West Bank, but a revolt by Israeli nationalist and religious opponents of the agreement brought down the government of Binyamin Netanyahu and scuppered the deal. At further negotiations at Camp David in 2000, the new Israeli prime minister Ehud Barak proposed that 20.5 per cent of the West Bank remain under Israel's control and the rest be handed over to the Palestinians. Yasser Arafat rejected the offer and the last best chance for a final peace was gone.

Subsequent developments, the growth of Israeli settlements in the West Bank, the eruption of a Second Intifada in 2000, the building by Israel of a security wall encircling the Palestinian enclaves, the coming to power of radical Islamist movements such as Hamas and Islamic Jihad in the West Bank and Gaza and an increasingly conservative, nationalist and religious Israeli electorate, have all meant that the 1995 Oslo Map and the peace process of which it formed a key part now seem the final, rather than the first stage in redrawing the map of Israel and Palestine.

Israeli-Palestinian Interim Agreement on the West Bank and the Gaza Strip

CONSOLIDATED MAP OF THE WEST BANK

superimposing map Nos. 1, 3, 4 and 7 of the Israeli-Palestinian Interim Agreement on the West Bank and the Gaza Strip

Note: For the purposes of this publication, the original map Nos. 1, 3, 4 and 7 of the Interim Agreement have been superimposed on a single map. In the Interim Agreement each of these four original maps consisted of four sections, each section being signed separately in identical signature boxes, a single copy of which has been reproduced here. The legends from each of the original maps have also been reproduced here. The scale of the maps has been reduced from 1:50,000 in the original to 1:100,000 here.

LEGEND

MAP NO. 1

AREA A
AREA B
SPECIAL CASE
JEWISH HOLY SITE IN AREA A
JORDANIAN BORDER
ROAD NO. 60

MAP NO. 3

GENERAL DIRECTORATE OF POLICE
DISTRICT HQ.
POLICE CENTER
POLICE STATION
POLICE POST
PUBLIC SECURITY CAMP
MOBILE POLICE UNIT
OBSERVATION POINT
PUBLIC SECURITY PATROL
SECURITY ACTIVITY PERIMETER
AREA A
AREA B
SPECIAL CASE

MAP NO. 4

JEWISH HOLY SITES IN AREA A
D.C.O.
JOINT MOBILE UNIT
JOINT PATROL
JOINT PATROL ROAD
JORDANIAN BORDER

MAP NO. 7

AREAS WITH ZONING ARRANGEMENTS
ROUTES WITH ZONING ARRANGEMENTS

NAMES AND BOUNDARY REPRESENTATIONS ARE NOT NECESSARILY AUTHORITATIVE

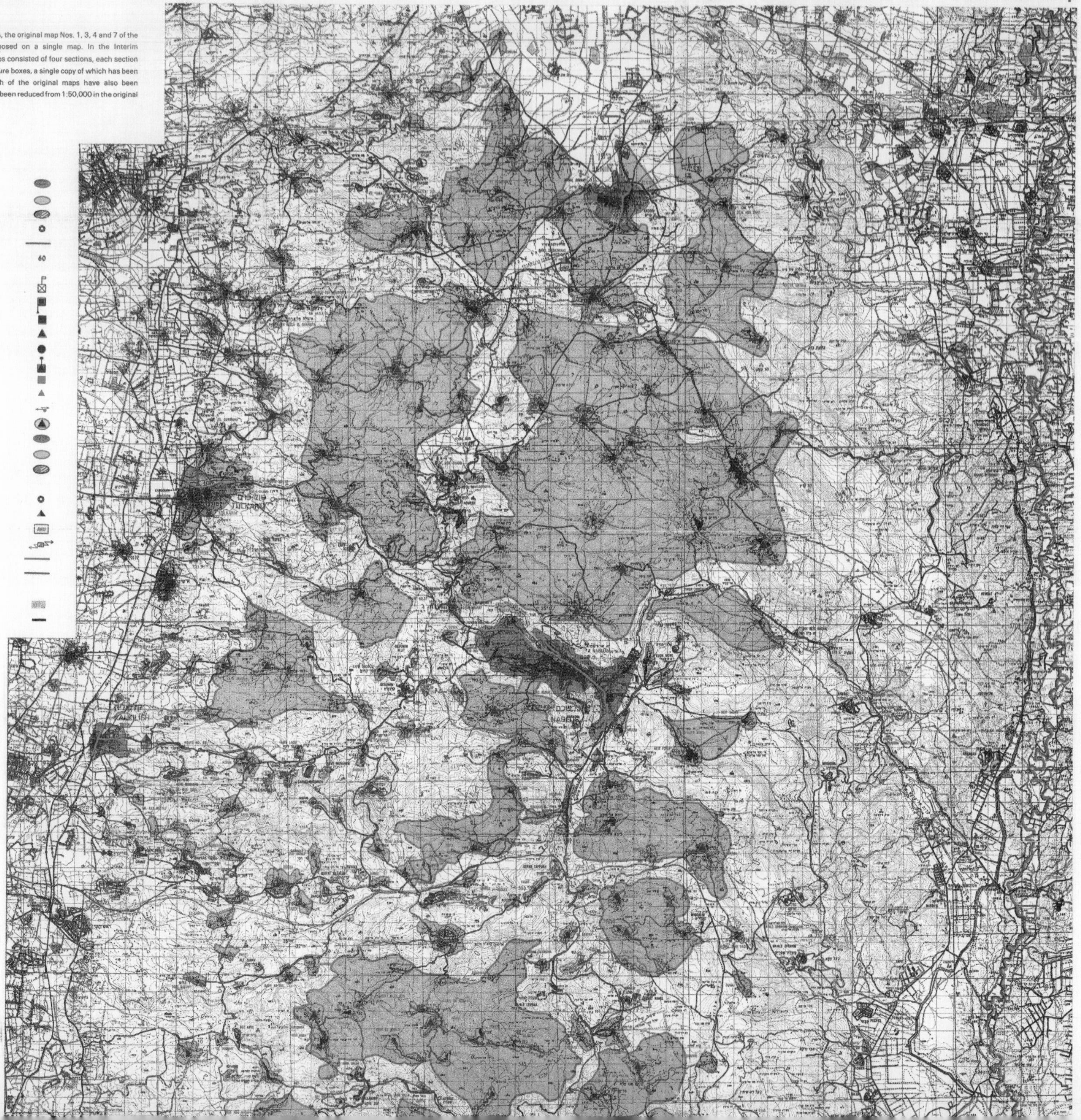

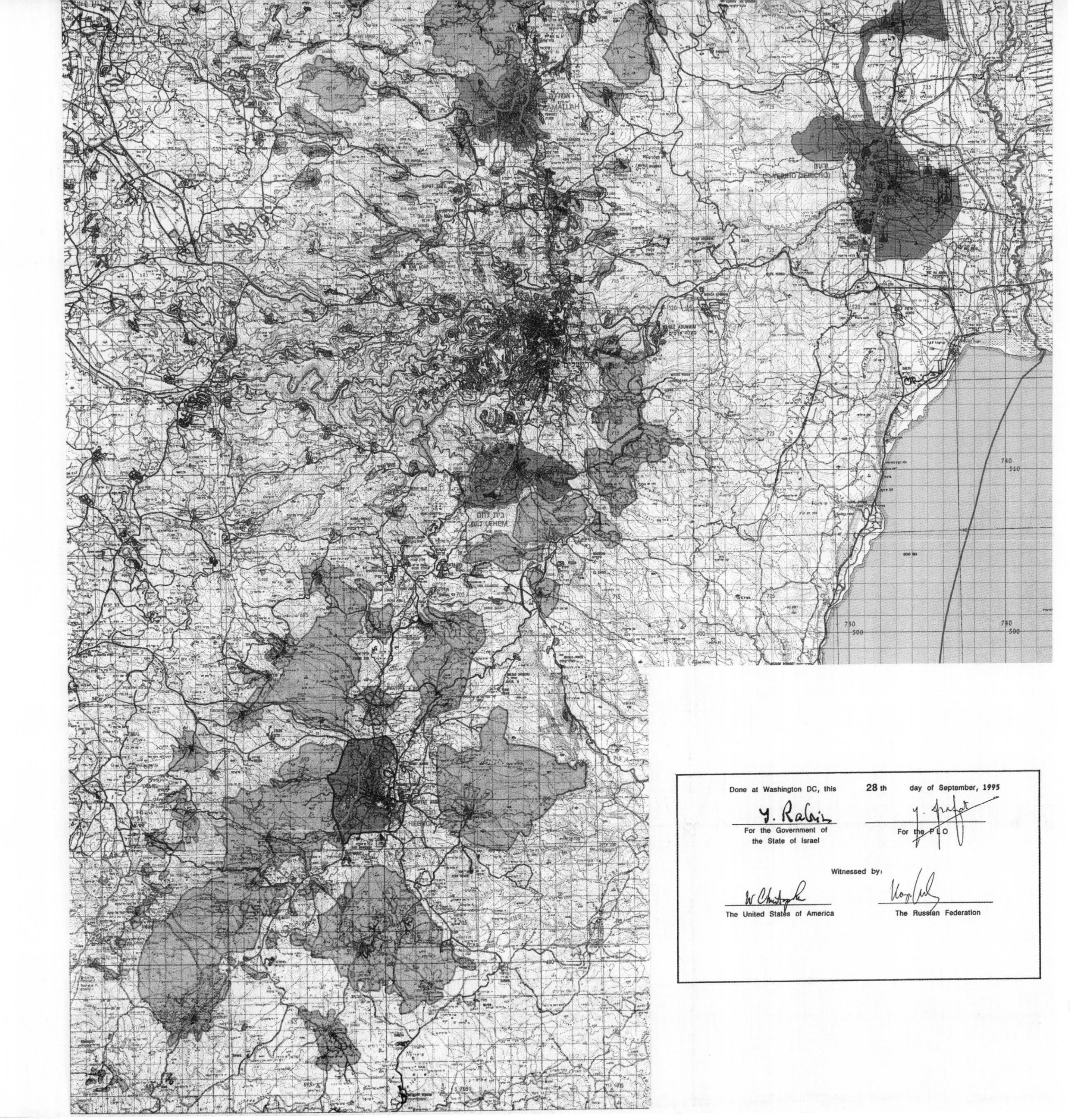

Done at Washington DC, this 28 th day of September, 1995

Y. Rabin
For the Government of
the State of Israel

Y. Arafat
For the PLO

Witnessed by:

W. Christopher
The United States of America

The Russian Federation

MAPPING TODAY

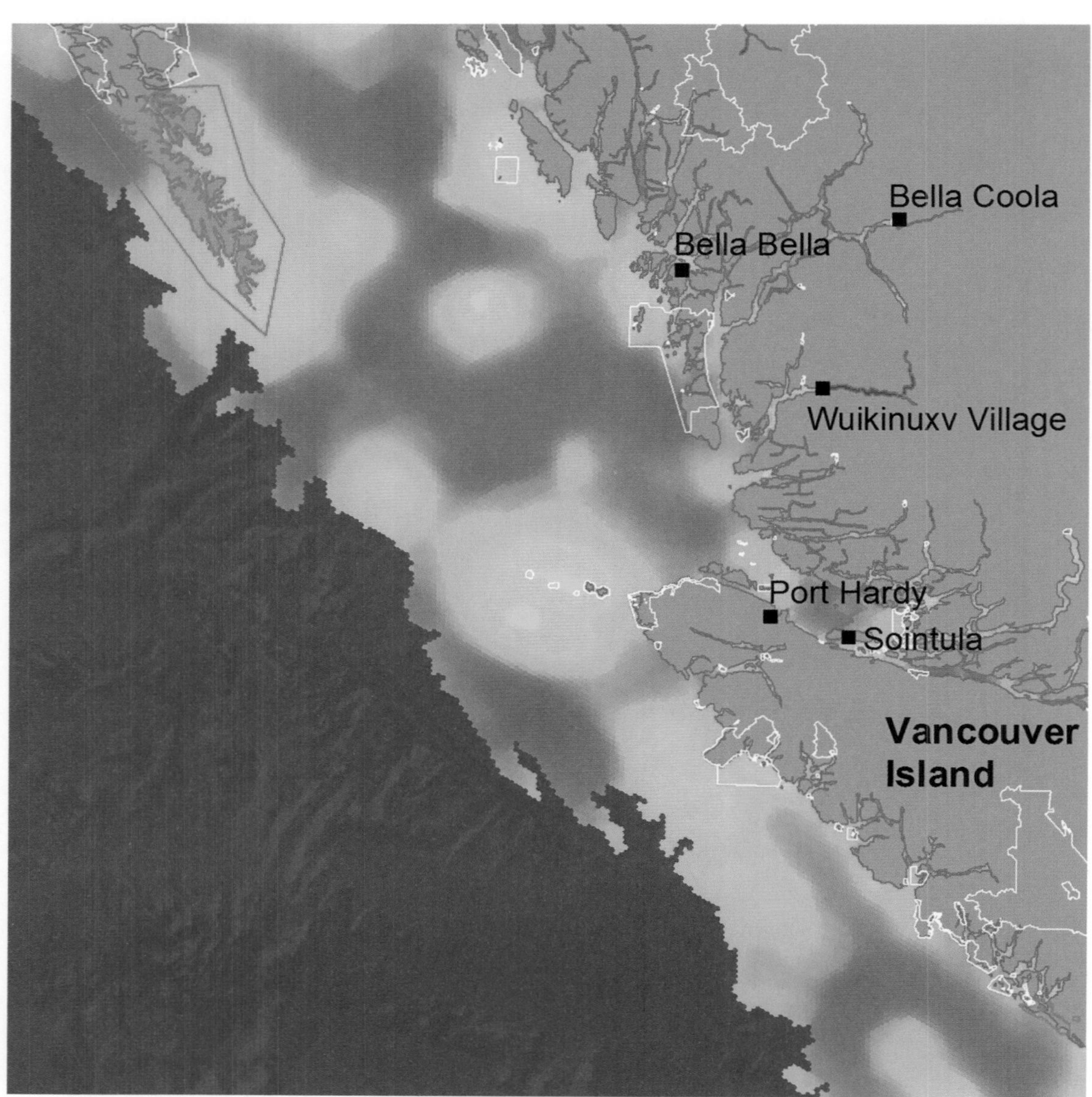

Google Maps can be manipulated and customized in many ways to show different types of geographic data for specific purposes. Examples include (overleaf): city plan with business locations (top left); hill-shaded relief map of the Himalaya (centre left); humanitarian crisis in Darfur, Sudan (bottom left); local air pollution in New York city (top right); Marine Protected Areas, British Columbia (centre right); historical map of eastern Russia and Alaska (bottom right).

Google Maps

Google, the company founded in the late 1990s by Larry Page and Sergey Brin has revolutionized the world of mapping. Google Maps, the web application which the company launched in 2005 has enabled maps to be accessed and used in ways never possible before – dramatically increasing the potential uses of geographic data.

In the late 1990s Larry Page and Sergey Brin, founded Google. Its mission was 'to organize the world's information and make it universally accessible and useful'. Google has since grown into a hugely successful multi-billion-dollar company, but perhaps more than that, it has had a profound effect on society and has become part of our daily lives. We use Google to find all kinds of information, to find out where things are and to discover and explore through our screens in a way which simply wasn't imaginable a few years ago. Google Maps is Google's browser-based product free at the point of use, which has grown into a desktop and mobile web mapping application which offers street maps, satellite imagery and 'Street View' perspectives, while delivering such things as live traffic information and route planning.

Google's 'geo' offerings have both disrupted and revolutionized the way people view, use and make maps and how they interact with their surroundings. Google's original mission statement has often been expanded: 'to organize the world's information, geographically'. What Google saw early on was the huge potential for showing people not only a search result but, crucially, where they could locate it. That may be a museum, a coffee shop, an archaeological dig or a school district. Most data have a spatial component, so basing searches on maps, thus allowing people to navigate their information geographically, was a masterstroke.

Google also grasped other converging technologies, in particular the internet and the rapid development and uptake of mobile devices. Global Positioning System (GPS) location data have also become ubiquitous in our devices so we can easily locate ourselves on the map, create tracks of our ski runs and tag our photos with locations which can be instantly uploaded to a map and shared with the world. In addition to becoming the default map of choice for finding places and navigating, the Google Maps application programming interface (API) has also underpinned the democratization of online mapping. This allows anyone to create, for example, 'mashups' of their own georeferenced data, to customize the Google base map into their own style, or to place other maps – for example historical maps – onto the Google Maps base to allow direct comparisons.

The design of Google's map has steadily improved so that it now delivers high quality online web mapping which is aesthetically pleasing, purposeful and globally consistent. And it is automatically modified depending on its use – for example, secondary roads widen at particular scales when traffic information and direction of traffic flow is shown, and the appearance of 3D buildings and moving shadows at large scales (in some cities) represent the built environment like never before. This is a single map, but one which is designed to work well at each map scale, across all scales and in different ways.

Of course, such disruptive technological change is not without consequence. Can we ignore the fact that Google brought the Mercator projection (see page 76) back to prominence when it is generally considered to be a poor choice for world mapping? Maybe not… but Web Mercator is a perfect technical solution for serving tiles of data efficiently and rapidly. Google have also come under significant attack over questions of privacy and security. For example, does the capture and embedding of Street View imagery impinge on basic human rights? While faces and car number plates are blurred, some countries have laws which restrict such image capture and publication. It is not universally popular.

Any company or organization which pushes the limits of technology and causes such changes is bound to face some difficulties and so society has to react to the obvious benefits which Google's map applications have brought, and address any concerns. Without disruptive technologies society cannot progress. What is perhaps unique about Google is their very rapid rise and the impact of their geo technologies not only on cartography but also on society. Given the pace of change in the first decade of Google's mapping activities, and the way in which it has shaped our use of technology in the twenty-first century, the map's evolution is sure to continue. And so are the multitude of ways in which it is used.

As we have seen, maps have reflected, influenced and sometimes even directly driven our shared history, and have commonly been significant factors in social, scientific and geopolitical developments in the past. As the technology used to create and use maps develops, and as the part they play in our lives through such things as mobile applications continues to increase, they can be expected to continue to play an important part in the future.

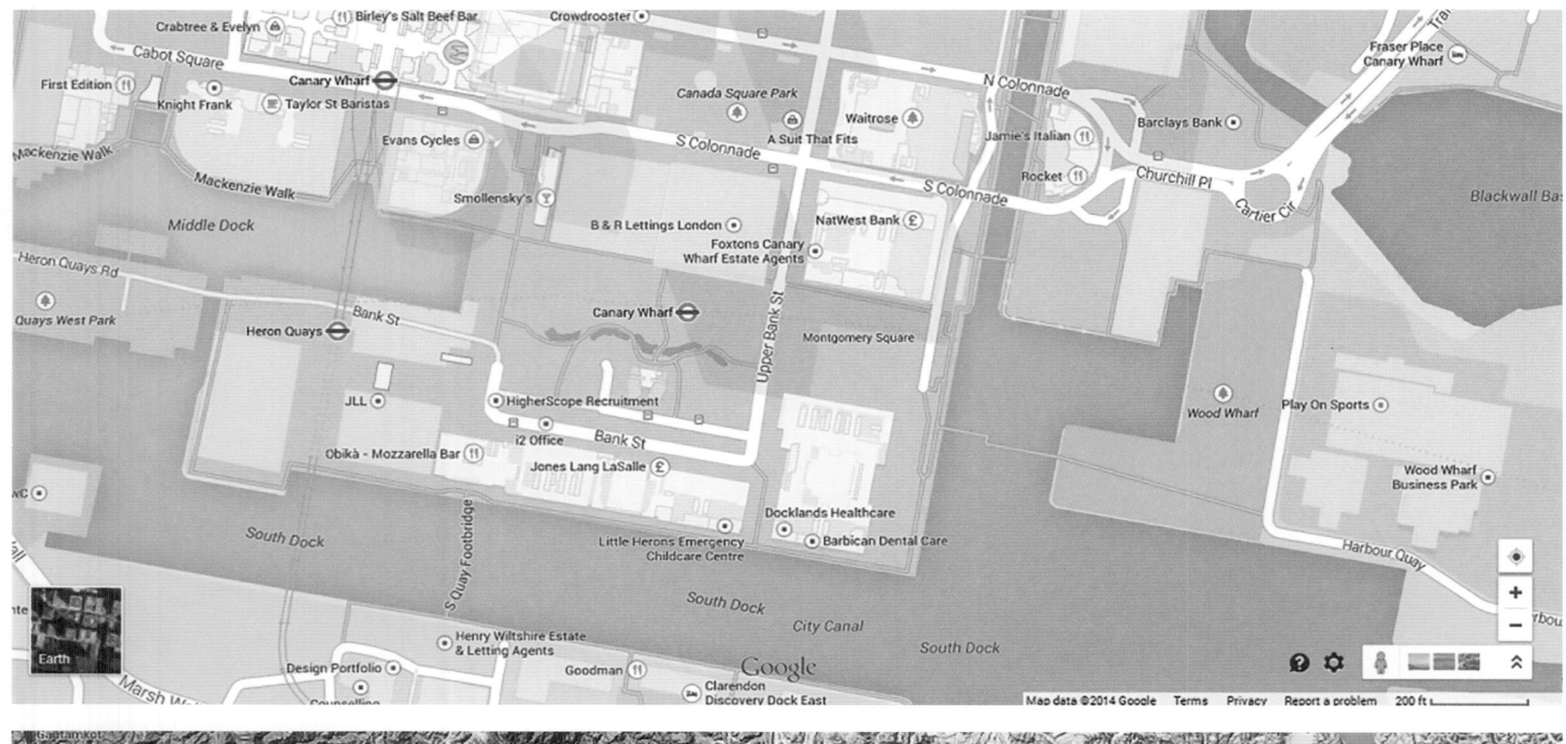

Crabtree & Evelyn
Birley's Salt Beef Bar
Crowdrooster
Cabot Square
First Edition
Canary Wharf
Knight Frank
Taylor St Baristas
Canada Square Park
Waitrose
N Colonnade
Fraser Place Canary Wharf
Evans Cycles
A Suit That Fits
Jamie's Italian
Barclays Bank
Mackenzie Walk
S Colonnade
Rocket
Churchill Pl
Smollensky's
Cartier Cir
Blackwall Ba
Middle Dock
B & R Lettings London
NatWest Bank
Foxtons Canary Wharf Estate Agents
Heron Quays Rd
Quays West Park
Bank St
Heron Quays
Upper Bank St
Montgomery Square
JLL
HigherScope Recruitment
i2 Office
Wood Wharf
Play On Sports
Obikà - Mozzarella Bar
Jones Lang LaSalle
Wood Wharf Business Park
South Dock
S Quay Footbridge
Docklands Healthcare
Little Herons Emergency Childcare Centre
Barbican Dental Care
Harbour Quay
Earth
South Dock
City Canal
Henry Wiltshire Estate & Letting Agents
South Dock
Design Portfolio
Goodman
Google
Clarendon Discovery Dock East
Map data ©2014 Google
Terms
Privacy
Report a problem
200 ft

Pokhara
Baglung
Tansen
Butwal
Bharatpur
Chitwan National Park
Kathmandu
Siddharthanagar
Kapilvastu
Rudrapur
Motipur
Chanai
Masina
Ramgram
Pragatinagar
Meghauli
Naya Belhani
Gorkha
Manaslu Conservation Area
Langtang National Park
Khairen
Earth
Google

Crisis in Darfur
West Darfur
South Darfur
Image Landsat
Google earth
Tour Guide
Imagery Date: 4/10/2013 12°10'35.28" N 23°49'49.56" E elev 903 m eye alt 706.73 km

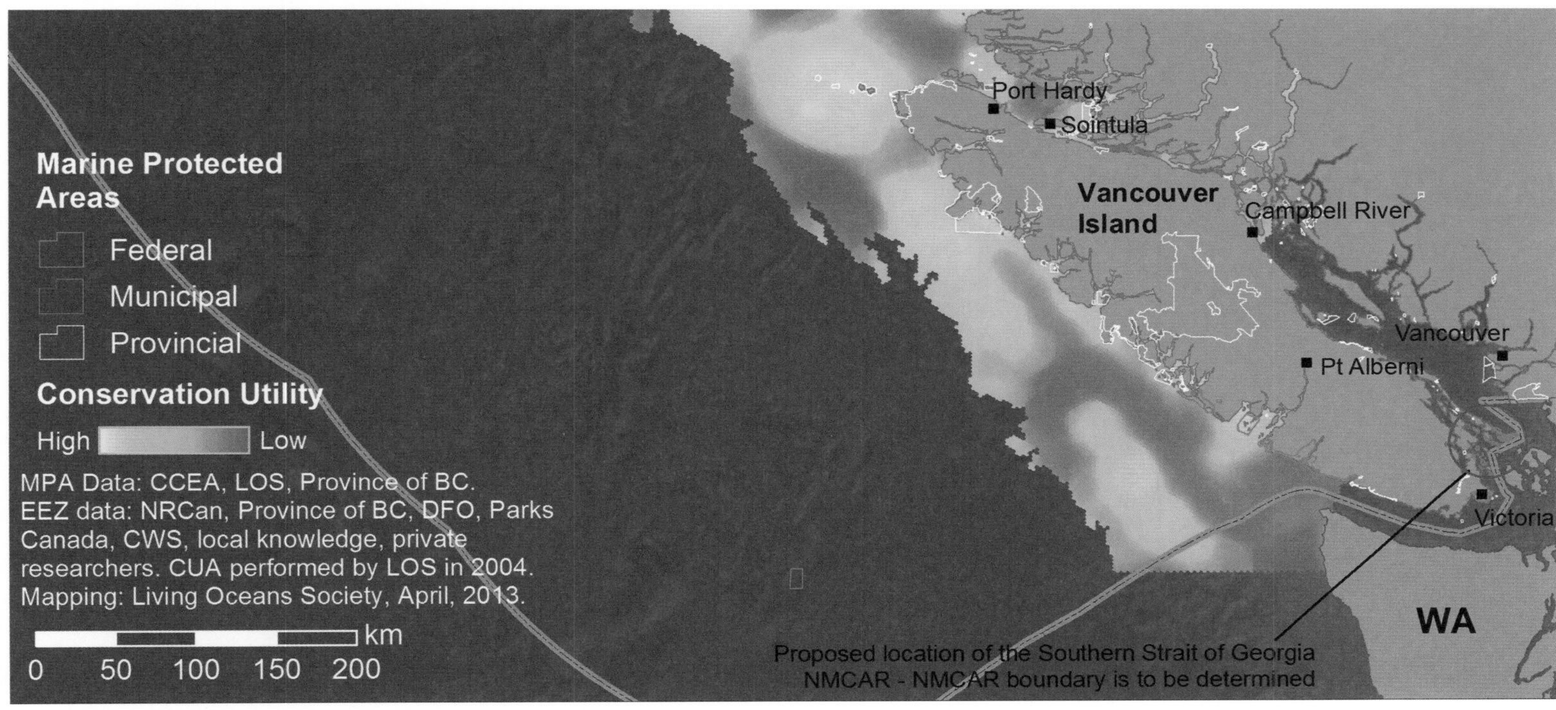

Marine Protected Areas
Federal
Municipal
Provincial
Conservation Utility
High
Low
MPA Data: CCEA, LOS, Province of BC.
EEZ data: NRCan, Province of BC, DFO, Parks Canada, CWS, local knowledge, private researchers. CUA performed by LOS in 2004.
Mapping: Living Oceans Society, April, 2013.
km
0 50 100 150 200
Port Hardy
Sointula
Vancouver Island
Campbell River
Vancouver
Pt Alberni
Victoria
WA
Proposed location of the Southern Strait of Georgia NMCAR - NMCAR boundary is to be determined

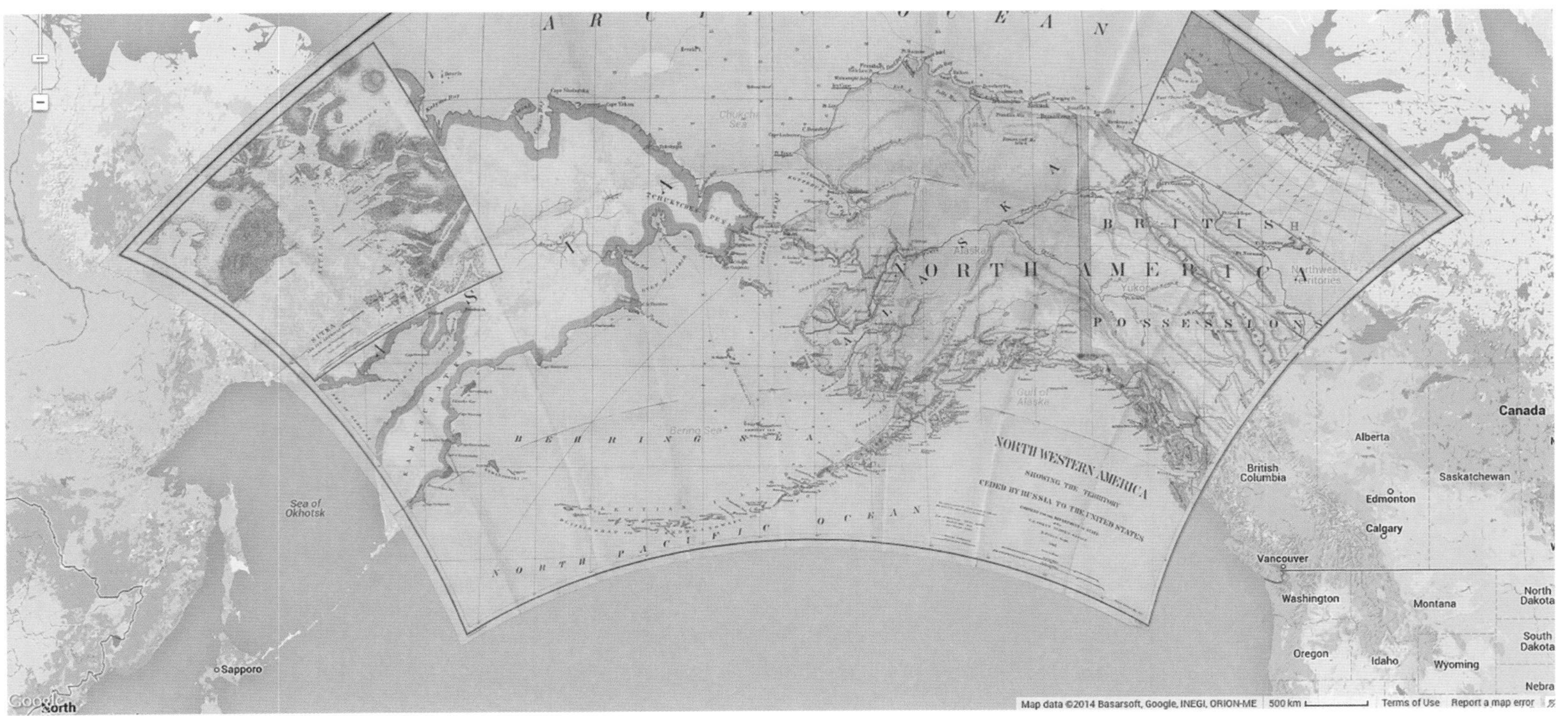

NORTH WESTERN AMERICA
SHOWING THE TERRITORY
CEDED BY RUSSIA TO THE UNITED STATES
Sea of Okhotsk
Sapporo
Canada
Alberta
Saskatchewan
British Columbia
Edmonton
Calgary
Vancouver
Washington
Montana
North Dakota
South Dakota
Oregon
Idaho
Wyoming
Map data ©2014 Basarsoft, Google, INEGI, ORION-ME
500 km
Terms of Use
Report a map error

INDEX

INDEX

Acknowledgements

Produced under the direction of Collins Learning, HarperCollins Publishers, Glasgow UK.

Project leader and chief editor: Mick Ashworth, Ashworth Maps and Interpretation Ltd, Bishopbriggs, Glasgow, UK.

www.ashworthmaps.co.uk

History editor and main contributor: Philip Parker.

DTP/layout: Gordon MacGilp.

Contributing authors:

Dr Peter Chasseaud, FRGS is a historian of military cartography, the founder of the Historical Military Mapping Group of the British Cartographic Society, and member of the Defence Surveyors' Association and the author of definitive works on trench mapping and toponymy. He is the author of *Mapping the First World War*, published by Collins in association with Imperial War Museums. (Pages 177, 189)

Nicholas Crane is a Fellow of the Royal Geographical Society and author of *Mercator, The Man Who Mapped the Planet.* (Page 77)

Catherine Delano-Smith, M.A., B.Litt., D.Phil. (Oxon), Senior Research Fellow, Institute of Historical Research, University of London. Her early research was on man-land relationships since the Neolithic in southern Europe, from which she moved into the history of early maps and mapping, with a particular interest in medieval and renaissance exegetical mapping. She is Editor of *Imago Mundi. The International Journal for the History of Cartography* and she organises the *Maps and Society lectures* at the Warburg Institute and teaches on map history in the London Rare Book School. (Page 37)

Dr Kenneth Field is a self-confessed carto-nerd. After twenty years in UK academia he now works for Esri Inc in cartographic research and development. He writes, teaches and blogs, is Editor of The Cartographic Journal, Chair of the ICA Map Design Commission, a Fellow of both the BCS and RGS and a Chartered Geographer (GIS). (Page 249)

Dr Adrian Fox, Head of Mapping and Geographic Information Centre, British Antarctic Survey.
Co-chief officer, SCAR Standing Committee on Antarctic Geographic Information.
Member of UK Antarctic Place-names Committee. (Page 213)

David Helliwell is Curator of Chinese Collections in the Bodleian Library, Oxford, UK. He is currently engaged in preparing an online catalogue of all the Library's pre-modern Chinese materials. His primary interest is in the Chinese book as an object, and he has produced a number of published and unpublished works in this area, with the primary aim of making the Bodleian's holdings better known. (Page 89)
http://serica.bodleian.ox.ac.uk/
http://oldchinesebooks.wordpress.com/

Manosi Lahiri holds a MSc degree from SOAS, University of London and a PhD from the University of Delhi. In 1993, she founded ML Infomap, a geographic information company. Her published books include *Here Be Yaks: Travels in Far West Tibet* and *Mapping India*, a five-hundred-year cartographic history of the country. (Pages 96, 117, 157, 169, 196)

Carla Lois. Geographer, PhD in History by the Universidad de Buenos Aires. Professor at the Universidad Nacional de La Plata, researcher at the Consejo Nacional de Investigaciones Científicas y Técnicas (CONICET, Argentina), Associate Editor for Volume 5 of the *History of Cartography* (University of Chicago Press). (Page 85)

David B. Miller, geography instructor at Northern Virginia Community College, specializes in cultural geography, physiography, geopolitics, and historical cartography. As a text and map editor for the National Geographic Society in Washington, D.C., David enjoys reading and making maps on all themes and regions. He holds a master's degree in geography from Western Illinois University, a bachelor's in geography from Miami University, Ohio, and is a proud Fellow of The British Cartographic Society. (Pages 93, 120, 141, 221, 224)

Philip Parker is a historian specializing in the classical and medieval world. He is the author of *The Empire Stops Here: A Journey Around the Frontiers of the Roman Empire* (Jonathan Cape 2009), the DK *Eyewitness Companion Guide to World History* (Dorling Kindersley 2010) and the Sunday Times best-seller, *The Northmen's Fury: A History of the Viking World* (Jonathan Cape 2014). He was General Editor of *The Great Trade Routes: A History of Cargoes and Commerce Over Land and Sea* (Anova 2012) and *Himalaya: The Exploration and Conquest of the Greatest Mountains on Earth* (Conway 2013). He previously worked as a diplomat and a publisher of historical atlases. (Pages 8, 10, 13, 21, 25, 29, 33, 45, 53, 57, 69, 101, 104, 106, 109, 125, 128, 132, 134, 137, 145, 148, 160, 173, 180, 184, 192, 208, 216, 228, 236, 240, 245)

Dick Pflederer, FRGS, has completed long term research projects on portolan charts while resident at several important institutions in the UK, the US and Italy. He is the author of several books on the subject of sea charts, navigation and exploration. (Pages 41, 48, 72)

Hance Smith, until 2011 Reader, Cardiff University School of Earth and Ocean Sciences, specializing in ocean development and management and marine geography. Over 120 academic papers and book chapters written. He was General Editor (and founder) of Routledge's Advances in Maritime Research, has written and co-authored academic papers on North Sea oil and contributed to *Scotland and Oil*, Edinburgh, Oliver & Boyd. (Page 200)

Albert Theberge is a retired officer from the commissioned service of the US National Oceanic and Atmospheric Administration (NOAA). He spent twenty-seven years primarily engaged in nautical charting work and seafloor mapping. He served twelve years on the US Board of Geographic Names Advisory Committee on Undersea Features and three years on its international GEBCO counterpart. The past eighteen years he have been spent as a research librarian at the NOAA Central Library and has written extensively on the history of hydrography, oceanography, and marine cartography. (Pages 80, 152)

Chet Van Duzer and is an Invited Research Scholar at the John Carter Brown Library in Providence, Rhode Island, and has published extensively on medieval and Renaissance maps. (Pages 17, 61, 65)

Dr Adrian Webb FSA FBCartS, holds a Master's and a PhD from the University of Exeter. He has written extensively on aspects of historic hydrography, and Somerset's local and economic history. He currently manages the Archive at the United Kingdom Hydrographic Office. (Pages 112, 114)

Additional thanks to:
Chris Fleet, Map Library, National Library of Scotland, Edinburgh, UK
Tom Harper, Map Library, British Library, London, UK
Gillian Hutchinson, Curator, History of Cartography, National Maritime Museum, Greenwich, UK
Peter Jones, Defence Geographic Centre, Feltham, UK
Nick Millea, Map Librarian, Bodleian Library, Oxford, UK
Rose Mitchell, Map archivist, The National Archives, Kew, UK
Kirsteen Valenti, University of Glasgow Library, UK

Image Credits

Dust jacket	Courtesy Bibioteca Nazionale di Napoli.
Cover	© National Maritime Museum, Greenwich, London, UK.
Endpapers	Courtesy Utrecht University Library.
9	De Agostini Picture Library/Bridgeman Images
11	Mary Evans/Iberfoto
12, 14–15	Courtesy James Harrell.
16, 18–19	Courtesy Bibioteca Nazionale di Napoli.
20, 22–3	Österreichische Nationalbibliothek, Vienna, Austria Cod.324.
24, 26–7	Giraudon/Bridgeman Images.
28, 30–1	Library of Congress, Geography and Map Division.
32, 34–5	The Art Archive/Bodleian Libraries, The University of Oxford.
36, 38–9	© British Library, MS Cotton Nero D 1, ff.183v-184r.
40, 42–3	Bibliothèque national de France.
44, 46–7	The Hereford Mappa Mundi Trust and the Dean and Chapter of Hereford Cathedral.
49–51	Bibliothèque national de France.
52, 54–5	Private Collection/Photo © Christie's Images Bridgeman Images.
56. 58–9	Beinecke Rare Book and Manuscript Library.
60, 62–3	De Agostini Picture Library/Bridgeman Images
64, 65–7	Library of Congress, Geography and Map Division.
68 (left), 70	Newberry Library, Chicago, Illinois, USA / Bridgeman Images
68 (right), 71	Museo Nacional de Antropologia, Mexico City, Mexico/Ian Mursell/Mexicolore Bridgeman Images.
73–5	© National Maritime Museum, Greenwich, London, UK.
76, 78–9	Bibliothèque Nationale de Cartes et Plans, Paris, France/Bridgeman Images.
81–3	Courtesy Utrecht University Library.
84, 86–7	Library of Congress, Geography and Map Division.
88, 90–1	The Bodleian Libraries, The University of Oxford
92, 94–5	© The British Library Board Maps.4.Tab.8.(1).
97–9	© National Maritime Museum, Greenwich, London, UK.
100, 102	Creative Commons Attribution/ Share-Alike License. Commons.wikimedia.org.
103	Collins Bartholomew
105	© National Maritime Museum, Greenwich, London, UK.
107	British Library/Science Photo Library.
108, 110–11	Library of Congress, Geography and Map Division Science Photo Library.
113	British Library/Science Photo Library.
115	Sourced from the UK Hydrographic Office (www.ukho.gov.uk).
116, 118–19	Library of Congress, Geography and Map Division.
121–3	© British Library Board. All Rights Reserved Bridgeman Images.
124, 126–7	Library of Congress, Geography and Map Division.
129–31	Courtesy of Mapseeker www.mapseeker.co.uk
133	Creative Commons Attribution/ Share-Alike License. Commons.wikimedia.org.
135	© Museum of London, UK/Bridgeman Images.
136, 138–9	Library of Congress, Geography and Map Division.
140, 142–3	Library of Congress, Geography and Map Division.
144, 146–7	Collins Bartholomew.
149–51	Library of Congress, Geography and Map Division.
153–5	NOAA Central Library Historical Collection.
156, 158–9	J.T.Walker/Royal Geographical Society.
161–3	Private Collection/© Look and Learn/ Bridgeman Images.
165–7	Collins Bartholomew.
168, 170–1	Collins Bartholomew.
172, 174–5	The National Archives.
176, 178–9	Imperial War Museum/Western Front Association IMW/M.5-000756.
181–3	Reproduced by permission of the National Library of Scotland, Edinburgh, UK.
185–7	CPA Media
188, 190–1	The National Archives, London, UK.
193–5	Creative Commons Attribution-Share Alike 3.0 Unported license.
197–9	© The British Library Board Maps MOD OR 6409.
201–3	Oxford University Press.
205–7	Collins Bartholomew.
209–211	Reproduced by permission of the National Library of Scotland, Edinburgh, UK.
212–15	Collins Bartholomew.
217–19	Collins Bartholomew.
220, 222–3	Collins Bartholomew.
225–7	Courtesy Lunar and Planetary Institute, Universities Space Research Association, Houston
229	© Bettman/CORBIS.
230-1	John F. Kennedy Presidential Library and Museum
232, 234–5	By kind permission of UNEP.
237–9	Library of Congress, Geography and Map Division.
241–3	© UK MOD Crown Copyright, 2014
244, 246–7	Image courtesy of the University of Glasgow Library.
248–51	(top left) Google Maps; (centre left) Google Maps; (bottom left) Google Maps/United States Holocaust Memorial Museum; (top right) Google Earth/Carbon Visuals www.carbonvisual.com; (centre right) Living Oceans Society; (bottom right) Google Maps/David Rumsey Map Collection www.davidrumsey.com

Every effort has been made to trace the copyright owners of any images used and to obtain permission for their reproduction. If any have been missed, the publishers apologize and invite the copyright owners to contact them to arrange any formal permissions required.

Selected further reading

Cartographia Mapping Civilizations, Vincent Virga, Little Brown 2007

The Golden Age of Maritime Maps, Hofmann, Richard and Vagnon, Firefly 2013

A History of the World in Twelve Maps, Jerry Brotton, Penguin 2013

The Image of the World, Peter Whitfield, British Library 1994

Lie of the Land – The Secret Life of Maps, Carlucci and Barber (Eds), British Library 2001

The Map Book, Peter Barber, Wiedenfeld & Nicolson 2005

Mapping India, Manosi Lahiri, Niyogi 2012

Mapping the First World War, Peter Chasseaud, Collins 2013

Mapping the World An Illustrated History of Cartography, Ehrenberg, National Geographic 2006

Maps That Made History, Lez Smart, TNA 2004

The Map that Changed the World, Simon Winchester, Penguin 2002

Mercator, Nicholas Crane, Phoenix 2002

On The Map, Simon Garfield, Profile 2013

The Oxford Map Companion, Patricia Seed, Oxford 2014

Remarkable Maps, John O.E. Clark and Jeremy Black, Conway 2005

The Sea Chart – The Illustrated History of Nautical Maps and Navigational Charts, John Blake, Conway 2009

Mr Selden's map of China, Timothy Brook, Profile Books 2013

The World Through Maps, John Rennie Short, Firefly 2003

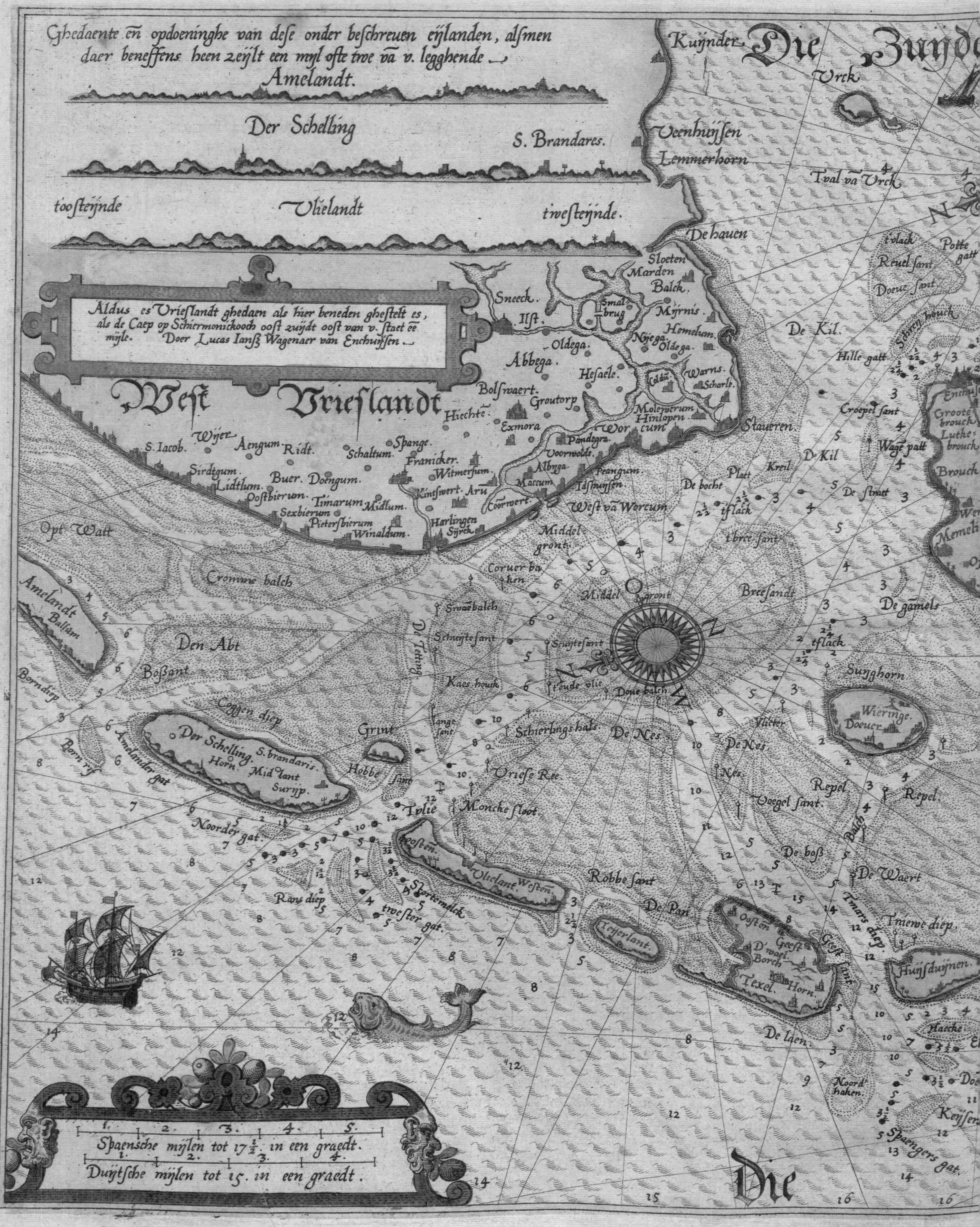
Ghedaente en opdoeninghe van dese onder beschreuen eylanden, alsmen daer beneffens heen zeylt een myl ofte twe va v. legghende
Amelandt
Der Schelling
S. Brandares.
toosteynde
Vlielandt
twesteynde.
Aldus es Vrieslandt ghedaen als hier beneden ghestelt es, als de Caep op Schirmonickooch oost zuydt oost van v. staet oe myle. Doer Lucas Ianß Wagenaer van Enchuyssen
West Vrieslandt
Kuynder
Die Zuyde
Vrck
Veenhuyssen
Lemmerhorn
Tval va Vrck
De hauen
Sloeten
Marden
Balck
Sneeck
Ilst
Smalbrug
Myrnis
Homelum
Nyega
Oldega
Abbega
Hesaele
Warns
Bolswaert
Groutorp
Hiechte
Exmora
Molquerum
Hinlopen
Staueren
Worcum
S. Iacob
Wyer
Aengum
Ridt
Schaltum
Spange
Franicker
Witmersum
Sirdtgum
Lidtlum
Buer
Doengum
Oostbierum
Timarum
Midlum
Sexbierum
Pietersbierum
Winaldum
Harlingen
Kintswert
Coorwert
West va Worcum
Opt Watt
Cromme balch
Amelandt
Ballum
Den Abt
Boßant
Borndiep
Coggen diep
Der Schelling
S. brandaris
Horn
Midlant
Surryp
Amelander gat
Grint
Hobbe sant
Tvlie
Noorder gat
Middel gront
Coruer bakken
Swaebalch
De Tetting
Schuytesant
Kaes houck
Schierlings hals
Vriese Ree
Monckesloot
Middel gront
Breesandt
Oude vlie
Doue balch
De Nes
De Kil
Hille gatt
Croepel sant
Wage patt
Reuel sant
Doeue sant
Potte gatt
Groote brouck
Luthe brouck
Brouch
De Kil
De bocht
t'flack
t'bree sant
De gamels
Surghorn
Wieringe
Doeuer
Vlitter
De Nes
Repel
Voegel sant
Vlielant
Westen
Robbe sant
Raus diep
twester gat
De Pan
Texel
Horn
D' wael
Borch
Oosten
Geest
Huysduynen
De Waert
Tmars diep
Tniewe diep
De laen
Noord hahen
Spaengers gat
Keyser
Spaensche mylen tot 17½. in een graedt.
Duytsche mylen tot 15. in een graedt.
Die